Michael Hülsmann · Katja Windt

Understanding Autonomous Cooperation and Control in Logistics

Michael Hülsmann · Katja Windt (Eds.)

Understanding Autonomous Cooperation and Control in Logistics

The Impact of Autonomy on Management, Information, Communication and Material Flow

With 91 Figures

Springer

Prof. Dr. Michael Hülsmann
University of Bremen
Faculty 07: Business Studies and Economics
Wilhelm-Herbst-Str. 12
28359 Bremen, Germany
michael.hülsmann@uni-bremen.de

Dr.-Ing. Katja Windt
University of Bremen
Bremen Institute of Industrial Technology
and Applied Work Science
Hochschulring 20
28359 Bremen, Germany
wnd@biba.uni-bremen.de

ISBN 978-3-642-07996-2 e-ISBN 978-3-540-47450-0

Springer is a part of Springer Science+Business Media
springer.com

Cover: WMXDesign, Heidelberg
Printed on acid-free paper 68/3180 YL - 5 4 3 2 1 0

Preface

The idea and results of the edited volume "Understanding Autonomous Cooperation and Control in Logistics – The Impact of Autonomy on Management, Information, Communication, and Material Flow" are based on the interdisciplinary research of the working group "Autonomous Cooperation" within the Collaborative Research Centre 637 (CRC 637) "Autonomous Cooperating Logistic Processes – A Paradigm Shift and its Limitations" at the University of Bremen.

The starting point of this research is to lay foundations for building a theory concerning the concept of autonomous cooperation and control (including technologies and instruments) in logistics. A further aim is to gain valid knowledge about the involved causal relations so as to apply the concept in practice. Therefore, the research of the CRC 637 tries to identify rules of the paradigm of autonomous cooperation and to find the means, whereby the degree of autonomous cooperation can be designed on all levels of logistic systems:

- On the decision making level;
- On the information and communication level;
- On the material flow level of logistics management.

It is expected that a higher degree of autonomous cooperation in logistic processes could be one approach to dealing with the increasing complexity and dynamics in logistic systems. This might be possible because on the one hand autonomous cooperation might lead to an increasing flexibility, which could further lead to positive emergency and improvement in process quality (i.e. robustness). On the other hand, autonomous cooperation could also have contradictious effects on productivity, which might be attributed to the immanent redundancy in resources as well as structures and the delegation of decision power. Thus, the CRC 637 is striving for the answer to the question what the optimal degree of autonomous cooperation might be.

In order to enable the implementation of self-organisation ideas as a principle of autonomous cooperation, control and organisation for logistic systems, it is the overarching aim of this edited volume to gain an interdisciplinary understanding of it. Therefore, the contributions in this edited volume try to develop an approach from different perspectives of production technology, electronics and communication engineering, informatics and mathematics, as well as business studies to determine how the concept of autonomous cooperation and control can be applied to logistics. This includes the individual description of the phenomena and principles of autonomous cooperation as well as an analysis of its implications for management, information, communication, and material flow. Therefore, the edited volume is to accomplish the following tasks:

- To collate various understandings of self-organisation, which have a comprehensive and differentiable description of its basic ideas and its adoption to logistics as an organisational principle;
- To identify and compare the scope and depth of autonomous cooperation and control resulting from various understandings of self-organisation, in order to summarise the commonness and differences and to allow development of an applicable understanding of autonomous cooperation and control for logistics;
- To establish an overarching conception of autonomous cooperation and control, which gives impulses for the research within different disciplines to answer the question, as to how logistics management can cope with complexity and dynamics in supply chains and networks in a better way;
- To develop a conceptual and terminological system for autonomous cooperation and control, but without a too detailed concretisation, which allows discipline-specific interpretation, functionalisation and application in the context of logistic systems.

Like most publications this edited volume is also based on the invaluable work and contributions of many helpful hands. Therefore, we, the editors, have the great honour and pleasure to thank everybody, who made this book possible. Firstly, we want to express our deep gratitude to the colleagues from the Collaborative Research Centre 637 (CRC 637) "Autonomous Cooperating Logistic Processes – A Paradigm Shift and its Limitations". Their contributions to this publication did not only shed light on our understanding of what autonomous cooperation and control is all about, what it implies for the design of logistics processes and systems, and what we can learn from other disciplines for the analysis of complexity and dynamics in logistics. It was also personally for us a real delight to work together with them on this edited volume. Secondly, we had the helpful support of the members of the board of the CRC 637, Prof. Dr. Carmelita Görg, Prof. Dr. Otthein Herzog, and Prof. Dr.-Ing. Bernd Scholz-Reiter. Their backing up was always an excellent motivation for us to proceed with our edited volume. They provided several inspiring ideas which helped us very much to realise this project. Thirdly, the always-courteous

Lore Zander handled many administrative duties. Many thanks for their inestimable help. Additionally, we could constantly rely on the cooperative coordination, careful editing, proof reading and accurate layout of Jan Tell, Dipl.-Ing. Thorsten Phillip, Ying Li, M.A., and Dan Smith. For this dependable support we, the editors, are greatly indebted. And of course, we want to express our appreciation to the publisher SpringerPhysica, represented by Thomas Lehnert. It was a constant source of stimulation to know, that we had been offered the occasion to publish our edited volume "Understanding Autonomous Cooperation and Control in Logistics – The Impact of Autonomy on Management, Information, Communication and Material Flow" at SpringerPhysica's. Finally, we would like to thank the German Research Foundation (DFG), which supported this research as part of the Collaborative Research Centre 637 "Autonomous Cooperating Logistic Processes – A Paradigm Shift and its Limitations".

Michael Hülsmann Katja Windt

Acknowledgement
This research was supported by the German Research Foundation (DFG) as part of the Collaborative Research Centre 637 "Autonomous Cooperating Logistic Processes – A Paradigm Shift and its Limitations" at the University of Bremen.

Contents

Associated Editors

Görg, Carmelita, Prof. Dr.
Fachbereich 1, Institute for Telecommunications and
High-Frequency Techniques (ITH)
Kommunikationsnetzwerke (ComNets)
University of Bremen
Otto-Hahn-Allee NW1
28359 Bremen, Germany

Gronau, Norbert, Prof. Dr.-Ing.
Chair of Business Information Systems
University of Potsdam
August-Bebel-Str. 89
14482 Potsdam, Germany

Habel, Annegret, Prof. Dr.
Department for Informatics,
Carl v. Ossietzky University
26111 Oldenburg, Germany

Herzog, Otthein, Prof. Dr.
TZI - FB 3
University of Bremen,
Am Fallturm 1
28334 Bremen, Germany

Kooten, Olaf van, Prof. Dr.
Horticultural Production Chains, Wageningen University,
Marijkeweg 22,
6709 PG Wageningen, The Netherlands

Maropoulos, Paul George, Prof. Dr.
Department for Mechanical Engineering
University of Bath, 4 East 2.11a, United Kingdom

Matiaske, Wenzel, Prof. Dr.
Universität Flensburg
Bahnhofstr. 38
24937 Flensburg, Germany

Mattfeld, Dirk C., Prof. Dr.
Institut für Wirtschaftswissenschaften
Technische Universität Braunschweig
Abteilung Wirtschaftsinformatik
Abt-Jerusalem-Str. 4
38106 Braunschweig, Germany

Pesch, Dirk, Dr.
Centre for Adaptive Wireless Systems
Department of Electronic Engineering
Cork Institute of Technology
Rossa Avenue
Cork, Ireland

Perera, Ranjit, Prof. H. Y.
Department of Electrical Engineering
University of Moratuwa
Katubedda
Moratuwa, Sri Lanka

Remer, Andreas, Prof. Dr.
Chair of Organization and Management (BWL VI)
University of Bayreuth
Universitätsstrasse 30
95440 Bayreuth, Germany

Scholz-Reiter, Bernd, Prof. Dr.-Ing.
Department of Planing and Control of Production Systems, BIBA
University of Bremen
Hochschulring 20
28359 Bremen, Germany

Schouten, Rob, Dr.
Horticultural Production Chains, Wageningen University,
Marijkeweg 22
6709 PG Wageningen,
The Netherlands

Spengler, Thomas, Prof. Dr.
Lehrstuhl für Produktion und Logistik
Institut für Wirtschaftswissenschaften, TU Braunschweig
Katharinenstr. 3
38106 Braunschweig, Germany

Tilebein, Meike, Prof. Dr.
DPD Endowed Assistant Professor of Innovation Management
EUROPEAN BUSINESS SCHOOL (EBS)
International University Schloss Reichartshausen
65375 Oestrich-Winkel, Germany

Wiendahl, Hans-Peter, Univ.-Prof. Dr.-Ing. Dr. mult. H.c.
Institut für Fabrikanlagen und Logistik IFA
Leibniz University of Hannover
An der Universität 2
30823 Garbsen, Germany

Authors

Agarwal, Robin, M.A.
BWL MNS, Universität Bremen
Wilhelm-Herbst-Str. 12, 28359 Bremen
ragarwal@uni-bremen.de

Arndt, Lars, Dipl. Oec
BWL NM, Universität Bremen
Wilhelm-Herbst-Str. 12, 28359 Bremen
larndt@uni-bremen.de

Becker, Markus, Dipl.-Ing.
TZI ComNets, Universität Bremen
Otto-Hahn-Allee, 28359 Bremen
mab@comnets.uni-bremen.de

Behrens, Christian, M.Sc.
ITEM, Universität Bremen
Otto-Hahn-Allee, 28359 Bremen
behrens@item.uni-bremen.de

Bemeleit, Boris, Dipl.Wi.-Ing.
ITAPT, Universität Bremen
Hochschulring 20, 28359 Bremen
bem@biba.uni-bremen.de

Böse, Felix, Dipl.-Wirtsch.-Inf.
PSPS, BIBA, Universität Bremen
Hochschulring 20, 28359 Bremen
boe@biba.uni-bremen.de

Dashkovskiy Sergey, Dr.
ZeTeM, Universität Bremen
Bibliothekstr.1, 28359 Bremen
dsn@math.uni-bremen.de

de Beer, Christoph, Dipl.-Phys.
PSPS, BIBA, Universität Bremen
Hochschulring 20, 28359 Bremen
ber@biba.uni-bremen.de

Freitag, Michael, Dr.-Ing.
c/o BIBA-IPS, Universität Bremen
Hochschulring 20, 28359 Bremen
fmt@biba.uni-bremen.de

Gehrke, Jan D., Dipl.-Inf.
TZI IS, Universität Bremen
Am Fallturm 1, 28359 Bremen
jgehrke@tzi.de

Görg, Carmelita, Prof. Dr.
TZI ComNets, Universität Bremen
Otto-Hahn-Allee, 28359 Bremen
cg@comnets.uni-bremen.de

Grapp, Jörn, Dipl.-Oec.
BWL MNS, Universität Bremen
Wilhelm-Herbst-Str. 12, 28359 Bremen
grapp@uni-bremen.de

Herzog, Otthein, Prof. Dr.
TZI IS, Universität Bremen
Am Fallturm 1, 28359 Bremen
herzog@tzi.de

Hildebrandt, Torsten, Dipl.-Wirtsch.-Inf.
PSPS, BIBA, Universität Bremen
Hochschulring 20, 28359 Bremen
hil@biba.uni-bremen.de

Hölscher, Karsten, Dipl.-Inf.
TZI TI, Universität Bremen
Linzer Str. 9a, 28359 Bremen
hoelsch@tzi.de

Hülsmann, Michael, Prof. Dr.
BWL MNS, Universität Bremen
Wilhelm-Herbst-Str. 12, 28359 Bremen
mhuels@uni-bremen.de

Jagalski Thomas, M.Sc.
PSPS, BIBA, Universität Bremen
Hochschulring 20, 28359 Bremen
jag@biba.uni-bremen.de

Jedermann, Reiner, Dipl.-Ing.
IMSAS, Universität Bremen
Otto-Hahn-Allee, 28359 Bremen
rjedermann@imsas.uni-bremen.de

Klempien-Hinrichs, Renate, Dr.
TZI TI, Universität Bremen
Linzer Str. 9a, 28359 Bremen
rena@informatik.uni-bremen.de

Knirsch, Peter, Dr.
TZI TI, Universität Bremen
Linzer Str. 9a, 28359 Bremen
knirsch@tzi.de

Kolditz, Jan, Dipl.-Wirtsch.-Ing.
PSPS, BIBA, Universität Bremen
Hochschulring 20, 28359 Bremen
kol@biba.uni-bremen.de

Kopfer, Herbert, Prof. Dr.
LfL, Universität Bremen
Wilhelm-Herbst-Str. 5, 28359 Bremen
kopfer@logistik.uni-bremen.de

Kreowski, Hans-Jörg, Prof. Dr.
TZI TI, Universität Bremen
Linzer Str. 9a, 28359 Bremen
kreo@tzi.de

Kuladinithi, Koojana, M.Sc.
TZI ComNets, Universität Bremen
Otto-Hahn-Allee, 28359 Bremen
koo@comnets.uni-bremen.de

Kuske, Sabine, Dr.
TZI TI, Universität Bremen
Linzer Str. 9a, 28359 Bremen
kuske@tzi.de

Lang, Walter, Prof. Dr.-Ing.
IMSAS, Universität Bremen
Otto-Hahn-Allee, 28359 Bremen
wlang@imsas.uni-bremen.de

Langer, Hagen, Dr. habil.
TZI IS, Universität Bremen
Am Fallturm 1, 28359 Bremen
hlanger@informatik.uni-bremen.de

Laur, Rainer, Prof. Dr.-Ing.
ITEM, Universität Bremen
Otto-Hahn-Allee, 28359 Bremen
rlaur@item.uni-bremen.de

Li, Ying, M.A.
BWL MNS, Universität Bremen
Wilhelm-Herbst-Str. 12, 28359 Bremen
linying@uni-bremen.de

Lorenz, Martin, Dipl.-Ing.
TZI IS, Universität Bremen
Am Fallturm 1, 28359 Bremen
mlo@tzi.de

Morales-Kluge, Ernesto, Dipl.-Wi.-Ing.
PSPS, BIBA, Universität Bremen
Hochschulring 20, 28359 Bremen
mer@biba.uni-bremen.de

Müller-Christ, Georg, Prof. Dr.
BWL NM, Universität Bremen
Wilhelm-Herbst-Str. 12, 28359 Bremen
gmc@uni-bremen.de

Philipp, Thorsten, Dipl.-Ing.
PSPS, BIBA, Universität Bremen
Hochschulring 20, 28359 Bremen
phi@biba.uni-bremen.de

Rekersbrink, Henning, Dipl.-Ing.
PSPS, BIBA, Universität Bremen
Hochschulring 20, 28359 Bremen
rek@biba.uni-bremen.de

Rüffer, Björn, M.Sc.
ZeTeM, Universität Bremen
Bibliothekstr. 1, 28359 Bremen
rueffer@math.uni-bremen.de

Scholz-Reiter, Bernd, Prof. Dr.-Ing.
PSPS, BIBA, Universität Bremen
Hochschulring 20, 28359 Bremen
bsr@biba.uni-bremen.de

Schönberger, Jörn, Dr.
LfL, Universität Bremen
Wilhelm-Herbst-Str. 5, 28359 Bremen
sberger@logistik.uni-bremen.de

Schumacher, Jens, Dr.-Ing.
ITAPT, Universität Bremen
Hochschulring 20, 28359 Bremen
jsr@biba.uni-bremen.de

Timm, Ingo J., Dr.-Ing.
TZI IS, Universität Bremen
Am Fallturm 1, 28359 Bremen
i.timm@tzi.uni-bremen.de

Timm-Giel, Andreas, Dr.-Ing.
TZI ComNets, Universität Bremen
Otto-Hahn-Allee, 28359 Bremen
atg@comnets.uni-bremen.de

Wenning, Bernd-Ludwig, Dipl.-Ing.
TZI ComNets, Universität Bremen
Otto-Hahn-Allee, 28359 Bremen
wenn@comnets.uni-bremen.de

Windt, Katja, Dr.-Ing.
PSPS, BIBA, Universität Bremen
Hochschulring 20, 28359 Bremen
wnd@biba.uni-bremen.de

Wirth, Fabian, PD Dr.
ZeTeM, Universität Bremen
Bibliothekstr. 1, 28359 Bremen
fabian@math.uni-bremen.de

Wycisk, Christine, Dipl.-Oec.
BWL MNS, Universität Bremen
Wilhelm-Herbst-Str. 12, 28359 Bremen
cwycisk@uni-bremen.de

1 Changing Paradigms in Logistics – Understanding the Shift from Conventional Control to Autonomous Cooperation and Control

Katja Windt[1], Michael Hülsmann[2]

[1] Department of Planning and Control of Production Systems, BIBA University of Bremen, Germany

[2] Management of Sustainable System Development, Institute for Strategic Competence-Management, Faculty of Business Studies and Economics, University of Bremen, Germany

1.1 Introduction

The understanding of logistics as the integrated planning, control, realization and monitoring of all internal and network-wide material-, part- and product flows including the necessary information flow along the complete value-added chain is still valid: but the logistic performance is becoming more and more dependent on technological innovations. One reason for this is increasing complexity in combination with a high incidence of potentially disruptive factors. The increasing number of part variants and their combination during the production process of automobiles, for instance, leads to a tremendous number of possible combinations. The resultant complexity can no longer be managed feasibly by means of centralized planning and control systems. In addition, today's customers expect a better accomplishment of the logistical targets, especially a higher due date reliability, and shorter delivery times. In order to cope with these requirements the integration of new technologies and control methods has become necessary. This is what characterizes the ongoing paradigm shift from a centralised control of "non-intelligent" items in hierarchical structures towards a decentralised control of "intelligent" items in heterarchical structures in logistic processes. Such intelligent items could include both raw materials, components or products, as well as transit equipment (e.g. pallets, packages) and transportation systems (e.g. conveyors, trucks).

The recent revolutionary developments within Information and Communication Technologies were marked by miniaturization, ubiquitous communications and digital convergence. The trend is towards embedded systems which are moving beyond local interfacing to globally connected systems and allow increased levels of "collective intelligence". These systems are based on recent IC technologies such as RFID and wireless communication networks, and intelligent items which can coordinate and communicate with each other. These new technological developments call for novel concepts and strategies designed to implement autonomy in logistic processes (Scholz-Reiter et al. 2004).

This anthology presents first approaches and results on autonomous cooperation and control methods for logistic processes. It is based on the research work within the Cooperative Research Center 637 "Autonomous Cooperating Logistic Processes – A Paradigm Shift and its Limitations" at the University of Bremen and it is supported by the German Research Foundation. The need for a better understanding of this new control paradigm in logistics will be explained in the second chapter of this introduction. Of equal importance is the analysis of the main drivers and the definition of autonomous cooperation and control, as well as the description of the major enablers which follows in the next chapter.

1.2 Drivers and enablers of autonomous cooperation and control in logistic processes

The drivers supporting the paradigm shift within logistics are categorised in fig. 1.1 as market, product, technologies and process drivers. The main change, which applies especially to logistic processes, is the significant reduction of time for the change of states, i.e. the time in between two different states of a system. The dynamics within logistic processes are increasing. This may be observed in the categories listed in fig. 1.1 A heterogeneous market with high demand fluctuations, products which incorpoprate a high number of variations and have short product lifecycles, new and fast developing information and communication technologies, as well as production on demand, characterise this situation. In parallel, the demands on logistic performance and logistic costs are increasing, too. This is indicated for instance by shorter delivery times, higher schedule reliability delivery flexibility and the use of reconfigurable technologies. As shown in the middle of fig. 1.1, besides the demands on shorter delivery time, higher schedule reliability, lower price and high quality, the complexity of all the internal and external influencing parameters of logistic systems is also increasing. Among other things, this increased complexity is due to production in global networks, an exponential increase in the amount of data with the use of new ICT, product structures with a high number of variations. In summary, logistic systems are confronted with increasing complexity in combination with many potentially disruptive factors. These impact factors are the drivers of change for a new control paradigm within logistic processes.

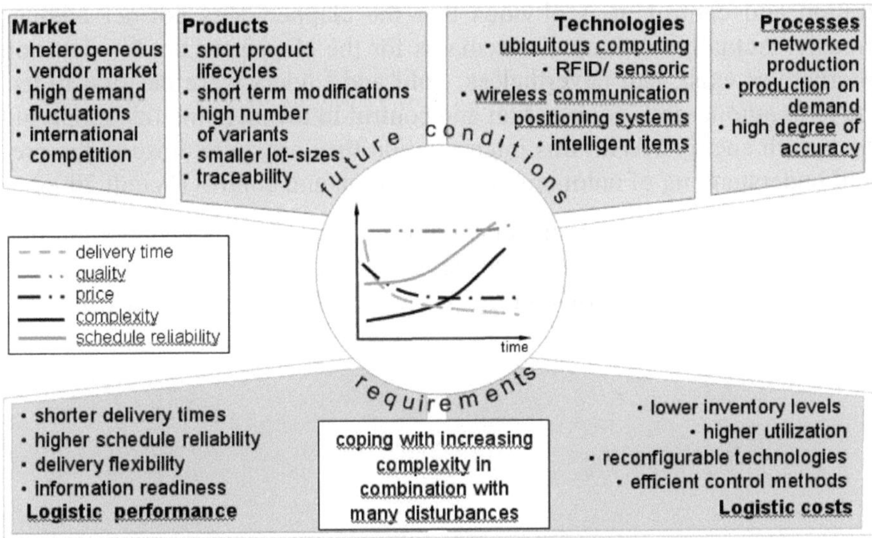

Fig. 1.1 Future conditions and requirements on logistic processes

The paradigm shift is based on the following hypothesis: The implementation of autonomous logistic processes provides a better accomplishment of logistic objectives in comparison to conventionally managed processes despite increasing complexity. In order to verify this thesis, it is necessary to characterize production systems with regard to their level of complexity during the development of an evaluation system.

Autonomous cooperation and control is one factor to guarantee the necessary changeability of logistic processes. Wiendahl et al. defines changeability as characteristics to accomplish early and foresighted adjustments of the factory's structures and processes on all levels to change impulses with small expenditure (Wiendahl et al. 2007).

Several similar terms exist besides autonomous cooperation and control e.g. self-organisation, self-management or self-regulation. The term autonomous control was initially used in the year 1930 by Pohl and Lüders (Pohl and Lüders 1930). The described example referred to the functionality of a door-bell. The clapper of the bell obtains quasi autonomously the energy for its oscillation by connecting the current to an electro-magnet via the use of a spring. Due to self-induction, the pendulum represented by the clapper is accelerated and consequently the electric circuit is disconnected. The task of the spring is to reconnect the electrical contact. Clearly, if there were a constant energy supply the ring tone would sound continuously. In the proper meaning of the aforementioned definition of autonomous cooperation and control, it is obvious that the clapper does not act autonomously. Actually, nothing else remains for the clapper to do. No decision alternatives exist. But nevertheless, Pohl and Lüders were the first to use the term autonomous cooperation and control in the meaning of "supplying itself with energy". With this interpretation they are quite close to the present understanding of autonomous cooperation and control (Windt 2006).

In order to get a better understanding of autonomous cooperation and control it is necessary to identify the enablers of autonomous cooperation and control which are shown in fig. 1.2 and explained in the following.

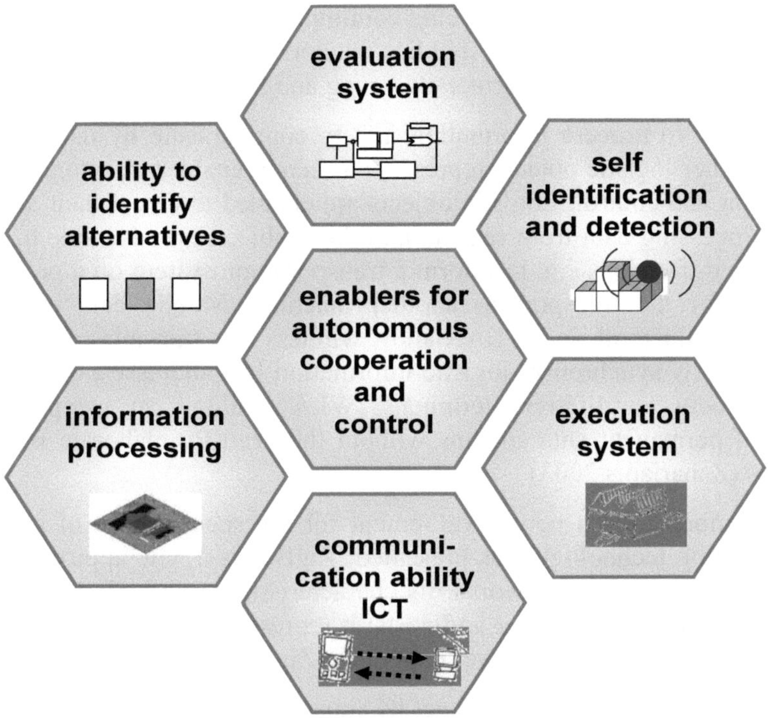

Fig. 1.2 Enablers of autonomous cooperation and control

In order to enable logistic objects (e.g. machine, transportation system, order, product, pallet) to act in an autonomous way the use of ICT is necessary. RFID technology plays a major role in autonomous logistic processes. While the current way of handling data in traditional logistic processes is by means of barcode, the information involved in autonomous processes is handled via RFID tag. Applications in logistics go from automatic stock control and pallet localisation, through automatic registration of goods inbound and outbound, to the saving of detailed information e.g. contents, destination or delivery date (Westkämper and Jendoubi 2003), (Finkenzeller 2002). Future systems will integrate sensors and processing units in embedded systems which will allow the use of a higher level of autonomous cooperation and control.

Positioning systems like the American GPS, the European Galileo or the Russian GLONASS, enable a complete localization of vehicles using a combination of satellite assisted positioning and mobile radio (Gebresenbet and Ljungberg 2001).

Network security systems are being continually enhanced and improved. Safe communication through public networks is an important pre-condition between logistics partners (Cheung and Misic 2002).

The ability to process information and to communicate by using new ICT with other logistic objects represents a second enabler of autonomous cooperation and control. Logistic objects are enabled to detect their situation by processing data from sensors and these objects are also able to assert rendered decisions e.g. to inform a transportation system on a production floor for the transport to another machine. Mobile data transfer systems like Bluetooth and WLAN allow wireless data transmission. Bluetooth can safely synchronize logistic information like addresses, dates and capacities between different terminals. WLAN allows an inexpensive transfer of permanent data streams without the need for elaborate wiring harnesses (Zahariadis 2003).

In December 2004 a new development for the specification of a new communication technology was presented: ZigBee is a new approach addressing wireless sensor networks. Its characteristics are a high density of nodes per network, low power and costs: it represents an optimized short-range wireless solution with lower data rates (ZigBee 2006).

If a logistic object is able to detect its situation on its own by the use of such new ICT, then one key characteristic of autonomous cooperation and control (in fig.1.2 self identification and detection) is attained. In order to acquire the other consecutive characteristics of autonomous cooperation and control it is necessary that the logistic object has the ability to identify alternatives in order to reach its target in a better way. This ability to identify alternatives is another enabler of logistic objects acting autonomously. Nevertheless, there is the need to decide between the identified and given alternatives. Consequently, an evaluation system has to provide methods to evaluate all alternatives. An evaluation system represents another of the enablers necessary for autonomous cooperation and control.

The typical job-shop-scheduling problem, which is characteristic for production logistics, leads to non-polynomial problems. One characteristic of such problems is that the solution space, meaning the range of possible alternative solutions, increases faster than the speed with which decision making takes place.

Without heuristic methods, central control methods are not able find an optimal solution, while nevertheless involving time-consuming arithmetic operations. These time-consuming arithmetic operations often have the effect that during this planning, processes are altered: this causes the elaborately made plan to be invalid even before the beginning of its implementation. Due to this, control systems need to cope with constantly changing plans and simultaneously occurring changes (which are neither visible nor can be influenced) during the process sequence. In addition to that, according to Wiesenthal the control system, has to "imagine" itself and its environment as different in the future. Due to the lack of reliable data and therefore an adequately accurate determination of the future system, the control system has to navigate into an undetermined future (Wiesenthal 2006). As a result of these circumstances, it is not purposeful to implement a complete planning for a longer period in a non-deterministic system. In fact, it appears that decentralized approaches cope in a better way with the previously described problems. Those decentralized control approaches reduce the number of necessary arithmetic operations, and in addition fewer parameters have to be taken into consideration. So decentralized or autonomous control approaches enable the use of conventional decision making methods, which need fewer computational efforts and are therefore time saving, thus reducing the chance of simultaneously occurring changes during processes or simultaneously appearing events. Autonomous cooperation and control hence is able to open new logistic potentials in interaction with complex and dynamically changing process structures. To utilise those potentials, first it is necessary to understand what the term of autonomous cooperation and control describes and what are the major characteristics of this phenomenon – which is the overarching aim of this anthology.

1.3 Autonomous cooperation and control – a general understanding

The basic foundations of autonomous co-operation and control reflect on the idea of self-organization, an interdisciplinary study which has been developing for about 35 years under the labels such as self-organization, autopoiesis, dissipative structures, emergency and complexity theory. The core of the self-organization concept is the formation and development of order in complex dynamic systems (Paslack 1999). In natural sciences, important representatives are Prigogine (Glansdorf and Priogine 1971), in chemistry (theory of dissipative structures), Peitgen and Richter (Peitgen and Richter 1986) in mathematics (chaos theory), Haken (Haken and Graham 1971; Haken 1973) and Foerster (Foerster 1960), in physics (synergetics and cybernetics), and Maturana and Varela (Maturana and Varela 1980), in Biology (autopoiesis). The last concept "autopoiesis" is also applied to other fields such as sociology (Luhman's (Luhman 1973) system theory), psychology concerning family therapy (Hoffmann 1984), jurisprudence regarding the theory of state (Tebner and Willke 1984), marketing (Schüppenhauer 1998) and management (Kirsch 1992). Such transference of research results to various scientific fields might be an indication of high relevance of self-organization for different sciences and its wide recognition. But it is still necessary to adopt the general idea of self-organisation to a capable understanding for logistics. That is why this anthology tries to develop such a definition, in which autonomous cooperation and control is regarded as the answer of a logistic system to complexity and dynamics. Therefore, autonomous cooperation and control is defined as:

> *Autonomous Control describes processes of decentralized decision-making in heterarchical structures. It presumes interacting elements in non-deterministic systems, which possess the capability and possibility to render decisions.*
>
> *The objective of Autonomous Control is the achievement of increased robustness and positive emergence of the total system due to distributed and flexible coping with dynamics and complexity.*

The given definition has been developed within the interdisciplinary working group autonomous cooperation and control of the Cooperative Research Centre (CRC) 637 "Autonomous Cooperating Logistic Processes – A paradigm Shift and its Limitations". Based on this global definition of the term autonomous cooperation and control, further developed definitions related to the relevant science fields will be presented within the articles included in this anthology.

What are the major general and constitutional characteristics of the definition of autonomous cooperation and control given before?

Decentralized Decision-making

Decision concerns the adoption of an action so that an object can reach a state (end state) from another state (starting state). Normally there are some alternative actions and the selection of one specific action has to be preceded by obtaining and processing of necessary information.

The goal-oriented selection between action alternatives is termed as decision-making. (Frese 1993; Laux 1998) Here actions could be either active (self-induced) or reactive (external-induced). Decentralization means the shift from the central point (Frese 1993). For the definition of autonomous cooperation, decentralization means the delegation of decision power, that is, individual system elements are allowed to make independent decisions and are capable of making such decisions by gaining access to necessary resources (e.g. relevant information)

Autonomy

An element of a larger system is autonomous when it is responsible for its own system design, direction and development. In other words, it can make decisions independent from the external entities (Probst 1987). The autonomy of a system or an individual is always measured according to certain criteria and the contextual conditions of the system (Varela 1979; Probst 1987). Criteria could be the scope and extent of decision power. Consequently, autonomy could be seen as the result of the processes of decentralization and delegation (Kappler 1992). In the context of autonomous cooperation, the concept of autonomy is understood as autonomous decision-making.

Interaction

Interaction describes the successful contact between elements (or systems, subsystems etc). "Being successful" means in this context that communication takes place. In other words, the intended contact is able to induce reactions (i.e. reciprocity)(Staehle 1999). Such interactions are central to the autonomous cooperating logistic systems and are realized through communication between system elements such as goods in transportation, vehicles and warehouses. During the interaction processes, information is exchanged in the form of specific data, which could assist in decision-making by the involved elements. With the use of advanced technology like RFID, elements of a logistic system could communicate with elements both inside and outside the system.

Heterarchy

Heterarchy describes the parataxis of system elements (Goldammer 2002). A Heterarchical system is featured by the absence of a permanently dominant entity (Probst 1992). In a heterarchical logistic system such as a production network, there are fewer superordinate and subordinate relationships between logistic elements. This means an increasing level of independence between single elements and a central logistic coordination entity.

Non-determinism

A system is non-deterministic if its behaviour cannot be predicted over a relatively long period despite precise measurement of system states and knowledge about all system laws (Flämmig 1998). For example, the exact output of the system cannot be predetermined based on the input in a non-deterministic system. With such observations, Prigogine brings forward the concept of bifurcation, which means that at this point there are various paths possible for system development. Neither the time point nor the development path to be selected could be predicted, as they follow no causal patterns (Prigogine 1996). With the characteristic of non-determinism, autonomous cooperation strives for higher efficiency in dealing with complexity and uncertainty within processes. The aim is to optimize production and improve order fulfilment. An example could be that components (meeting technological prerequisites such as with imbedded chips) seek the optimal processing path and thus control the production line by themselves. Disruption of the whole or a large part of the process could be prevented, as components could react to disturbance flexibly with alternative actions in hand.

To understand autonomous co-operation and control in logistics on has to delimitate the concept of "Autogenous Processes" vs. the concept of "Autonomous Processes". Generally speaking, autonomous cooperation could be divided into autogenous processes and autonomous processes (Bea and Göbel 1999). **Autogenous processes** refer to formation of spontaneous order as a result of dynamics and complexity of systems. Such an order is the result of human actions but not human designs (Hayek 1967). In **autonomous processes**, all system members could influence the system order, which could in turn better adapt to system needs and environmental challenges and consequently become more efficient (Bea and Göbel 1999). Here autonomous cooperation is understood as autonomous processes with decentralized intelligence and decision-making. System elements will be given tasks, meta-structures and methods in a general way by external entities, which embody a certain degree of external control. However, the situational concretization of processes within the established framework will be left to the knowledge and capability of elements.

A second delimitation is necessary, which gives an ordered understanding of "Autonomous Cooperation" vs. "Self-organization" vs. Self-management". The three concepts all describe a system's ability of creating order with its own resources. Nevertheless there exist differences concerning the form and degree of such an ability. Therefore, a differentiation between these three concepts will be carried out here. **Self-management** is a broad concept, describing the fully autonomous development of a system, which means that the system can formulate its own objectives and plans as well as deciding its own organization forms and necessary resources (Manz and Sims 1980). As a component of management, **self-organization** depicts the way how a system arranges its own structure and processes through its own abilities (Probst 1992). **Autonomous cooperation** has a narrow meaning and refers to only the selection freedom of system members. Regarding the actual situations, system elements could choose among alternatives, which are principally predefined by external entities (i.e. management) (Bea and Göbel 1999).

1.4 Aims of the edited volume

In the preface, the major objectives of this anthology were mentioned:

- To collect **various understandings of self-organization**, which had a comprehensive and differentiable description of the basic ideas about the concept;
- To identify and compare **the scope and depth of autonomous cooperation and control** resulting from various understandings of self-organization, and to summarize the commonness and differences for the terminological purpose;
- To establish **a common conception of autonomous cooperation and control**, which stimulated the cooperation in the research through reflecting various perspectives from different disciplines;
- To develop **a concept system for autonomous cooperation and control** but without concretization, which allowed discipline-specific interpretations in the context of logistic systems.

Concretely, those overarching aims of the anthology set up its focus, which consists of tasks like:

- To define and characterize autonomous cooperation and control;
- To outline the history of autonomous cooperation and control;
- To model autonomous cooperation and control;
- To show the impacts and necessary changes for the management;
- To sketch concepts of autonomous cooperation and control methods;
- To present the use of ICT for autonomous cooperation and control;
- To give first examples of the implementation of autonomous cooperation and control.

1.5 Structure of the edited volume

To answer those questions lying behind the tasks described above, the starting point for this anthology was a now more than three years lasting research within the working group "Autonomous cooperation" of Collaborative Research Center 637 (CRC 637) "Autonomous Cooperating Logistic Processes – A Paradigm Shift and its limitations". As explained before, the overarching aim of CRC637 is to lay foundations for theory building concerning the concept of autonomous cooperation and control (including technologies and instruments) in logistics and to gain extensive knowledge about the involved causal relations so as to apply the concept in practice. In order to achieve these objectives, the research of the CRC 637 tries to identify rules of the paradigm of "autonomous cooperation and control" and to find the means to influence the degree of autonomous cooperation and control on all levels of logistic systems (decision level, information and communication level, and material flow level). The research expects that a higher degree of autonomous cooperation in logistic processes could be one approach to handling complexity and dynamics in logistic systems by increasing flexibility, which could further lead to positive emergence and robustness (i.e. improvement in process quality and achievement of logistical targets). Meanwhile, autonomous cooperation and control could also have negative effects on productivity, which might be attributed to the immanent redundancy in resources as well as structures and the delegation of decision power. Thus, CRC 637 is striving for the solution to the problem of finding the optimal degree of autonomous cooperation and control. Therefore, it was the aim of the working group to set up a common understanding of autonomous cooperation and control, which can be adapted to the individual research aims, contexts, and terminological frameworks of the single subprojects of the CRC 637.

In order to fulfil its objective, the working group "Autonomous Cooperation and Control" first tried to get an **overview of existing ideas about autonomous cooperation and control**. Subprojects each introduced their own understandings of autonomous cooperation and elaborated those characteristics they considered as constitutive. The commonness and differences of the understandings were then discussed.

Next all subprojects of the working group "Autonomous Cooperation and Control" worked out a **catalogue of criteria**, which were used to develop an overarching definition shared by the whole CRC637. Such a catalogue ensured that the conception process conformed to the academic quality criteria regarding definition formulation. Besides, this catalogue also included those criteria that ensure the connectivity between the common definition to be developed and the specific research requirements of the four individual disciplines working together in the CRC637 (production engineering, communication and electrical engineering, computer science and mathematics, economics and business administration). In addition, criteria in this catalogue allowed the global definition to be adapted to the research questions specific to the subprojects, to the application scenarios and to the theory conception for analysis within individual tasks.

Based on this catalogue of criteria and the existing ideas of autonomous cooperation in the subprojects, the subprojects first **redefined their individual understandings** according to those criteria. The new definitions specific to respective subprojects were then again compared so that an oriented and systematic canalization of various understandings could be achieved and the scope of constitutive characteristics could be narrowed down.

In this way the working group "Autonomous Cooperation and Control" deduced **a global definition of autonomous cooperation and control**. On the one hand, this definition reflected the essential understandings of individual subprojects through the procedure outlined above. On the other hand, it satisfied the main terminological interests (in a common understanding) as well as the rules for a transdisciplinary language, and requirements for theory development and practical application.

Next, the necessary **transformation and adaptation of the global definition** was carried out in individual subprojects to better satisfy the individual interests in research without undermining the whole terminological system and the agreed language rules. Consequently, in-depth ideas about autonomous cooperation could be obtained for specific problems, which complement a collectively developed as well as shared and consistent terminology of CRC637.

In order to get a profound understanding of autonomous cooperation and control it is necessary to distinguish between the three main layers referring to Ropohl management, information and communication and the material flow layer (Ropohl 1979). Therefore, the anthology is structured in three main categories.

The **second chapter** "Fundamental Basics and Concepts of Autonomous Control and Cooperation" following this introduction focus on the fundamental basics and the description of autonomous cooperation and control concepts. The historical development of autonomous cooperation and control as well as the main criteria are presented. Furthermore, the modelling problem of autonomous cooperation and control is addressed in several articles.

The **third chapter** "Autonomous Control Methods for the Managment, Information and Communication Layer" picks up the ICT developments and how the management processes have to be changed if autonomous co-operation and control is to be integrated in logistic processes. Besides the management view, also knowledge management and knowledge-based risk-management plays an important role and is addressed in this chapter.

The **fourth chapter** "Autonomous Control Methods and Examples for the Material Flow Layer" concentrates on the material flow layer where the developed autonomous cooperation and control methods need to be implemented and executed. Therefore, one enabler of autonomous cooperation and control – an evaluation system for autonomous logistic processes – is presented. Other articles describe scenarios, the implementation and first results of autonomous cooperation and control in practice or on the basis of simulation studies.

References

Bea FX, Göbel E (1999) Organisation: Theorie und Gestaltung, Stuttgart

Cheung KH, Misic J (2002) On virtual private networks security design issues, in: Akyildiz, I., Rudin, H. (Eds.), Computer Networks, Volume 38, Issue 2, 165-179

Chmielewicz K (1979) Forschungskonzeptionen der Wirtschaftswissenschaft, 2nd edn., Stuttgart

Finkenzeller K (2002) RFID-Handbuch – Grundlagen und praktische Anwendungen induktiver Funkanlagen, Transponder und kontaktloser Chipkarten. 3. Aufl., Carl Hanser Verlag, München

Flämig M (1998) Naturwissenschaftliche Weltbilder in Managementtheorien: Chaostheorie, Selbstorganisation, Autopoiesis, Frankfurt am Main

Foerster vH (1960) On Self-Organizing Systems and their Environment, in: Yovitis, M. C. / Cameron, S. (Ed): Self-Organizing Systems, London

Frese E (1993) Grundlagen der Organisation: Konzept - Prinzipien – Strukturen, 5th revised edn., Wiesbaden

Gebresenbet G, Ljungberg D (2001) Coordination and Route Optimization of Agricultural Goods Transport to Attenuate Environmental Impact. In: Journal of Agricultural Engineering Research. Academic Press, Volume 80, Issue 4, 329-342

Glansdorff P, Prigogine I (1971) Thermodynamic theory of structure, stability and fluctuations, London

Goldammer Ev (2002) Heterarchy and Hierarchy – Two Complementary Categories of Description, in: Vordenker

Haken H(1973) Synergetics: cooperative phenomena in multi-component systems, Stuttgart
Haken H, Graham R (1971) Synergetik - Die Lehre vom Zusammenwirken, in: Umschau in Wissenschaft und Technik 6, p. 191-195.
Hayek FA (1967) Studies in philosophy, politics and economics, London
Hill W, Fehlbaum R, Ulrich P(1994) Ziele, Instrumente und Bedingungen der Organisation sozialer Systeme, 5th edn., Bern
Hoffman L (1984) Grundlagen der Familientherapie, 2nd edn.
Kappler E (1992) Autonomie, in: Frese, E. (Ed): Handwörterbuch der Organisation, 3th edn., Stuttgart, p. 272-280.
Kirsch W(1992) Kommunikatives Handeln, Autopoiese, Rationalität: Sondierungen zu einer evolutionären Führungslehre, München
Laux H (1998) Entscheidungstheorie, 4th edn., Berlin
Luhmann N (1973) Zweckbegriff und Systemrationalität: Über die Funktion von Zwecken in sozialen Systemen, Frankfurt am Main
Manz C, Sims H (1980) Self-Management as a Substitute for Leadership: A Social Learning Theory Perspective, in: American Manager Review 5, p. 361-367.
Maturana HR, Varela F 1980) Autopoiesis and cognition: the realization of living, Reidel
Paslack R (1991) Urgeschichte der Selbstorganisation: zur Archäologie eines wissenschaftlichen Paradigmas, Braunschweig
Peitgen H, Richter PH (1986) The beauty of fractals: images of complex dynamical systems, Berlin
Prigogine I (1996) The End of Certainty: Time, Chaos, and the New Laws of Nature, New York
Probst GJB (1992) Organisation: Strukturen, Lenkungsinstrumente und Entwicklungsperspektiven, Landsberg/Lech
Probst GJB (1987) Selbst-Organisation: Ordnungsprozesse in sozialen Systemen aus ganzheitlicher Sicht, Berlin
Pohl RO, Lüders K (1930) Pohls Einführung in die Physik, Berlin/Heidelberg
Ropohl GJB (1979) Eine Systemtheorie der Technik – Grundlegung der Allgemeinen Theorie. Carl Hanser Verlag. München
Scholz-Reiter B, Windt K, Freitag M (2004) Autonomous Logistic Processes – New Demands and First Approaches. In: Monostori, L (eds.): Proceedings of 37th CIRP International Seminar on Manufacturing Systems. Hungarian Academy of Science, Budapest (Hungary)
Schüppenhauer A (1998) Multioptionales Konsumentenverhalten und Marketing: Erklärungen und Empfehlungen auf Basis der Autopoiesetheorie, Wiesbaden
Staehle WH (1999) Management: eine verhaltenswissenschaftliche Perspektive, 8th edn., München
Teubner G, Willke H (1984) Kontext und Autonomie: gesellschaftliche Selbststeuerung durch reflexives Recht, Florence
Varela F J(1979) Principles of biological autonomy, New York
Westkämper E, Jendoubi L (2003) Smart Factories – Manufacturing Environments and Systems of the Future. In: Bley, H (Eds.): Proceedings of the 36th CIRP International Seminar on Manufacturing Systems. pp 13-16

Wiendahl HP, ElMaraghy HA, Nyhuis P, Zäh M, Wiendahl HH, Duffie N, Kolakowski M (2007) Changeable Manufacturing: Classification, Design, Operation. In: Annals of the CIRPVol. 56/2/2007 Revision Jan 15, 2007

Wiesenthal H (2006) Gesellschaftssteuerung und gesellschaftliche Selbststeuerung. VS Verlag für Sozialwissenschaften. Wiesbaden

Windt K (2006) Selbststeuerung intelligenter Objekte in der Logistik. In Hütt, M, Vec, M, Freund, A (Eds): Selbstorganisation: Ein Denksystem für Natur und Gesellschaft. Böhlau Verlag, Köln, Weimar, Wien, 2006, 271-314

Zahariadis Th (2003) Evolution of the Wireless PAN and LAN standards. In: Schumny, H (Eds.): Computer Standards and Interfaces. Volume 26, Issue 3, 175-185

ZigBee (2006) Alliance, Spezifikation Dezember 2006, verfügbar unter: http://www.zigbee.org. Letzter Abruf am 16.02.2007

2 Fundamental Basics and Concepts of Autonomous Control and Cooperation

2.1 Perspectives on Initial Ideas and Conceptual Components of Autonomous Cooperation and Control

Katja Windt[1], Michael Hülsmann[2]

[1] Department of Planning and Control of Production Systems, BIBA University of Bremen, Germany

[2] Management of Sustainable System Development, Institute for Strategic Competence-Management, Faculty of Business Studies and Economics, University of Bremen, Germany

In order to enable the implementation of self-organisation ideas for logistics – concretised as control and organisation principles –, one has to understand the fundamental basics and characteristics of autonomous cooperation and control as well as its foundations. In this respect, the basic underlying idea is the concept of self-organisation like shown above. It is an interdisciplinary concept that has been developing for more than 35 years under labels such as self-organisation, autopoiesis, dissipative structures as well as emergency and complexity theory. The core of the self-organisation concept is the formation and development of order in complex dynamic systems (Paslack 1991). In natural sciences, important exponents are Prigogine (Glansdorff and Prigogine 1971) in chemistry (theory of dissipative structures), Peitgen and Richter (Peitgen and Richter 1986) in mathematics (chaos theory), Haken (Haken and Graham 1971,Haken 1973) and Foerster (Foerster 1960) in physics (synergetics and cybernetics), and Maturana and Varela(Maturana 1973) in biology (autopoiesis).

Those ideas still exert a great influence on other disciplines working on questions of self-order creation. The last concept "autopoiesis" is for example applied to other fields such as sociology Luhmann's system theory (Luhmann 1973).

Consequently, the idea of implementing the concept of autonomous cooperation and control into the organisation of supply chains and supply networks sees double interdisciplinarity: On the one hand, the fundamental ideas of self-organisation – which is the principle lying behind autonomous cooperation and control – come from sources of various disciplines which could be intertwined; on the other hand, these different interdisciplinary perspectives on its application could lead to different or even diverged interpretations. Therefore the first general task of a scientific process, namely the terminological task (Hill et al.1994), is more important for the research on autonomous cooperation than for other research fields where the objects and approaches are mono-disciplinary. That is why representatives from production engineering, communication technology, electrical engineering, computer science and mathematics, as well as from business studies and management science were invited to contribute to this chapter "Fundamental Basics and Concepts of Autonomous Cooperation and Control" and to explain their individual perspectives on initial ideas and conceptual components of this specific organisational principle for logistics. All the articles in this chapter are intended to contribute towards an overarching conception of the application of autonomous cooperation from an interdisciplinary perspective and to identifying the basics for managing, measuring, and modelling autonomous cooperating logistic processes. This chapter would like to establish a differentiated and multi-usable overview to enable an interdisciplinary understanding of what autonomous cooperation and control is all about. This furnishes the terminological basis for all further research on models, methods, and applications.

The first article **"Prologue to Autonomous Cooperation — the Idea of Self-Organisation as its Basic Concepts"** – written by Michael Hülsmann, Christine Wycisk, Robin Agarwal, and Jörn Grapp – deals with self-organisation, the origin of autonomous cooperation, by exploring different understandings of self-organisation and common characteristics underlying these concepts. Autonomous cooperation describes processes of decentralized decision-making in heterarchical structures. The implementation of autonomous cooperation aims at a flexible self-organizing system structure that is able to cope with dynamics and complexity while maintaining a stable status. The basic idea of the concept of autonomous cooperation derives from concepts of self-organisation, which analyze the emergence of ordered and robust structures in complex systems in general. For transferring the idea of self-organizing systems into the concept of autonomous cooperation, a first step would be to understand the roots and principles of self-organisation. In this chapter, the core aspects of selected concepts of self-organisation are presented with a brief description of each. Next, to give a clear picture of the idea of self-organisation, the characteristics which form the basis of self-organizing systems contained in the selected concepts are extracted and juxtaposed by means of the general criteria of system structure, system behaviour and system abilities.

In the second contribution **"Historical Development of the Idea of Self-Organisation in Information and Communication Technology"**, Markus Becker, Koojana Kuladinithi, Andreas Timm-Giel, and Carmelita Görg summarize how the idea of self-organisation has been applied in ad hoc networks (including mesh and sensor networks), peer-to-peer networks, autonomic computing and autonomic communication. The constituting features of autonomous control (non-centralized design and operation, heterarchy, interaction, autonomy, decision process) have been used and enhanced since the beginnings of Information and Communication Technology. In this chapter, proactive and reactive routings, autonomic address assignment and mobile agents in ad hoc networks are described. Then specific applications of peer-to-peer networks are introduced. Next, examples of autonomic computing with its "self-" principles and examples of autonomic communication as well as related issues concerning self-organisation (i.e. controllability, reliability and security) are presented.

In the article **"Business Process Modelling of Autonomously Controlled Production Systems"** written by Felix Böse and Katja Windt, a specification of the main criteria of autonomous cooperation and control is introduced. Based on this, the ARIS concept (Architecture of Integrated Information Systems) as an integrated method for the modelling of processes and information systems is analysed regarding its suitability for describing autonomous control in production systems. Furthermore, changes in order processing are exemplarily illustrated in several views of a business process model using the ARIS concept.

The next chapter **"Catalogue of Criteria for Autonomous Control in Logistics"**, contributed by Felix Böse and Katja Windt, tries to explain the concept of autonomous control and describes its main criteria in contrast to conventional controlling methods in logistics systems. Over the years there has been an increase in the complexity of production and logistics systems regarding organisational, time-related and systemic aspects. As a result, it is often impossible to make all necessary information available to a central entity in real time and to perform appropriate measures of control in terms of a defined target system. Therefore, demands were placed on new control methods. Autonomous control seems to be an appropriate alternative, whose idea is to develop decentralised and heterarchical planning and controlling methods. In this chapter, a definition of autonomous control is introduced. The constituent characteristics of this definition are considered in a developed catalogue of criteria in the form of an operationalized morphological characteristic schema in order to describe autonomous logistic processes and emphasize how conventionally managed and autonomous logistic processes differ. The criteria and their properties are explained in a concrete way by investigating a production logistics scenario of a job shop manufacturing system.

Lars Arndt and Georg Müller-Christ deal with **"Strategic Decisions for Autonomous Logistics Systems"** and intend to explain decision issues involved in the application of autonomous cooperation. Autonomous cooperation in logistics is based on the capability of logistics objects to decide and coordinate among themselves. Though the role of new technologies, especially multi-agent technology in enabling local self-coordination has been addressed by several authors, the underlying decision problem remains unclear. Therefore, this chapter elaborates the strategic nature of decision in autonomous cooperating logistics processes. More specifically, it describes autonomous cooperation in logistics as a particular form of delegation of decision making, attributes the strategic character of this delegation process to the necessity for organisations to open their boundaries, and outlines a concept of boundary management in order to foster and regulate the boundary opening and thus to provide the appropriate organisational context for the decision to implement autonomous cooperation.

The following article, which describes **"Autonomous Units: Basic Concepts and Semantic Foundation"** – written by Karsten Hölscher, Renate Klempien-Hinrichs, Peter Knirsch, Hans-Jörg Kreowski, and Sabine Kuske – proposes the concept of autonomous units for modelling logistics objects acting autonomously, while interacting with each other for the purpose of accomplishing certain tasks. The guiding principle of autonomous units is the possibility to integrate autonomous control into the model of processes. This provides a framework for a semantically sound investigation and comparison of different mechanisms of autonomous control. Concretely speaking, this chapter describes algorithmic and particularly logistic processes in a general and uniform way, portrays the range of applications and their according methods, introduce the rule-based approach and elaborate autonomous units on different levels.

Bernd Scholz-Reiter, Fabian Wirth, Michael Freitag, Sergey Dashkovskiy, Thomas Jagalski, Christoph de Beer, and Björn Rüffer discuss in their contribution **"Mathematical Models of Autonomous Logistics Processes"** fundamental concepts of autonomy within a logistic network and mathematical tools which can be used to model this property. Autonomous control in a logistic network describes a decentralised coordination of intelligent logistic objects (parts, machines etc.) and allocation of jobs to machines by intelligent parts themselves. To develop and analyze such autonomous control strategies, dynamic models are required. This chapter describes and compares several possible models for autonomous logistic processes (discrete models and fluid approximations, partial differential equations and ordinary differential equations) and discusses how autonomous control enters these models and what its effects on the dynamics and stability of the processes are. By means of an example, this chapter further presents the advantages of autonomous control and points out the related stability problem.

In the chapter **"Autonomous Decision Model Adaptation and the Vehicle Routing Problem with Time Windows and Uncertain Demand"** Jörn Schönberger and Herbert Kopfer investigate generic procedures and rules for an automatic feedback controlled adaptation of decision models for a variant of the well-known Vehicle Routing Problem with Time Windows. This task is driven by the realization that static decision models fail to work at times of changes in the real world. This chapter presents the considered decision problem in more detail, introduces the algorithmic framework for autonomous adaptation of the decision model, and proves the framework's general applicability within numerical simulation experiments.

References

Foerster v H (1960) On Self-Organizing Systems and their Environment. In: Yovits M C, Cameron S (eds) Self Organizing Systems. London

Glansdorff P, Prigogine I, (1971) Thermodynamic theory of structure, stability and fluctuations. Wiley, New York

Haken H, Graham R (1971) Synergetik - Die Lehre vom Zusammenwirken. Umschau in Wissenschaft und Technik 6: 191-195

Hill W, Fehlbaum R, Ulrich P (1994): Organisationslehre 1: Ziele, Instrumente und Bedingungen der Organisation sozialer Systeme, 5. Auflage, Bern et al

Luhmann N (1973): Zweckbegriff und Systemrationalität. Reihe: Suhrkamp-Taschenbuch Wissenschaft. Band 12. Frankfurt am Main

Maturana HR, Varela FJ (1973) Autopoiesis and Cognition: the Realization of the Living. Reidel, Dordecht

Paslack R (1991) Urgeschichte der Selbstorganisation: zur Archäologie eines wissenschaftlichen Paradigmas. Vieweg, Braunschweig

Peitgen HO, Richter RH (1986): The Beauty of Fractals: Images on Complex Dynamical Systems. Berlin

Teubner G, Willke H (1984) Kontext und Autonomie. Gesellschaftliche Selbststeuerung durch reflexives Recht. Zeitschrift für Rechtssoziologie 6: 4-35

2.2 Prologue to Autonomous Cooperation – the Idea of Self-Organisation as its Basic Concepts

Michael Hülsmann, Christine Wycisk, Robin Agarwal, Jörn Grapp

Management of Sustainable System Development, Institute for Strategic Competence-Management, Faculty of Business Studies and Economics, University of Bremen, Germany

2.2.1 Introduction

Autonomous cooperation describes processes of decentralized decision-making in heterarchical structures. The implementation of autonomous cooperation aims at a flexible self-organizing system structure that is able to cope with dynamics and complexity while maintaining a stable status (Hülsmann and Windt 2005). The basic idea of the concept of autonomous cooperation derives from concepts of self-organisation, which analyze the emergence of ordered and robust structures in complex systems in general (Paslack 1991). The idea of self-organisation has its historical roots in different academic fields such as Physics, Biology and Chemistry and dates back to at least 500 BC of the pre-Socratic Heraclites and Aristotle who identified self-organized processes in natural phenomena (Paslack and Knost 1990; Paslack 1991). An increasing number of literature written by different scientists from different disciplines concern explicitly with self-organizing systems can be found from the 1970's, as for example in Cybernetics von Foerster (1960), in Chemistry Prigogine and Glansdorff (1971), in Physics Haken (1973) and in Biology Maturana and Varela (1980).

It does not seem feasible to apply a concept of natural sciences (the idea of self-organizing systems) cent per cent into social sciences, since there are essential differences between those systems in nature, constitution. There may exist attempts of its application to business, for instance to logistics in terms of autonomous cooperation which is believed to incorporate the self-organizing principles (Hülsmann and Windt 2005). Transferring the idea of self-organizing systems into the concept of autonomous cooperation a first step would be to understand the roots and principles of self-organisation.

The aim of this paper is to unlock via its primal foundation concepts the understanding of self-organisation and its different common characteristics underlying these concepts. This shall serve as a platform to get introduced into the working principles of self-organizing systems. These concepts are seen as the foundation for explaining the underlying principles as to how complex systems autonomously create ordered structures. It may be presumed that these concepts shall set the trajectory and common ground for understanding processes of autonomous order creation, which in turn forms the basis for autonomous cooperation.

Therefore, the core aspects of selected concepts of self-organisation are presented with a brief description of each in the subsequent section of this paper. Later to give a clear picture of the idea of self-organisation, the characteristics which form the basis of self-organizing systems out of the selected concepts shall be extracted and juxtaposed by means of the general criteria of system structure, system behavior and system abilities. Finally, a conclusion is drawn about the general understanding of the concept of self-organisation with emphasis on its potential application and further areas of research.

2.2.2 Concepts of self-organisation

In this section, the so called "primal concepts" of self-organisation out of which the main ideas of autonomous order creation have emerged are introduced. (Paslack and Knost 1990) mention the approaches Synergetic (Haken 1973), Dissipative Structures (Prigogine 1969), Autopoiesis (Maturana and Varela 1973), Cybernetics (von Foerster 1960), Ecosystems (e.g. Bick 1973) and Chaos Theory (e.g. Mandelbrot 1977 and Lorenz 1963) among those primal self-organisation concepts (see also Grapp et al. 2005).

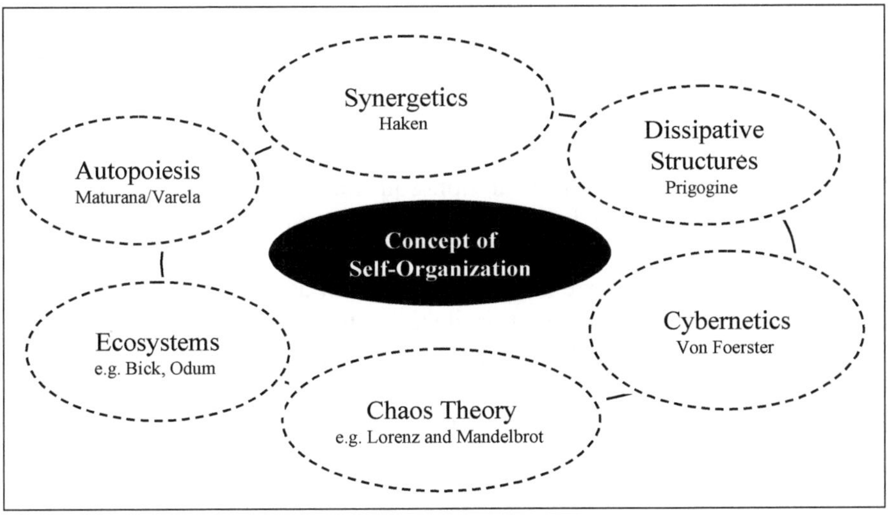

Fig. 2.1 Primal concepts of the idea of self-organisation

Synergetics

Self-organisation of systems has been the subject of central discussions of Synergetics in several research disciplines since its inception. The concept of Synergetics was invented by Haken in 1969 who for the first time saw this subarea of Physics as a new field of interdisciplinary research (Ulrich 1984). Though it originated from Physics (e.g. lasers, fluid instabilities, plasmas) it found applications not only in the natural sciences, such as Chemistry (e.g. chemical reactions resulting in pattern formation, including flames), Biology (e.g. morphogenesis, evolution theory), Meteorology, Neurobiology, Computer Sciences (e.g. synergetic computer), Movement Science, but also in the Humanities such as Sociology (e.g. city growth), Psychology and Psychiatry (including Gestalt Psychology). Several other authors who contributed to this field are Buckminster (1975), Ulrich (1984), Probst (1984), Kriz (1990), Tschacher (1992), Tschacher, Schiepek and Brunner (1992), Stadler and Kruse (1995), Dauwalder and Tschacher (1999), Malik (2000).

According to Buckminster, Synergetics can be applied to all aspects of human endeavor because it is capable of providing a method, a design and a philosophy for problem solving. It involves the integration of geometry and philosophy and accounts for both physical and metaphysical understanding of several methods and processes (Buckminster 1975).

Haken defines the core aspect of Synergetics as the cooperation of individual parts of a complex system that interact with each other and thereby autonomously produce macroscopic spatial, temporal or functional structures. The concept attempts to explain that these structures develop spontaneously in nature by virtue of self-organisation. In physical systems, Synergetics studies the nonlinear non-equilibrium process, where – after energy is being pumped into a system – macroscopic structures emerge from disorder in behavior of large number of microscopic particles. The functioning of a laser, for example, can be seen as such a synergetic process. A laser is a light source that produces light with properties, which vary from conventional lamps. For instance in the case of a gas discharge lamp, individual atoms are excited by means of electric current. Each excited atom then emits a light wave track making their transitions entirely independent from one another, i.e. the light emission is entirely irregular. On the contrary, in case of a laser a transformation of energy occurs where the random motion of electrons of electrical current is transformed into highly ordered energy of the light field, i.e. a beam of coherent light is emitted out of the chaotic movement of particles exhibiting harmony among them (Haken 1978).

The coherent process in Synergetics as described above exhibits a process of self-organisation. Seen from a thermodynamic point of view it seems to contradict with the second law of thermodynamics, which states that no system can convert energy from one form to another useful form with hundred percent efficiency and all systems tend towards disorder (Kuhn 1978). However this contradiction gets resolved by the fact that the laser is an open system through which permanently energy is pumped, while the thermodynamics second law deals with closed systems. As chaos turns into order, Synergetics makes use of probabilities (to describe uncertainty) and information (to describe approximation) and therefore deals with stochastic (chance) and deterministic (necessity) processes. This transition from disorder to order is found to be related with the concept of entropy (degree of disorder). But Synergetics has replaced the entropy principle by a dynamic principle which refers to open systems through which energy (and matter) can be pumped into the system (Haken 1981).

In this open system, competition sets in between different forms of collective modes. Those modes which win the competition slave the whole system (known as "slaving principle") and thus determine the macroscopic order (known as "order parameters"). Here, Haken (1981) states that neither the elements of the system nor the order parameters determine the state of order but rather that order parameters and elements determine each other. He further explains that despite the different nature of individual disciplines, the corresponding order parameters obey the same equations which describe logical processes. These logical processes can take place in different substrates or in different systems. More high ordered states can arise in different disciplines due to change in external conditions. In this way more and more complex structures arise in a self-organized way i.e. evolution of new structures internally and not from external sources (Haken 1981). These complex non-equilibrium systems are studied by Synergetics and self-organisation theory (Tschacher et al. 2003).

According to Fuchs (2002), Haken's work infers self-organisation differently as Haken has tried to transfer the synergetic principle of slaving directly from Physics to Sociology. Moreover, Fuchs argues that the term 'slaving' does not seem to be proper wording in social contexts and he views slaving as a terminus technicus which has no ethical or other implication.

Cybernetics

The term "Cybernetics" is derived from the Greek word kybernetes which means steersman, governor, or pilot (Drosdowski 1990). The Oxford Dictionary defines 'Cybernetics' as "the science of communications and control in machines (e.g. computers) and living things (e.g. by the nervous system)" (Oxford Dictionary 2002). The term was first coined in 1948 by Wiener to address the study of "teleological mechanisms" (Wiener 1948). Cybernetics is an interdisciplinary field being studied in Philosophy, Biology and Medical Sciences, Engineering as well as in Business Studies. Authors who have made major contributions are McCulloch and Pitts (1943), Wiener (1948), von Foerster (1960) and others such as Ashby (1970), Pask (1979), Probst (1984), Walter (1996), Heylighen and Joslyn (2001).

Speaking in general terms, the influence of Cybernetics may be seen in several contemporary disciplines such as computer science, information theory, control theory, automata theory, artificial neural networks, cognitive science, dynamical systems, artificial intelligence and artificial life. The main feature of Cybernetics which differentiates between Cybernetics on the one hand and information theory and control theory is its emphasis on communication and control. Not only in artificial or engineered systems but also in evolved and natural systems, which behave by setting their own goals rather than being controlled by their creators (Heylighen a. Joslyn 2001). Cybernetics has extended its application in various concepts like self-organisation (von Foerster 1960; Ashby 1970), computer architectures, cellular automata, and game theory (Aspray 1990), autonomous robots (Braitenberg 1984), and artificial neural networks (McCulloch and Pitts 1943).

Referring to Probst, Cybernetics takes into account the research on the mechanisms of control in its broadest sense. Using cybernetic principles, it might be possible to help managers in finding other and perhaps more adequate solutions for design, control and development of purposeful social systems. This may be achieved by deriving rules of action or confirming or rejecting the prevailing and accepted managerial rules of action (Probst 1984).

The theory of observing design or discovery in general and the science of communication were seen as 'Cybernetics of the first order' by von Foerster (1979). Through considering the whole domain as a system, he found necessary requirements and functions for observing this system. He termed this understanding as 'Cybernetics of the second order' or 'Cybernetics of observing systems'. Second-order Cybernetics explores the construction of models of systems. It studies Cybernetics with an increased awareness that the observers are a part of the system as well, i.e. the examiner (the observer) and the examination are part of the system being observed. Von Foerster also referred to this as 'Cybernetics of Cybernetics' (von Foerster 1979). The proceedings of the Macy Conference edited by von Foerster found that Cybernetics manages itself based on the notion of circular causality (von Foerster 1960). Following this, two generalizations were drawn by von Foerster. First, recursion that is implicit in Cybernetics of Cybernetics and tends to stabilize at a particular value (or a self-function generating a self-value), which he thought was a manifestation of an object, and therefore presents a model for the appearance of stability. Second, since each one of us is our own observer, every individual has its own unique way of understanding and observing things, which might vary from observer to observer (von Foerster 1979). This is in conjunction with Pask's conversation theory, which provides common means of communication in case understanding of individuals vary (Pask 1979).

Each dynamical system that belongs to attractors (which may have any type of shape or dimension within the system) finally results in having one of the attractors, thereby losing its independence to visit any other system's state space. This is what Ashby (1970) referred to as the principle of self-organisation. He also pointed out that if the system is composed of several subsystems, then the constraint generated by self-organisation implies that the subsystems have either become mutually dependent or mutually adapted. For example, in case of magnetization, initially the assembly of magnetic spins point in random directions (maximum entropy), but later end up being aligned in the same direction (minimum entropy, or mutual adaptation) (Ashby 1962). Self-organisation according to von Foerster can be enhanced by stochastic perturbations ('noise') of the system's state, in which the descent of the system gains momentum and forces shallow attractors to exit the system. This is referred to as order from noise principle (von Foerster 1960).

Hence it can be seen that early work on Cybernetics focused on defining and applying principles through which systems may be controlled. However recent work has endeavored to understand how systems organize and control themselves. Cybernetics – though not developed as an individual discipline yet – has developed as an emerging concept among varied processes involving people as active organizers, sharing communicators, and as autonomous, responsible individuals (Umpleby 1999).

Dissipative structures

The term 'Dissipative Structures' was coined by the physicist Prigogine in order to explain the phenomena of non-equilibrium thermodynamics (Prigogine 1969). The application of the concept can be found not only in Physics and Chemistry but also in Biology and Sociology. Authors who work in cooperation with Prigogine on this subject are Glansdorff (Glansdorff and Prigogine 1971), Balescu (1975), Nicolis (Nicolis and Prigogine 1977), Lefever (1978), Stengers (Prigogine and Stengers 1984), Goldbeter (1997), and Herschkowitz (2001).

Prigogine was awarded the Nobel Prize for his contribution to non-equilibrium thermodynamics, which is seen as a source of order in a system, and particularly for the theory of dissipative structures, which results from dynamic states of matter caused by irreversible processes (Prigogine 1980). Prigogine describes the world as evolving from order to disorder, and considers thermodynamics as the science of 'becoming' from 'being' (Prigogine 1980). He has shown that the behavior of matter under non-equilibrium conditions can be radically different from its behavior at, or near equilibrium condition. This difference introduces different alternatives such as self-organisation and complex dynamics (Thore 1995).

Near equilibrium, the description of the temporal evolution of a system can be expressed by linear equations. Far from equilibrium one deals with nonlinear equations, which may result in bifurcations and the spontaneous appearance and evolution of organized states of matter of the so called Dissipative Structures. As an example of a dissipative structure consider a pan of liquid heated from below. When the temperature is low, heat passes through the liquid by conduction. As the heating is intensified, regular convection cells appear spontaneously. The liquid boils. Energy is transferred from thermal motion to convection currents. The boiling dissipative structure is radically different from the equilibrium structure of the liquid. However, the order can be maintained in this boiling dissipative structure far from equilibrium conditions only through a sufficient flow of energy. According to Prigogine, the world can be seen as subject to self-organisation and evolution. He views energy dissipation as the driving force of evolution. Despite the increase in organisation and complexity of living systems, the biological evolution has accelerated over a period of time. Each new step increasing the functional organisation has in itself the germs for further evolution. For instance, mathematical relations describing the evolution of thermodynamical systems can be adapted to understand the notion of survival of the fittest in predator and preys. On the one hand, the prey evolves as to exploit available resources more efficiently and tries to prevent itself from being caught by the predator. On the other hand, the predator evolves as to increase the frequency of capturing the prey and to decrease its death rate. The ratio of the biomass of predator to prey can be seen as gradually increasing with evolution (Prigogine 1969).

According to the second law of thermodynamics the world can be seen as evolving from order to disorder while biological evolution is about the complex emerging from the simple i.e. order arising from disorder (Scaruffi 2003). Though both views being contradictory show that irreversible processes and non-equilibrium states are an integral part of the real world. Nicolis and Prigogine stress the need for a system composed of independent units that interact with each other, in which flow of energy drives the system away from equilibrium and nonlinearity. This non-equilibrium and nonlinearity excels the spontaneous development of self-organizing systems of ordered structure and behavior in open systems regardless of the general increase in entropy by ejecting matter and energy in the environment (Nicolis a. Prigogine 1977).

Autopoiesis

The origin of the term "Autopoiesis" lies in its Greek meaning, wherein 'Auto' means self and 'poiesis' means creation or production (Drosdowski 1989). Put together, it means self-creation or self-production i.e. a process where an organisation produces itself (Maturana a. Pörksen 2002). The biologists Varela and Maturana introduced the concept of Autopoiesis in 1973, which is concerned with the question "What is life?" or more precisely what differentiates living systems from non-living systems

(Maturana a. Varela 1973). They explained Autopoiesis as follows: "An autopoietic machine is a machine organized (defined as a unity) as a network of processes of production (transformation and destruction) of components which: (i) through their interactions and transformations continuously regenerate and realize the network of processes (relations) that produced them; and (ii) constitute it (the machine) as a concrete unity in space in which they (the components) exist by specifying the topological domain of its realization as such a network." (Maturana a. Varela 1973). The main objective of Maturana and Varela is to explain the totality of living systems through an entire conceptual theory (Maturana a. Pörksen 2002). This concept has diffused into several other disciplines of study like Psychology (Walter 1996), Law (Teubner 1995; Teubner a. Willke 1984), Politics (Beyerle 1994) and social sciences (Luhmann 1984). Several other authors who have made contribution to the study of Autopoiesis are Uribe (Varela, Maturana a. Uribe 1974), Goguen (Goguen a. Varela 1979), Kauffman (Kauffman a. Varela 1980), Winograd and Flores (1986), Dyke (1988), Mingers (1989), Luisi (Luisi a. Varela 1989), Capra (1996).

Maturana and Varela examined Autopoiesis or self-production as a key to understand biological phenomena, which express that the mechanism of self-production explains both the diversity and the uniqueness of living systems. Autopoiesis endows living systems with the property of being autonomous. A typical autopoietic system is a biological cell. For example, the eukaryotic cell, which is made of various biochemical components like proteins and nucleic acids, is organized into bounded structures such as the cell nucleus, a cell membrane and cytoskeleton. On the basis of external flow of molecules and energy these structures produce components which in turn continue to retain the organized bounded structure. Hence, it can be seen that the concept of Autopoiesis lays emphasis on reproduction, evolution, and cognitive aspects (Maturana and Varela 1980).

The process of Autopoiesis explains the dynamics of living systems. Dyke refers to it as the dynamics of non-equilibrium thermodynamic system, or organized states what may also be understood as dissipative structures, which remain stable despite the continuous flow of matter and energy through them (Dyke 1988).

Chaos theory

Chaos and complexity can be represented by a mathematical model of phenomena of emergence of order out of chaos. Lorenz was the one who – while making experiments for weather predictions – came up with a theory which is well known as Chaos Theory. Lorenz found that even small and minor changes in initial stages can lead to a severe change in the long term behavior of a system (Lorenz 1963). Poincaré advocated this theory as well much earlier as Lorenz's work (Poincaré 1890). This behavior of changes may be seen as masquerading with the flapping of the wings of a butterfly, also known as Butterfly Effect. This phenomenon may demonstrate the Chaos Theory as it has high sensitive dependence on initial conditions. For example, two variables in flipping of a coin may be seen as sensitive dependence on initial conditions. First, how high the coin flips, and second, when the coin will hit the ground (Lorenz 1963). Apart from Poincaré and Lorenz, Chaos Theory has been worked upon by other scholars. They are for example Birkhoff (1923), Cartwright (1965), Prigogine (1969), May (1976), Derrida (1976), Mandelbrot (1977), Gleick (1987), Littlewood (1988), Kolmogorov (1991), Ruelle (1991), Binnig and Feigenbaum (1995), Smale and Hirsch (2004).

The phenomenon of emergence shows how structure arises from the interaction of many independent units. In physical and mathematical terms, it can be described as nonlinear equations out of which unpredictable solutions emerge. Based on sensitivity to initial conditions as discussed above, every system follows its laws of motion and traces some trajectory in phase space. 'Phase Space' is the space in which all possible states of a system are represented, with each possible state of the system corresponding to one unique point in the phase space. The different shapes that chaotic systems produce in this phase space are known as "strange attractors" (Lorenz 1963). These strange attractors can occur in both discrete as well as in continuous dynamical systems. An example of continuous dynamical systems could be the equations used by Lorenz to make weather predictions, while an example for discrete dynamical systems could be the Hénon Map (Dickau 1992).

Chaos Theory can be said to be an interdisciplinary field of research. The application of this theory could be seen in ecology and biological population predictions. The changes in growth rates make it even more difficult to make such predictions. May (1976) found out that after a certain point in growth rate it becomes impossible to forecast the growth behavior using equations. However, with a closer look some order could be traced in form of white strips on the graph, wherein the equation passed through bifurcations before returning to chaos. It can be interpreted that the graph has an exact replica of itself within. This exhibits self-similarity (May 1976). Mandelbrot studied this self similarity by taking into account 100 years cotton price fluctuations. On examining the data he noticed the following fact: each particular price change was random and unpredictable. But the sequence of changes was independent on scale, where curves for daily price changes and monthly price changes matched perfectly (Mandelbrot 1977). These findings reflect a common thing which is self-organisation i.e. how interaction among independent parts produces structures.

Hypercycles

Eigen, a German biophysicist and chemist, won the Nobel Prize in 1967 for his discovery that very short pulses of energy can induce extremely fast chemical reactions. Together with Schuster he came up with the model of "Hypercycles" (Eigen a. Schuster 1977). Hypercycles can be understood as self-replicating entities that integrate several autocatalytic elements into an organized unit by helping each other in a cyclic way. The main contributions to this concept were given by Eigen and Schuster (1979), but some other authors like Kuhn (1978), Smith (1979), Winkler (Eigen et al. 1981), Hofbauer and Sigmund (1988), Mallet-Paret (1993), Vespalcova, Holden and Brindley (1995) also contributed to this field of research. Theoretical and practical applications of hypercycles may be found in Biology, Chemistry, as well as in Physics, for example on hypercircuits in hypergraphs, molecular Biology, and in cellular automata.

Hypercycles are a network of cyclic reactions i.e. cyclic linkage of chemical reactions. This network gets formed with the help of combination of catalytic reactions. It stays in equilibrium when there is an adequate flow of energy and may contain closed loops known as catalytic cycles. A higher flow of energy drives the system far away from equilibrium, thereby influencing the combination of catalytic cycles to form closed loops of higher order, known as hypercycles. The production of enzymes within these hypercycles acts as a catalyst for its subsequent cycle in the loop turning each link in the loop into catalytic cycle of its own. Life is the product of a hierarchy of hypercycles in which basic catalytic cycles may get organized into an autocatalytic cycle i.e. a cycle which is capable of self-reproducing. A set of autocatalytic cycles in turn may get organized in a catalytic hypercycle. This catalytic hypercycle represents the basics of life (Eigen and Schuster 1979).

Eigen views hypercycles as a self-reproducing hypothetical stage of macromolecular evolution, which could follow quasispecies. Each specie acts as a catalyst for the replication of next either directly (ribozymes) or via intermediary enzymes (Hofbauer a. Sigmund 1988). The dual process of unity (due to the use of a universal genetic code) and diversity (due to the trial and error approach of natural selection) in evolution started even before the existence of life. Evolution of species may be seen as a precedent in parallel to process of molecular evolution. The difference between hypercycles and living systems may be seen in a way that hypercycles define no boundaries (boundary is understood as a container where chemical reaction takes place), while living organisms have a boundary as part of the living system, for example skin (Scaruffi 2003). In short it can be said that given a set of self-reproducing entities, which nourishes itself through common and limited resources like energy and material supply, natural selection is inevitable (Eigen 1971).

Ecosystems

The term "Ecology" was coined by the German zoologist Haeckel (1875). It has its origin in the Greek word oikos, which means "household" (Drosdowski 1990). Haeckel defines ecology as the science of relations between organisms and their environment. The concept of Ecosystems makes it possible to preserve, conserve, or protect both biotic and abiotic existing natural resources (Innis 1979).

Odum places energy as the central focus of his attention. He considers organisms and their physical environment as a single integral system and stresses that the flow of energy and nutrient cycling are rather more important than the entities that perform the function (Odum 1999). The fundamental goal of ecology, however, may be seen as identifying mechanisms that generate pattern. The spatial attributes of habitat, and individuals occupying habitat greatly influences the dynamics of biological systems, and thereby influences patterns in abundance, distribution, behavior, functioning, and evolution of organisms (Johnson 1997). The main authors contributing to the idea of ecosystems are Haeckel (1875), Bick (1973), and others like May (1976), Boerlijst and Hogeweg (1991), Camazine (1991), Nowak (1992), Karsai and Penzes (1993), Odum (1999).

A different approach to ecosystems is to study the dynamics of systems in which the spatial factor of interacting individuals or sub populations matter, wherein self-organisation which refers to the spontaneous emergence of global structure comes into play. The individuals or beings in the system are greatly influenced by their local environment. This biological phenomenon is as diverse as evolution of pre-biotic self-replicating molecules (Boerlijst a. Hogeweg 1991), evolution of cooperative behavior (Nowak and May 1992, 1993), co existence in fungal communities (Halley et al. 1994), and organisation in social insects (Camazine 1991; Karsai a. Penzes 1993). These models are equitable of the fact that spatial factors of individuals are crucial to the dynamics of system in terms of density, frequency, and population size. They affect the process which in turn affects the behavior of individuals. Hence, there can be seen a feedback between self-organizing behavior, system dynamics, and evolution of individuals within the system (Solé a. Bascompte 2006).

2.2.3 Characteristics of self-organizing systems

At this point it may be seen that it is the organisation of systems which plays a major role in the patterns of interaction and overall behavior, structure and abilities. For example, if all organs of a living organism are put together, a body cannot be expected to become alive. A body must necessarily self-organize in order to function, sense, grow, develop, react or respond (Mishra 1994). Hence, it can be said that the importance of self-organizing systems focuses on the relationships of their components and not on the components itself. Interaction among the components of systems may be seen as a necessary condition for setting a path for its future courses of action.

Having introduced the primal concepts of self-organisation, the potent factors of self-organizing may be seen in the principles and conditions that govern those systems. In order to outline the major principles and conditions of self-organisation, the characteristics forming the base of self-organizing systems with reference to the selected foundation concepts shall be discussed below. Therefore, criteria like system structure, system behavior and system abilities shall be used. In using those criteria, from a system theoretical point of view (Bertalanffy 1951), it can be ensured that all necessary perspectives are taken into consideration to gain an overall and clear understanding of self-organisation.

Characteristics concerning the system structure

It may be seen that all introduced concepts deal with complex systems. Thereby, what is more central to the issue is not what kind of nature they are attributed to (e.g. living or non-living systems), but the extent of occurrence of existing interrelations between the elements of the system as well as between the system and its environment (Dörner 2001; Malik 2000). Probst and Gomez particularly emphasize the aspect of dynamics in their understanding of complex circumstances, which differentiates complex systems from complicated systems. Dynamics is described as the rate of modification of a system over a specific period of time. A system can be described as complicated if it features various internal elements and links as in a functional description of a major machine. Complexity is not reached until high dynamics between the system elements is identifiable

(Probst and Gomez 1989). This interaction of the system elements is one precondition for the process of self-organizing. Haken introduced in this context the term of emergence, which describes a result of self-organisation. Through the process of interaction of the individual elements new qualitative characteristics of the system arise – so called emergences – which cannot be related to individual system components, but result from the complex synergy effects of the interacting elements (Haken 1993).

Self-organizing systems are open systems that means that they are open to absorb information and resources. The more information and resources absorbed by the system, the more changes of its status are assumed thereby influencing the internal dynamics of the system. However, the system openness enables self-organizing systems to adapt to significant changes in the environment (Varela 1979; Malik 2000).

Characteristics concerning the system behavior

Self-organizing ordered structures do evolve autonomously from the interaction of individual elements. Haken's study of self-organisation by investigating laser light provides an instructive example of this. He observed individual light waves. After supplying them with a certain mass of energy, they autonomously arrange themselves through interactions from a chaotic system state to a profoundly structured state the laser (Haken 1987). Prigogine and Glansdorff (1971) could observe similar results when they fed a liquid with energy. It displayed autonomous patterns in the form of dissipative structures. The concept of self-organisation presumes that through interaction of the systems elements an ordered structure evolves autonomously, which enables the system to cope with complexity and dynamics.

This implies that self-organizing systems contain autonomous system elements. A system's or an individual's autonomy can be identified if they form, guide or develop themselves, meaning that their decisions, relations and interactions are not dependent on external instances (Probst 1987). In doing so, a complete independence of the system from other systems cannot be assumed however (Varela 1979; Malik 2000). Each system only represents a part of a wide-ranging total system (environment) which it is in some way dependent on and influenced by. Therefore, it has to be understood as a relative autonomy of the individual or the system in relation to certain criteria (Varela 1979; Probst 1987). Regarding autonomously cooperating processes within a company, these criteria are defined by the given scope of action and decision making of the autonomous subject. For this reason autonomy manifests itself in the company as a result of processes of decentralization and delegation (Kappler 1992). Additionally, the autonomously acting systems are operationally-closed, which is termed as self-reference. It implies that the system defines its actions and borders by itself (Luhmann 1984). The system only induces actions which are essential for further survivability.

The characteristic of non-linearity can be found in all self-organizing systems. Non-linearity could be understood as a non-deterministic behavior referring to a system whose behavior is not causally predetermined and hence not predictable (Haken 1987; Prigogine 1996). In social autonomously cooperating systems, a framework of general rules of decision-making is predetermined (Hülsmann a. Windt 2005) and the desired final state of the system may be predictable, but not the mode of achieving it. Based on the ability of autonomous decision-making and autonomous acting of the individual system elements, the system behavior is not casually predetermined and thus not predictable. However, an organisation's way of acting is not completely non-linear. In general, a reason may be found in corporate history. According to the theory of path dependency a grown system is always predetermined by its former decisions. Thus, the amount of acting alternatives is always limited by former irreversible decisions (Schreyögg, Sydow and Koch 2003).

Characteristics concerning the system abilities

Complex systems are defined as systems being in a state far from equilibrium (Prigogine a. Glansdorff 1971). This may be seen in a way that complex systems are permanently open to absorb information and resources that are essential for it to sustain and survive. The system openness results in an everlasting change of the system status, which forces the system to stabilize its ordered structure permanently. When two reversible processes occur at the same rate, it manifests a dynamic equilibrium. Equally, Maturana and Varela (1980) as well as Odum (1999) found that natural systems – unhindered by human interference – also seek stability and balance through the capability of self-control mechanism, e.g. ecosystems are able to restore stable status within its system until a certain degree if necessary (Odum 1999).

Within an autopoietic system, like a biological cell for example, the components of the system are permanently involved in the production of new system elements. The cell possesses the ability of self-replication. Processes of self-replication may play an important role in self-organizing systems. The cell for example produces its own borders through this process which distinguishes the cell from its environment.

Flexibility could be seen as a competence from a system viewpoint as it supports the system with the level of adaptiveness required for it to sustain and survive in a dynamic, complex and highly competitive environment (Hülsmann and Wycisk 2005). The ability of being flexible by the components of the system helps them in self-organizing and forming, communicating and establishing desired relationships. Being flexible also aides the process in how complex systems autonomously create ordered structures because of its ability to adapt flexibly to the demanding complex and dynamic situation. Moving from a self-management perspective to a more abstract level of system perspective, it can be said that self-organisation creates the ability within the elements of the system to organize itself autonomously i.e. the system determines its own goals, autonomously chooses its strategies and organisational structure and also raises the necessary resources itself (Manz and Sims 1980).

2.2.4 Conclusions

The aim of the paper as reflected throughout was to develop a general understanding of the basic principles underlying autonomous cooperation. Therefore, it is necessary to understand the sources of the basic idea, which lay in concepts of self-organisation. Having seen a glimpse above of the origin of primal foundation concepts of the idea of self-organisation, it may be realized that concepts like entropy, Synergetics, Cybernetics, dissipative structures, autopoiesis and chaos theory have made an imprint in academia. What can be seen as an area of core shift today is towards self-reference, self-similarity, self-organisation and autonomy. Autonomous systems derive their autonomy from their intrinsic self-organisation (Vernon and Furlong 1992). The multitude of the facets of self-organisation seems to span boundaries across the ability of systems and maintain its identity and autonomy.

The phenomena of self-organisation may be considered to serve as explanations of the adaptive, intentional, and purposive functioning of many complex systems, especially of cognitive, biological, and social systems (Tschacher et al. 2003). As Bremermann puts it: "Self-organisation is creation without a creator attending to details" (Bremermann 1994), "Self" in this context may be seen as a result of internal mutual or reciprocal relations. Self-organisation may not only mean that it constitutes the idea of one science or idea of several sciences but the underlying basis or unifying substructure of various sciences (Zwierlein 1994). From the characteristics of self-organizing systems as discussed in Section 3 above, it can be said that the patterns of interaction among the elements of the system plays an important role in shaping the system's structure, behavior, and abilities. The concept of self-organisation may be recognized as a potential field capable of having its application in business processes as it increases the organisational ability and provides the flexibility to self-organize and cope with complex situations in a dynamic environment. There are attempts, however, to transfer and integrate the idea of self-organisation in autonomous co-operating logistics processes using modern technologies like RFID, sensors, etc.

Hence a general understanding of self-organisation that has been developed through this work is presumed to be helpful to management practice as a first step towards its application and transfer into autonomous cooperating business processes, for instance in logistics. However, the question that still persists is whether self-organisation is a sequel, progression or succession to autonomous cooperation. What remains to be answered in future research is to what extent the idea of self-organisation can be transferred to or used in the concept of autonomous cooperation and how they can be applied to obtain optimum performance in business processes.

References

Ashby WR (1962) Principles of the self-organizing system. In: Foerster v H, Zopf GW (eds) Principles of self-organisation. Illinois Symposium. Pergamon Press, London, pp 255-278

Ashby WR (1970) Introduction to Cybernetics. Chapman and Hall, London

Aspray W (1990) John von Neumann and the Origins of Modern Computing. MIT Press, Mass.

Balescu R (1975) Equilibrium and nonequilibrium statistical mechanics. Wiley, New York

Bertalanffy Lv (1951) General systems Theory – A New Approach to the Unity of Science. Human Biology 23: 302-361

Beyerle M (1994) Staatstheorie und Autopoiesis über die Auflösung der modernen Staatsidee im nachmodernen Denken durch die Theorie autopoietischer Systeme und der Entwurf eines nachmodernen Staatskonzepts. Reihe, Beiträge zur Politikwissenschaft, Band 59, Berlin, zugl. Dissertation, Heidelberg

Bick H (1973) Population Dynamics of Protozoa Associated with the Decay of organic Materials in Fresh Water. American Zoologist 13(1): 149-160

Binnig G, Feigenbaum M (1995) Chaos und Kreativität: rigorous chaos. Internationales Zermatter Symposium: Kreativität in Wirtschaft, Kunst und Wissenschaft 3

Birkhoff GD (1923) Relativity and modern physics. Harvard Univ. Press, Cambridge

Boerlijst MC, Hogeweg P (1991) Spiral wave structure in prebiotic evolution: hypercycles stable against parasites. Pysica D 48: 17-28

Braitenberg V (1984) Vehicles, experiments in synthetic psychology. MIT Press, Mass.
Bremermann HJ (1994) Self-organisation in Evolution, Immune Systems, Economics, Neural Nets, and Brains. In: Mishra RK, Maass D, Zwierlein E (eds) On Self-oganization. Springer, Berlin
Buckminster FR (1975) Synergetics. Macmillan, Vol 1, pp 164-189
Camazine S (1991) Self-organizing pattern formation on the combs of honey bee colonies. Behavioral ecology and sociobiology 28: 61-76
Capra F (1996) The Web of Life: A New Scientific Understanding of Living Systems. Anchor Books, New York
Cartwright ML (1965) Contributions to the theory of nonlinear oscillations. Princeton University Press, Princeton, NJ
Dauwalder JP, Tschacher W (1999) Dynamics, Synergetics, Autonomous Agents. World Scientific, Singapore
Derrida J (1976) Of Grammatology. Trans. Gayatri Spivak. Johns Hopkins University Press, Baltimore
Dickau RM (1992) The Hénon Attractor.
Dörner D (2001) Die Logik des Misslingens: Strategisches Denken in komplexen Situationen. 14th edn. Hamburg
Drosdowski G (1990) Duden Fremdwörterbuch Dictionary, 5th edn
Drosdowski G (1989) Duden Herkunftswörterbuch Dictionary, 2nd edn
Dyke C (1988) The Evolutionary Dynamics of Complex Systems: A Study in Biosocial Complexity. OUP, New York
Eigen M (1971) Molekulare Selbstorganisation und Evolution. Naturwissenschaften 58 (10): 465-523
Eigen M, Schuster P (1977) The hypercycle. A principle of natural self-organisation. Part A: Emergence of the hypercycle. Naturwissenschaften, 64: 541-565
Eigen M, Schuster P (1979) The Hypercycle: A principle of natural self-organisation. Springer, Berlin
Eigen M, Gardiner W, Schuster P, Winkler OR (1981) The origin of genetic information. Sci Am 244(4): 78-94
Fuchs, C. (2002) Concepts of Social Self-Organisation. INTAS-Project: Human Strategies in Complexity - Research Report. Vienna University of Technology. Available Online at: http://www.self-organisation.org
Gleick J (1987) Chaos: Making a New Science. Viking, New York
Goguen JA, Varela FJ (1979) Systems and distinctions: Duality and complementarily. International Journal of General Systems 5: 31-43
Goldbeter A (1997) Biochemical oscillations and cellular rhythms: the molecular bases of periodic and chaotic behavior. Cambridge University Press, Cambridge
Glansdorff P, Prigogine I (1971) Thermodynamic theory of structure, stability and fluctuations. Wiley, New York
Grapp J, Wycisk C, Dursun M, Hülsmann M (2005) Ideengeschichtliche Entwicklung der Selbstorganisation – Die Diffusion eines interdisziplinären For-

schungskonzeptes. Reihe: Forschungsbeiträge zum Strategischen Management, hrsg v Hülsmann M, Bd 8. Bremen

Haeckel E (1875) Ziele und Wege der heutigen Entwicklungsgeschichte. Dufft, Jena

Haken H (1973) Synergetics: cooperative phenomena in multi-component systems. Symposium on Synergetics, 30 April to 6 May 1972. Schloß Elmau, Stuttgart

Haken H (1978) Synergetics — An Introduction. Springer Verlag, Berlin

Haken H (1981) Synergetics and the problem of self organisation. In: Roth G, Schwegler H (eds) Self Organizing Systems. Campus Verlag, Frankfurt/M, p 12

Haken H (1987) Die Selbstorganisation der Information in biologischen Systemen aus Sicht der Synergetik. In: Küppers BO (ed) Ordnung aus dem Chaos. München, pp 35-60

Haken H (1993) Synergetik: Eine Zauberformel für das Management? In: Rehm W (ed) Synergetik: Selbstorganisation als Erfolgsrezept für Unternehmen. Symposium IBM. Stuttgart, pp 15-43

Halley JM, Comins HM, Lawton JH, Hassell MP (1994) Competition, succession, and pattern in fungal communities: towards a cellular automation model. Oikos 70: 435-442

Herschkowitz N (2001) Das vernetzte Gehirn: seine lebenslange Entwicklung. Huber, Bern

Heylighen F, Joslyn C (2001) Cybernetics and Second-Order Cybernetics. In: Meyers RA (ed) Encyclopedia of Physical Science and Technology (3rd ed). Academic Press, New York

Hofbauer J, Sigmund K (1988) Evolutionary Games and Population dynamics. Cambridge Univ Press, Cambridge

http://mathforum.org/advanced/robertd/henon.html

Hülsmann M, Windt K (2005) Selbststeuerung – Entwicklung eines terminologischen Systems. Bremen

Hülsmann M, Wycisk C (2005) Contributions of the concept of self-organisation for a strategic competence-management. The Seventh International Conference on Competence-Based Management: Value Creation through Competence-Building and Leveraging. 2 - 4 June 2005. Antwerpen, Belgien

Innis GS (1979) Systems analysis of ecosystems. Intern. Co-operative Publ. House, Fairland

Johnson CR (1997) Self Organizing in spatial competition systems. Frontiers in Ecology 20: 245-263

Kappler E (1992) Autonomie. In: Frese E (ed) Handwörterbuch der Organisation. 3rd edn, Stuttgart. pp 272-280

Karsai I, Penzes Z (1993) Comb building in social wasps: Self-organisation and stigmeric script. Journal of theoretical biology 161: 505-525

Kauffman LH, Varela FJ (1980) Form dynamics. Journal of Social Biological Structures 3: 171-206

Kolmogorov AN (1991) Selected works of A.N. Kolmogorov, edited by Tikhomirov VM, translated by Volosov VM- 3 volumes. Kluwer Academic Publishers, Dordrecht, Boston

Kriz J (1990) Synergetics in clinical psychology. In: Haken H, Stadler M (eds) Synergetics of cognition. Springer, Berlin, New York, pp 393-404

Kuhn T (1978) Black-Body Theory and the Quantum Discontinuity 1894-1912. The University of Chicago Press, Chicago, p 13

Lefever R (1978) Molecular movements and chemical reactivity as conditioned by membranes enzymes and other macromolecules. Wiley, New York

Littlewood JE (1988) Littlewood's miscellany, edited by Béla Bollobás. Cambridge Univ Press, Cambridge

Lorenz EN (1963) Deterministic Nonperiodic Flow. Journal of the Atmospheric Sciences 20: 130 –141

Luhmann N (1984) Soziale Systeme: Grundriß einer allgemeinen Theorie. Suhrkamp, Frankfurt/M

Luisi P, Varela FJ (1989) Self-replicating micelles - a chemical version of a minimal autopoietic system. Origins of Life and Evolution of the Biosphere 19: 633-643

Malik F (2000) Strategie des Managements komplexer Systeme: Ein Beitrag zur Management-Kybernetik evolutionärer Systeme. 6th edn, Bern

Mallet-Paret J (1993) Obstructions to the existence of normally hyperbolic inertial manifolds. Providence, RI: Div of Applied Math, Brown University

Mandelbrot B (1977) The Fractal Geometry of Nature. Freeman, New York

Manz C, Sims HP (1980) Self-management as a substitute for leadership: A social learning theory perspective. In: AMR 5, pp 361-367

Maturana HR, Pörksen B (2002) Vom Sein zum Tun: Die Ursprünge der Biologie des Erkennens. Carl-Auer-Systeme-Verl, Heidelberg

Maturana HR, Varela FJ (1973) Autopoiesis and Cognition: the Realization of the Living. Reidel, Dordecht

Maturana HR, Varela F (1980) Autopoiesis and cognition: the realization of living. Reidel, Dordrecht

May RM (1976) Theoretical Ecology: Principles and Applications. Blackwell Scientific Pub, Oxford

McCulloch WS, Pitts W (1943) A logical calculus of the ideas immanent in nervous activity. Bulletin of Mathematical Biophysics 5: 115–133

Mingers J (1989) An introduction to autopoiesis - implications and applications. Systems Practice 2(2): 159-180

Mishra RK (1994) Living State of Self-oganization. In: Mishra RK, Maass D, Zwierlein E (eds) On Self-oganization. Springer Verlag, Berlin

Nicolis G, Prigogine I (1977) Self organisation in non equilibrium systems. Wiley, New York

Nowak MA, May RM (1992) Evolutionary games and spatial chaos. Nature 359: 826-829

Nowak MA, May RM (1993) The spatial dilemmas of evolution. International Journal of bifurcation and chaos 3: 35-78

Odum, EP (1999) Grundlagen, Standorte, Anwendung. übersetzt von Overbeck, 3. Aufl. Thieme, Stuttgart
Oxford Dictionary (2002), 8th ed
Pask, G (1979) A conversation theoretic approach to social systems. In: Geyer F, van der Zouwen J (eds) Sociocybernetics. Martin Nijholf, Amsterdam, pp 15-26
Paslack R (1991) Urgeschichte der Selbstorganisation: zur Archäologie eines wissenschaftlichen Paradigmas. Vieweg, Braunschweig
Paslack R, Knost P (1990) Zur Urgeschichte der Selbstorganisationsforschung, Ideengeschichtliche Einführung und Bibliographie (1940 – 1990). Kleine, Bielefeld
Poincaré HJ (1890) Sur le problème des trois corps et les équations de la dynamique. Acta Mathematica 13:1
Prigogine I (1969) Structure, dissipation and life. In: Marois M (ed) Theoretical Physics and Biology. Amsterdam, pp 23 – 52
Prigogine I (1980) From Being to Becoming. Freeman, San Francisco
Prigogine I (1996) The End of Certainty: Time, Chaos, and the New Laws of Nature. The Free Press, New York
Prigogine I, Glansdorff P (1971) Thermodynamic Theory of Structure, Stability and Fluctuation. Wiley, London
Prigogine I, Stengers I (1984) Order out of Chaos: Man's New Dialogue with Nature. Heinemann, London
Probst GJB (1984) Cybernetic Principles for the design, control and development of Social Systems. In: Ulrich H, Probst GJB (eds) Self Organisation and Management of Social Systems. Springer, Berlin, pp 127-131
Probst GBJ (1987) Selbstorganisation: Ordnungsprozesse in sozialen Systemen aus ganzheitlicher Sicht. Parey, Berlin
Probst GJB, Gomez P (1989) Vernetztes Denken: Unternehmen ganzheitlich führen. Gabler, Wiesbaden
Ruelle D (1991) Chance and Chaos. Princeton University Press, Princeton, NJ
Scaruffi P (2003) Thinking About Thought: Towards a Unified Understanding of Life, Mind and Matter. Writers Club Press, Lincoln
Schreyögg G, Sydow J, Koch J (2003) Organisatorische Pfade – Von der Pfadabhängigkeit zur Pfadkreation? In: Schreyögg G, Sydow J (eds) Strategische Prozesse und Pfade. Managementforschung 13, Wiesbaden
Smale S, Hirsch MW, Devaney RL (2004) Differential equations, dynamical systems, and an introduction to chaos. Elsevier Acad. Press, Amsterdam
Smith JM (1979) Hypercycles and the origin of life. Nature 280: 445-446
Solé RV, Bascompte J (2006) Self-organisation in complex ecosystems. Princeton Univ. Press, Princeton, NJ
Stadler M, Kruse P (1995) Ambiguity in mind and nature: Multistable cognitive phenomena. Springer, Berlin
Teubner G (1995) Wie empirisch ist die Autopoiese des Rechts? In: Martinsen (ed) Das Auge der Wissenschaft: zur Emergenz von Realität. Baden-Baden, pp 137-155

Teubner G, Willke H (1984) Kontext und Autonomie: Gesellschaftliche Selbststeuerung durch reflexives Recht. In: Zeitschrift für Rechtssoziologie 5
Thore S (1995) The diversity, complexity and evolution of high tech capitalism. IC² Institute, The University of Texas at Austin. Kluwer Academic Publishers, Boston Dordrecht London
Tschacher W (1992) Self-organisation and clinical psychology: empirical approaches to Synergetics in psychology. Springer, Berlin
Tschacher W, Dauwalder JP, Haken H (2003) Self-organizing systems show apparent intentionality. In: Tschacher W, Dauwalder JP (eds) The Dynamical Systems Approach to Cognition. World Scientific, pp 183-200, Singapore
Tschacher W, Schiepek G, Brunner EJ (1992) Self-organisation and clinical psychology: empirical approaches to Synergetics in psychology. Springer, Berlin New York
Ulrich H (1984) Self-organisation and management of social systems: insights, promises, doubts, and questions. Springer, Berlin
Umpleby SA (1999) Papers from the 1997 meeting of the American Society for Cybernetics. Taylor and Francis, Philadelphia
Varela FJ (1979) Principles of biological autonomy. North Holland, New York
Varela FJ (1981) Autonomy and Autopoiesis. In: Roth G, Schwegler H (eds) Self Organizing Systems. Campus Verlag, Frankfurt/M New York, p 18
Varela FJ, Maturana HR, Uribe R (1974) Autopoiesis: the organisation of living systems, its characterization and a model. Biosystems 5: 187–196
Vespalcova Z, Holden AV, Brindley J (1995) The effect of inhibitory connections in a hypercycle: A study of the spatio temporal evolution, Phys. Lett. A 197: 147-156
Vernon D, Furlong D (1992): Relativistic Ontologies, Self-Organisation, Autopoiesis, and Artificial Life: A Progression in the Science of the Autonomous. Proceedings of the Workshop: Autopoiesis and Perception, edited by Barry McMullin, Dublin City University
von Foerster H (1960) On Self-Organizing Systems and their Environment. In: Yovits MC, Cameron S (eds) Self Organizing Systems. London
von Foerster H (1979) Cybernetics of Cybernetics. In: Krippendorff K (ed) Communication and Control in Society. Gordon and Breach, New York, pp 5-8
Walter HJ (1996) Angewandte Gestalttheorie in Psychotherapie und Psychohygiene. Westdt. Verlag, Opladen
Wiener N (1948) Cybernetics or Control and Communication in the Animal and the Machine. MIT Press, New York
Winograd T, Flores F (1986) Understanding Computers and Cognition. Ablex, Norwood NJ
Zwierlein E (1994) The Paradigm of Self-organisation. In: Mishra RK, Maass D, Zwierlein E (eds) On Self-oganization. Springer Verlag, Berlin

2.3 Historical Development of the Idea of Self-Organisation in Information and Communication Technology

Markus Becker, Koojana Kuladinithi, Andreas Timm-Giel, Carmelita Görg

Communication Networks, NW1, University Bremen, Germany

Information and Communication Technology includes many different concepts, implementations and usages of Self-Organisation (Serugendo et al. 2004; Czap et al. 2005; Brueckner et al. 2005). These are among others: ad hoc routing, autonomic communication, Self-Star and peer-to-peer networks.

The constituting features of autonomous control, as already mentioned in Chapter 3.4, 3.7, 4.3 and 4.4, have been used and enhanced since the beginnings of Information and Communication Technology. **Non-centralised** or **distributed** design and operation is naturally present in ICT systems: The components of the ICT networks are distributed, e.g. the base stations of cellular networks, are distributed over the coverage area. **Heterarchy** is present in non-hierarchical networks, e.g. peer-to-peer networks, as explained later in this chapter. User-Network-Interaction and Network-Network-Interaction, as specified for example by the Border Gateway Protocol (BGP), make up the constituting property of **interaction**. **Non-determinism** exists, e.g., in the Internet for packets taking different routes to reach the same destination. Each Internet router acts **autonomously**, which is another constituting property of autonomous control. Finally, the **decision process** is also found in ICT systems: for example in policy-based decision processes. Those constituting features can be found in the following examples of ad hoc routing, peer-to-peer networks, autonomic computing and autonomic communication.

2.3.1 Ad hoc networks

Definition of ad hoc networks

An ad hoc network is a self-configuring network of hosts that have equal or similar functionalities and equal or similar responsibilities (Blazevic et al. 2001; Garbinato a. Rupp 2003). Especially important for the functioning of the network is the routing functionality that has to be present in all nodes. Usually this functionality is implemented only in a subset of network nodes, called routers, which provide this service also to those nodes which do not have this functionality. The functionality of ad hoc networks is especially challenging in wireless and mobile environments. The term ad hoc network usually implies a wireless ad hoc network. In this kind of network the nodes communicate by means of radio frequency transmission. A mobile ad hoc network, abbreviated MANET, is a wireless ad hoc network, in which the nodes are free to move. The links between the nodes are created by the routing functionality. The geometric arrangement of the nodes together with the links is called the topology of the network. Ad hoc networks have a dynamic topology due to the movement of the nodes. The routing protocol adapts the topology to the physically possible communication links.

An ad hoc network can include nodes from non-ad hoc networks, e.g. the Internet. Such nodes may provide access to the Internet for the other nodes of the ad hoc network. This extends the area covered by the Internet-node – usually called Access Point – without the need for installation of further Access Points or infrastructure cabling.

Ad hoc networks have several advantages over usual infrastructure networks:

- Ease of deployment: Ad hoc networks do not need the elaborate setup of Access Points, e.g. cabling, addressing, setup. Although in this case the usage is limited to the ad hoc network with no access to the Internet;
- Speed of deployment: As setup is easier, the deployment is also faster and cheaper (no network and power cabling);
- Decreased dependency on infrastructure: Single points of failure are eliminated, e.g., Access Points.

Characteristic properties of ad hoc networks are:

- Decentralized: Each component has the same functionality, rights and responsibilities. There is no central instance;
- Self-organized: Routes are found without manual or central interaction;
- Self-deployed: Except for physically placing the nodes and switching them on, no setting up needs to be done;
- Dynamic network topology: Depending on the propagation conditions the topology of the network can be changing and is handled by the ad-hoc network;
- Local knowledge: There is no central instance in the network that has knowledge of the complete network. All components of the network only have local knowledge;
- Interaction and cooperation of the elements/nodes of the network: The components work together to find routes to other components;
- Adding/removing nodes is dynamic: Once a new node is added, it announces itself and answers requests for routes in the same way that all other nodes are functioning as part of the network. When a node is removed, the routes using this node break. This break is detected and a new route is set up by the remaining nodes.

The dynamism in wireless systems is very high compared to wired networks. The attachment and detachment of nodes to the network can be more frequent, as there is no physical attachment necessary via cables.

Routing in ad hoc networks

The main functionality of ad hoc networks is the routing protocol. There are two main families of routing protocols: reactive and proactive routing protocols. An extensive list of routing protocols can be found in (Various Authors 1 2006). Hybrid versions of the two different routing approaches are a natural extension.

Reactive routing protocols

Protocols that create routes, only if requested by the user of the network are called reactive routing protocols. Examples are: Ad-hoc on Demand Distance Vector (AODV), Dynamic Source Routing (DSR) and Dynamic MANET On-demand (DYMO). Reactive Protocols are more appropriate, when the topology is highly dynamic. New routes, which appear frequently, need not be propagated through the whole network, as they are not needed by the hosts most of the time.

Proactive routing protocols
Protocols that maintain a list of routes to other nodes are called proactive routing protocols. Destination-Sequenced Distance Vector (DSDV), Optimized Link State Routing (OLSR) and Source Tree Adaptive Routing (STAR) are examples for such protocols. Proactive protocols are advantageous over reactive ones, when the topology is only slowly changing. These protocols do not require generation of routes, when a node wants to communicate, thus the initial delay is shorter.

Autonomic address assignment

A very important aspect of ad-hoc networks, which is ideally suited to highlighting the issues associated with self-organisation, is the area of address auto-configuration in ad-hoc networks. Address auto-configuration selects the Internet Protocol Address of devices autonomously without the need for a central instance (e.g. a Dynamic Host Configuration Protocol server). In a static network autonomous configuration of IP addresses can be done by a mechanism called link-local addressing. In dynamic mobile ad hoc networks, however, the situation is more complex. First, not all stations are within a distance of one hop of each other (i.e. not having a direct link). Additionally, there is a possibility of two MANETs joining to form a new MANET with members with the same assigned addresses. These circumstances need to be handled in a self-organized fashion by MANET protocols. A comparison of different techniques can be found e.g. in (O'Grady et al. 2004).

History of ad hoc networks

Mobile ad hoc networks are derived from so called packet radio networks of the 1970s. These projects were sponsored by the American Defense Advanced Research Projects Agency (DARPA). In 1983 the Survivable Adaptive Network (SURAN) project supported a larger scale network. With the common use of IEEE 802.11 components, an increased academic interest could be observed starting in the mid 1990s.

An Internet Engineering Task Force (IETF) working group was established, called MANET (Mobile Ad Hoc Networks). The term MANET was introduced by the IETF MANET charter. A variety of ad hoc network routing protocols have been discussed and promoted by this working group (Various Authors 2 2006, Wikipedia Authors 1 2006).

The development of AODV (Perkins et al. 2003) is based on Destination-Sequenced Distance Vector (DSDV), which is a protocol for static networks. AODV is an improvement over DSR (Johnson et al. 2004) by reducing the overhead needed for the routing. OLSR (Clausen et al. 2003) as a proactive protocol is derived from Link State Routing (LSR). AODV, DSR and OLSR are currently experimental Request for Comments (RFCs) of the Internet Engineering Task Force (IETF). The integration into standard track RFCs is done by merging DSR and AODV to a protocol called DYMO (Chakeres et al. 2006) and by enriching OLSR with ideas from other protocols to a protocol called OLSRv2 (Clausen et al. 2006). There are further efforts in unifying protocols from the two domains – reactive and proactive – into a common protocol with extensions specific to each domain.

Mesh and sensor networks

Specific kinds of ad hoc networks are mesh networks and sensor networks. Mesh networks are specific MANETs that consist of mostly static mesh routers and try to supply a backhaul service to mesh clients. In the past there have been several community initiatives to build such systems in urban areas, cf. (Aguayo et al. 2003; Various Authors 6 2006; Various Authors 7 2006). Similar solutions as in MANETs are used in mesh networks, additionally self-organisation is exploited with regard to Dynamic Channel Allocation (Akyildiz et al. 2005; Subramaniam 2006).

Wireless Sensor Networks (WSN) are a recent research field (Karl and Willig 2005; Akyildiz et al. 2002). WSNs combine MANETs with low-power design and the ability to sense and/or actuate. Self-organisation in the area of WSNs is focused on enabling lower energy consumption and thus a prolonged life time of the battery-powered devices. An example of this is the adaptation of the duty cycle (the ratio of time awake and time asleep) to the context of the WSN as done for example in (Neugebauer et al. 2005).

Active networks and mobile agents

Various aspects of autonomous control in data communication can be identified in all 7 layers of the ISO/OSI reference models (1: physical, 2: link, 3: network, 4: transport, 5: session, 6: presentation, 7: application) under different names.

It is called Active Networks in the lower layers. The transmitted data packets are accompanied by code components, which are executed on the transit nodes (routers). This leads to a certain degree of independence of the version of the router. More important is the possibility of autonomous control, i.e. each data packet chooses its actions individually (Lededza et al. 1998).

In the higher layers autonomous control is known as Mobile Agents, which are moving autonomously and cooperatively in the network and aim for individual goals, representing the user. Mobile agents extend the concept of agent technology as described in (Jennings and Wooldridge 1995) by the ability to move the agent to the location of the data. The applicability and performance of mobile agents has been studied by (Straßer and Schwehm 1997; Helin et al. 1999; Farjami et al. 1999; Hartmann et al. 1999; Yang et al. 2002). The general aspects of Agent technology are described in Chapter 3.7.

2.3.2 Peer to peer networks

Peer to Peer Networks (P2P) are another incarnation of the self-organisation idea in the information and communication technology field. Contrary to the traditional Client-Server-Architecture, all computers have the same or at least similar functionality like in ad hoc networks (Various Authors 3 2006). The nodes are called peers or "servents" to represent a combination of server and client. The predominant purpose of P2P networks is the retrieval and distribution of content, such as multimedia files. These networks have to handle the addition and removal of nodes to and from the network smoothly.

There are two families of P2P networks. Hybrid P2P or centralized P2P networks have peers that act as servers and client-peers are connected in a star-like fashion to a single super-peer. These super-peers handle special functionalities, such as the indexing of the content or the distribution of search requests. Pure P2P or decentralised P2P does not have super-peers, and all the peers have identical functionalities and responsibilities.

The oldest Peer to Peer Networks are Usenet and FidoNet. The most well-known, recent ones are Napster, Kazaa, Gnutella, eDonkey, JXTA and Bittorrent, (Androutsellis-Theotokis and Spinellis 2004).

Peer to Peer Networks currently also find applications in the context of Voice over Internet Protocol (VoIP), where the voice data of the telephone calls is transported by a P2P network (Baset and Schulzrinne 2004).

2.3.3 Autonomic computing

Technological systems are growing rapidly in size and complexity. In order to enable further growth of information technology systems, the Autonomic Computing Initiative was started in 2001 by IBM, (IBM Press 2003; Kephart 2003; Ganek 2003). The aim is to allow control of the growing complexity.

Increasing complexity necessitates more specialists, if there is no change to the way ICT systems are currently being handled. Those specialists might not be available or affordable. Usually these specialists have to maintain systems that have been created by a different set of specialists. Maintenance specialists cannot know everything about the system and the side effects of actions.

Furthermore, dependency on information and communication technology (ICT) systems and the monetary losses due to their failures are increasing. Many companies and organisations from many different economic sectors such as banks, IT companies, electrical power plants, police, and military organisations are highly dependent on the availability of their ICT systems. Autonomic Computing has therefore become a topic of interest for academia as well as major companies such as IBM, Sun, Daimler-Chrysler and Fujitsu-Siemens (Gu et al. 2005).

Autonomic Computing is a concept of self-managed computing systems with minimum human conscious awareness or involvement, derived from the human autonomic nervous system – a sophisticated autonomic entity.

Autonomic Elements are supposed to be the composing elements of autonomic computing. An autonomic element consists of the managed element and an autonomic manager. The autonomic manager consists of components that monitor, analyze, plan and execute based on knowledge that is available or has been gathered. The autonomic manager therefore acts as a control loop. The control loop describes how the resource and control interact with each other. The resource is measured and based on the measurements a decision is taken and the decision controls the resource.

The building blocks of Autonomic Computing are the self-* principles, that is self-configuring, self-optimising, self-healing, and self-protecting (Babaoglu et al. 2004; Wikipedia Authors 2 2006).

There are different levels of Autonomic Operations stated by (Ganek and Corbi 2003), starting from basic, managed, predictive, adaptive to autonomic systems. Gradually the manual handling involved decreases and the autonomic handling increases over these levels. The autonomy will be visible in processes, tools, skills as well as in benchmarks.

Autonomic Computing principles are already applied in middleware, database systems, and software engineering.

2.3.4 Autonomic communication

Closely related to Autonomic Computing is Autonomic Communication (Smirnov 2004, Various Authors 4 2006; Various Authors 5 2006), also called AutoComm. Autonomic Communication tries to solve the same set of problems for the Communication area, that Autonomic Computing is tackling largely for the Information Technology. The Autonomic Communication Forum initiative is founded on the belief that a radical paradigm shift towards a self-organising, self-managing and context-aware autonomous networks, considered in a technological, social and economic context, is the only adequate response to the increasingly high complexity and demands now being placed on the Internet (Wikipedia Authors 2 2006).

In Communication Technology the situation is similar and closely connected to the situation in Information Technology. The high demand for specialists in these fields cannot be satisfied (Bitkom 2003). The product innovation cycle is accelerating in such a way that the integration of older systems cannot be satisfied in time. Additionally the usage of and number of components of communication system is increasing. This requires distributed and self-organising structures, relying on simple and dependable elements that are capable of collaborating to produce a sophisticated behaviour of the system.

The technologies enabling this are so called self-* technologies, namely self-configuration, self-healing, self-optimisation and self-protection. Self-configuration describes the ability to automatically (re)configure components of the network, self-healing the detection and treatment of errors. The automatic surveillance and control of the usage of resources for an optimal usage of those resources is called self-optimisation. Self-protection is characterized by the ability to identify and prohibit attacks on components.

One key aim of Autonomic Communication is to enable zero-effort deployment. This describes the deployment of a network of communication units without having to do any configuration steps other than putting the units into place. The units will configure themselves in cooperation with the other units.

Key research areas of self-management are currently to answer the questions related to:

- Controllability: How is the ownership reflected in the process, when autonomous elements are negotiating with each other? What happens if an autonomous element cannot be controlled because of its ownership?
- Reliability: Does reliability emerge when autonomous elements are collaborating or does the unreliability increase?
- Security: How are the autonomous elements secured against unwanted control by other elements?

The history of autonomous control in Information and Communication Technology goes back to the 1950s. It started with research on what is now known as the Internet. Today's applications of ICT already heavily depend on autonomous control and this will increase in future as networks are growing.

2.3.5 Conclusions and future directions

This chapter introduced the development and the application of self-organisation in Information and Communication Technology. It summarizes how the idea of self-organisation has been applied in ad hoc networks (including mesh and sensor networks), peer to peer networks, autonomic computing and autonomic communication.

As shown in this chapter self-organisation has been a continuous theme during the past evolution of ICT. In future more sophisticated self-organisation ideas need to be included in ICT to cope with the increasing complexity. As one example context adaptivity can be named. This topic is of relevance in all communication areas from sensor networks to satellite networks. The context is spread over all functional layers of a communication system. A self-organized context adaptation taking into account information from all layers is one of many research topics. Additionally, research is needed with respect to stability issues of self-organisation in ICT (Dolev 2000). In the CRC 637 self-organisation concepts are being used in the demonstrator "Intelligent Transportation System" based on Radio Frequency Identification, Wireless Sensor Networks and Agent Technology, as introduced in Chapter 4.6. The demonstrator is continuously being enhanced based on the self-organisation principle.

References

Aguayo D, Bicket J, Biswas S, De Couto D (2003) MIT Roofnet: Construction of a Production Quality Ad-Hoc Network

Akyildiz I, Su W, Sankarasubramaniam Y, Cayirci E (2002) A survey on sensor networks, IEEE Commun. Mag. 40(8) 102-114

Akyildiz I, Wang X, Wang W (2005) Wireless mesh networks: a survey, Computer Networks 47: 445-487

Androutsellis-Theotokis St, Diomidis Spinellis D (2004) A survey of peer-to-peer content distribution technologies. ACM Computing Surveys 36(4):335–371 (doi:10.1145/1041680.1041681)

Babaoglu O, Jelasity M, Montresor A, Fetzer Ch, Leonardi St, van Moorsel A, van Steen M (2004) Self-star Properties in Complex Information Systems: Conceptual and Practical Foundations (Lecture Notes in Computer Science)

Baset S A, Schulzrinne H (2004) An Analysis of the Skype Peer-to-Peer Internet Telephony Protocol

Bitkom (2003) Innovationen für Wachstum und Beschäftigung - Das 10-Punkte-Programm der ITK-Wirtschaft 2003/2004, Bundesverband Informationswirtschaft Telekommuinikation und neue Medien e.V. http://www.bitkom.org/files/documents/BITKOM_10_Punkte_Programm_2003_23.09.03.pdf

Blazevic L, Buttyan L, Capkun S, Giordono S, Hubaux J-P, Boudec JY (2001) Self-Organisation in Mobile Ad Hoc Networks: The Approach to Terminodes, IEEE Communications Magazine, pp 166 -173

Brueckner S, Serugendo G, Hales D, Zambonelli F (2005) Engineering Self-Organising Systems, Third International Workshop, ESOA 2005, Utrecht, The Netherlands, July 25, 2005, Revised Selected Papers. Lecture Notes in Computer Science 3910 Springer 2006, ISBN 3-540-33342-8

Chakeres I, Perkins C (2006) IETF Draft: Dynamic MANET On-demand (DYMO) Routing

Clausen T, Jacquet P (2003) IETF RFC 3626: Optimized Link State Routing Protocol (OLSR)

Clausen T, Dearlove C, Jacquet P (2006) IETF Draft: The Optimized Link-State Routing Protocol version 2

Czap H, Unland R, Branki C (2005) Self-Organisation and Autonomic Informatics

Dolev Sh (2000) Self-Stabilization, MIT Press

Dressler F (2006) Self-Organisation in Autonomous Sensor/Actuator Networks. 19th IEEE/ACM/GI/ITG International Conference on Architecture of Computing Systems - System Aspects in Organic Computing (ARCS'06), Frankfurt, Germany, Tutorial

Farjami P, Görg C, Bell F (1999) A Mobile Agent-based Approach for the UMTS/VHE Concept. In: Proc. Smartnet'99 - The Fifth IFIP Conference on Intelligence in Networks, pp 149-162

Farjami P, Görg C, Bell F (1999) Advanced Service Provisioning based on Mobile Agents. In: Proc. MATA'99 - First International Workshop on Mobile Agents for Telecommunication Applications, pp 259-272.

Ganek AG and Corbi TA (2003) The dawning of the autonomic computing era. IBM Systems Journal 42(1): 5-18. ISSN 0018-8670. Publ. IBM Corp., Riverton, NJ, USA

Garbinato B, Rupp Ph (2003) From Ad Hoc Networks to Ad Hoc Applications. ERCIM News No. 54, July 2003. SPECIAL THEME: Applications and Service Platforms for the Mobile User

O'Grady JP, McDonald A, Pesch D (2004) Network Merger and its Influence on Address Assignment Strategies for Mobile Ad Hoc Networks. In Proc. of IEEE Vehicular Technology Conference Fall 2004, Los Angeles, CA, USA

Gu X, Fu X, Tschofenig H, Wolf L (2005) Towards Self-Optimizing Protocol Stack for Autonomic Communications: Initial Experience. I. Stavrakakis and M. Smirnov (Eds.), Proceedings of the 2nd IFIP International Workshop on Autonomic Communication (WAC'05), Springer Lecture Notes in Computer Science Vol. 3854 (LNCS), pp 186-201. Athens, Greece

Hartmann J, Evensen R, Görg C, Farjami P, Long H (1999) Agent-based Banking Transaction and Information Retrieval – What About Performance Issues? In: Proc. European Wireless'99, pp 205-210

Helin H, Laamanen H, Raatikainen K (1999) Mobile Agent Communication in Wireless Networks. In Proc. of European Wireless'99/ITG'99, pp 211-216

Wooldridge M, Jennings NR (1995) Intelligent Agents: Theory and Practice. Knowledge Engineering Review 10(2): 115-152

Johnson DB, Maltz DA, Hu YC (2004) IETF Draft: The Dynamic Source Routing Protocol for Mobile Ad Hoc Networks (DSR)

Karl H, Willig A (2005) Protocols and Architectures for Wireless Sensor Networks, Wiley

Kephart J, Chess D (2003) The Vision of Autonomic Computing. IEEE Computer 36(1): 41–50

Ledezda U, Wetherall D, Guttag JV (1998) Improving the Performance of Distributed Applications Using Active Networks. In: Proc. of INFOCOM, pp 590-599

IBM Press (2003), Autonomic Computing Initiative. http://www.autonomic-computing.org

Neugebauer M, Ploennigs J, Kabitzsch K (2005) Duty Cycle Adaptation with Respect to Traffic

Perkins C, Belding-Royer E, Das S (2003) IETF RFC3561: Ad hoc On Demand Distance Vector (AODV) Routing

Serugendo G, Karageorgos A, Rana OF, Zambonelli F (2004) Engineering Self-Organising Systems: Nature-Inspired Approaches to Software Engineering (Lecture Notes in Computer Science)

Smirnov M (2004) Report on FET consultation meeting on Communication paradigms for 2020, Brussels, 3-4 March 2004, Area: Autonomic Communication

Straßer M, Schwehm M (1997) A Performance Model for Mobile Agent Systems. In: H. R. Arabnia (Ed.): 'Int. Conf Parallel and Distributed Processing Techniques and Applications (PDPTA'97)', CSREA 1997 Volume II, pp 1132-1140

Subramaniam AP, Gupta H, Das S (2006) Minimum-Interference Channel Assignment in Multi-Radio Wireless Mesh Networks, Technical Report

Various Authors 2 (2006) http://www.ietf.org/html.charters/manet-charter.html

Various Authors 3 (2006) http://en.wikipedia.org/wiki/P2P

Various Authors 4 (2006) Autonomic Communication Forum. http://www.autonomic-communication-forum.org

Various Authors 5 (2006) Autonomic Communication Initiative. http://www.autonomic-communication.org

Various Authors 6 (2006) http://www.freifunk.net/

Various Authors 7 (2006) http://www.freenetworks.org/

Wikipedia Authors 1 (2006) http://en.wikipedia.org/wiki/Ad_hoc_routing_ protocol_ list

Wikipedia Authors 2 (2006) http://en.wikipedia.org/wiki/Autonomic_computing

Yang B, Liu D, Yang K (2002) Communication Performance Optimization for Mobile Agent System. In Proc. of the IEEE First International Conference on Machine Learning and Cybernetics (ICMLC 2002), pp 327-335 4-5 November, 2002, Beijing, China

2.4 Catalogue of Criteria for Autonomous Control in Logistics

Felix Böse, Katja Windt

Department of Planning and Control of Production Systems, BIBA
University of Bremen, Bremen, Germany

2.4.1 Introduction

Over the past years an increase in complexity of production and logistics systems regarding organisational, time-related and systemic aspects could be observed (Philipp et al. 2006). As a result, it is often impossible to make all necessary information available to a central entity in real time and to perform appropriate measures of control in terms of a defined target system. This development is caused by diverse changes, for example, short product life cycles as well as a decreasing number of lots with a simultaneously rising number of product variants and higher product complexity (Scherer 1998). Hence, new demands were placed on competitive companies, which cannot be fulfilled with conventional control methods. Conventional production systems are characterized by central planning and control processes, which do not allow fast and flexible adaptation to changing environmental influences. Establishing autonomous control seems to be an appropriate method to meet these requirements. The major aim of establishing autonomous logistics processes is to improve the logistics system's performance. The basis for achievement of this objective is a comprehensive understanding of the term autonomy in the context of logistics processes. The idea of autonomous control is to develop decentralised and heterarchical planning and controlling methods in contrast to existing central and hierarchical planning and controlling approaches (Scholz-Reiter et al. 2006). Autonomous decision functions are shifted to logistic objects. In the context of autonomous control, logistic objects are defined as material items (e.g. part, machine and conveyor) or immaterial items (e.g. production order) of a networked logistic system, which have the ability to interact with other logistic objects of the considered system. Autonomous logistic objects are able to act independently according to their own objectives and navigate through the production network themselves. The autonomy of logistic objects is possible due to recent developments by ICT (information and communication technologies), for example RFID technology (Radio Frequency Identification) for identification, GPS (Global Positioning System) for positioning or UMTS (Universal Mobile Telecommunications System) and WLAN (Wireless Local Area Network) for communication tasks (Böse and Lampe 2005). Furthermore comprehensive research in the field of agent-based computation in manufacturing (Monostori et al. 2006) is of particular importance for the implementation of autonomously controlled logistics systems.

These new approaches of autonomously controlled logistics systems are currently being investigated within the Collaborative Research Center 637 "Autonomous Cooperating Logistic Processes – A Paradigm Shift and its Limitations" at the University of Bremen, which deals with the implementation of autonomous control as a new paradigm for logistic processes (Scholz-Reiter et al. 2004).

The intention of this article is to explain what is meant by autonomous control and describe its main criteria in contrast to conventional controlling methods in logistic systems. Therefore, a definition of autonomous control is introduced. The constituent characteristics of this definition are considered in a developed catalogue of criteria in the form of an operationalised morphological characteristic schema in order to describe autonomous logistic processes and emphasize how conventionally managed and autonomous logistic processes differ. The catalogue of criteria represents an instrument that allows characterising a considered logistic system concerning its level of autonomous control. The criteria and their properties are explained in a concrete way by investigating a production logistics scenario of a job shop manufacturing system. In conclusion, further research activities concerning evaluation of autonomous control are presented.

2.4.2 Definition of autonomous control

The vision of autonomous control emphasizes the transfer of qualified capabilities to logistic objects as explained above. According to the system theory, there is a shift of capabilities from the total system to its system elements (Krallmann 2004). By using new technologies and methods, logistic objects are enabled to render decisions by themselves in a complex and dynamically changing environment. Based on the results of the work in the context of the CRC 637, autonomous control can be defined as follows:

"Autonomous Control describes processes of decentralized decision-making in heterarchical structures. It presumes interacting elements in non-deterministic systems, which possess the capability and possibility to render decisions independently.

The objective of Autonomous Control is the achievement of increased robustness and positive emergence of the total system due to distributed and flexible coping with dynamics and complexity." (Chapter 1 in this edited volume)

Based on this global definition of the term Autonomous Control, a definition in the context of engineering science was developed, which is focussed on the main tasks of logistic objects in autonomously controlled logistics systems:

"Autonomous Control in logistic systems is characterised by the ability of logistic objects to process information, to render and toexecute decisions on their own."

For a better understanding, terms in the given definitions of autonomous control such as decentralised decision-making in heterarchical systems, system elements ability of interaction as well as non-deterministic systems and positive emergence are described and discussed below.

Decentralised decision-making in heterarchical systems

One feature of autonomous control is the capability of system elements to render decisions independently. Autonomy in decision-making is enabled by the alignment of the system elements in the form of a heterarchical organisational structure (Goldammer 2006). Therefore, decentralisation of the decision-making process from the total system to the individual system elements is a specific criterion of autonomous control. Each system element represents a decision unit which is equipped with decision-making competence according to the current task (Frese et al. 1996). Due to the fact that decision-making processes are purposeful, according to the decision theory, each system element in an autonomously controlled system is characterised by target-oriented behaviour. Global objectives, for example, provided by the corporate management, can be modified independently by the system elements in compliance with their own prioritisation. For example, the objective low work in process can be replaced in favour of high machine utilization by the machine itself. Thus the objective system of single elements is dynamic because of ability to modify prioritisation of the objectives over time, i.e. during the production process.

System element's ability of interaction

Decentralized decision-making processes require the availability of relevant information for the system elements. Consequently, the capability of system elements to interact with other is a mandatory condition and thus one constitutive characteristic of autonomous control. The ability of interaction can accomplish different values depending on the level of autonomous control. The allocation of data, which other autonomous logistic objects can access, represents a low level of autonomous control. Communication, i.e. bi-directional data exchange between autonomous logistic objects, and coordination, i.e. the ability of autonomous logistic objects to cooperate and coordinate activities of other objects, represents higher level of autonomous control.

Non-deterministic system behaviour and positive emergence

In accordance with the above mentioned definition, the main objective of autonomous control is the achievement of increased robustness and positive emergence of the total system due to a distributed and flexible coping with dynamics and complexity. Non-determinism means that despite precise measurement of the system status and knowledge on all influencing variables of the system, no forecast of the system status can be made. Knowledge of all single steps between primary status and following status is not sufficient to describe the transformation completely (Flämig 1998). Thus a fundamental criterion of autonomous control is that for the same input and values, there are different possibilities for transition to a following status. As already explained, decentralisation of decision-making processes to the system elements leads to a higher flexibility of the total system because of the ability to react immediately to unforeseeable, dynamic influencing variables. In this way, autonomous control can lead to a higher robustness of the overall logistic system. Furthermore positive emergence is a main objective of autonomous control. Emergence stands for development of new structures or characteristics by concurrence of simple elements in a complex system. Positive emergence means that the concurrence of single elements leads to a better achievement of objectives of the total system than it is explicable by considering the behaviour of every single system element. That means, related to the context of autonomously controlled logistic processes, that autonomous control of individual logistic objects (e.g. machines, parts, orders) enables a better achievement of objectives of the total system than can be explained by individual consideration of the decentralised achievement of objectives (e.g. higher rate of on-time delivery, lower delivery times) of each single logistic object.

2.4.3 System layers of autonomous control

Based on the definition of autonomous control in the context of engineering science, its main characteristics are the ability of logistic objects to process information and render and execute decisions. Each characteristic can be assigned to different layers of work in an enterprise. In accordance with Ropohl (Ropohl 1979), different layers of work can be classified in organisation and management, informatics methods and information and communication technologies as well as in flow of material and logistics. These layers relate to decision, information and execution systems. Figure 2.2 presents the assignment of the characteristics to the system layers, illustrates their correlations and introduces the main criteria of autonomous control.

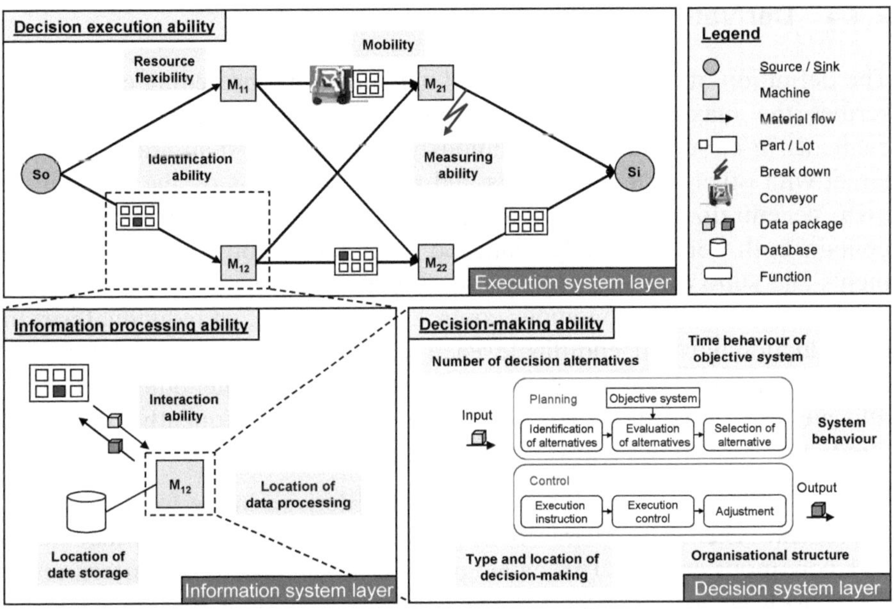

Fig. 2.2 System layers and criteria of autonomous control

The decision system is characterised by the decision-making ability. As mentioned before, in autonomously controlled production systems decision functions are shifted to logistic objects, which are aligned in a heterarchical organisational structure. These functions contain planning and control tasks and enable logistic objects to assign their progression. The decision-making process includes the identification and evaluation of decision alternatives on the basis of a decentralised objective system, the selection, instruction and execution of the best rated alternative, as well as possible adjustments.

The basis for decision-making is the information processing ability on the information system layer. In autonomously controlled production systems, logistic objects must be able to interact with each other as well as to store and to process data.

The execution system layer is characterised by the decision execution ability of logistic objects. Autonomous logistic objects are able to measure their current state and react flexibly to unforeseeable, dynamic influencing variables. Mobility and high flexibility of the resources are other main criteria of autonomous control in production systems.

2.4.4 Derivation of a catalogue of criteria

The definition of autonomous control explained in a preceding chapter describes the maximum level of autonomous control. Thus, all system-elements in an absolutely autonomously controlled system are able to interact with other system elements and render decisions on the basis of an own, decentralized target system. In general, logistics systems probably contain both conventionally managed and autonomously controlled elements and sub-systems. Furthermore, it is assumed that there are different degrees or levels of autonomous control. For example, an individual part in a production lot can coordinate each production step of the lot which represents a high level of autonomous control; meanwhile, other parts only allocate data regarding their processing states. Consequently, the latter mentioned case shows a lower level of autonomous control.

In the following, a catalogue of criteria is derived in the form of a morphological scheme for characterising logistic systems based on their level of autonomous control. This catalogue of criteria consists of thirteen criteria as well as corresponding properties, which allow a first approximate analysis of autonomously controlled logistic order processing. With respect to the derivation and definition of the constituent criteria, there was no predetermination concerning dedicated domains of corporate logistics (Wiendahl 2005). On the contrary, each criterion was defined with a very high degree of abstraction to enable a universal application in different fields of logistics, for example in production logistics as well as transportation logistics.

According to the morphological scheme for characterising structures of order processing (Luczak et al. 1998) several demands regarding selection and description of criteria are defined as follows:

- Each criterion must concern the organisation as well as the planning and control functions of a logistic system;
- Each criterion must sufficiently describe the field from conventional control to autonomous control in logistic systems in the form of its properties;
- Each criterion must allow measuring and evaluating of its properties with adequate accuracy;
- The application of each criterion must be possible with an appropriate effort.

Criteria category	Criteria	Properties			
Decision-making criteria	Time behaviour of objective system	static	mostly static	mostly dynamic	dynamic
	Organisational structure	hierarchical	mostly hierarchical	mostly heterarchical	heterarchical
	Number of decision alternatives	none	some	many	unlimited
	Type of decision making	static	rule-based		learning
	Location of decision making	system layer	subsystem layer		system-element layer
	System behaviour	elements and system deterministic	elements non-/ system deterministic	system non-/ elements deterministic	elements and system non-deterministic
Information processing criteria	Location of data storage	central	mostly central	mostly decentralised	decentralised
	Location of data processing	central	mostly central	mostly decentralised	decentralised
	Interaction ability	none	data allocation	communication	coordination
Decision execution criteria	Resource flexibility	inflexible	less flexible	flexible	highly flexible
	Identification ability	no elements identifiable	some elements identifiable	many elements identifiable	all elements identifiable
	Measuring ability	none	others	self	self and others
	Mobility	immobile	less mobile	mobile	highly mobile

increasing level of autonomous control →

Fig. 2.3 Criteria and properties

For the purpose of structuring of the catalogue of criteria, three categories are introduced based on the system layer of autonomously controlled logistics systems described in the preceding chapter. These categories are decision-making criteria, information processing criteria and decision execution criteria. In figure 2.3 the criteria and their properties for autonomously controlled systems are illustrated in the form of a morphological scheme that contains the main criteria of autonomous control and its properties, which represent the different levels of autonomous control.

The vision of autonomous control encompasses transferring qualified capabilities (e.g. decision-making, data processing, measuring) from the total system to the system elements. So the visualized criteria relate both to the total system and the system elements. Each criterion has a series of properties, with an increasing level of autonomous control in their order from left to right. For example, a logistic system with decentralised decision-making by its elements has a higher level of autonomous control than a system rendering central decisions.

2.4.5 Operationalisation of the catalogue of criteria

The catalogue of criteria as described above allows a qualitative determination of the level of autonomous control of a considered logistic system. So it is possible to describe a logistic system as mainly autonomously controlled or rather conventionally controlled by means of the property allocation with an increasing level of autonomous control in their order from left to right in figure 2.3 The catalogue of criteria allows basically a comparison of different logistics systems regarding their level of autonomous control. The remarks concerning the definition and description of the term autonomous control in the context of logistics explained in the chapters before suggest that the criteria do not all have the same influence on the determination of the level of autonomous control. For example the criterion location of decision-making seems to be a more important characteristic for autonomously controlled logistic systems than the criterion resource flexibility. For this reason an operationalisation of the catalogue of criteria seems necessary to ensure a precise determination of the level of autonomous control and allow an accurate comparison of logistic systems regarding their level of autonomous control.

For the purpose of evaluating the level of autonomous control of a considered logistics system the method of the value-benefit analysis, a frequently used evaluation method in practise, seems to be suited. Subject matter of the value-benefit analysis is the investigation of a number of complex alternatives in order to arrange these options according to the preferences of the decision maker by a multidimensional system of objectives in terms of values of benefit (Zangemeister 1976). In the present investigation the aim of the application of this method is not the determination of the top-rated alternative by means of a multidimensional system of objectives, but rather the evaluation of the level of autonomous control of a considered logistics system on the basis of constitutive criteria of autonomous control. However, the methodological procedure is the same except for the comparison of the total evaluation values of different alternatives which is not done in the case of the catalogue of criteria.

As a first step, each criterion of autonomous control is defined and assigned to the criteria categories: decision-making criteria, information processing criteria and decision execution criteria. After that, the weight of each criterion is ascertained. These weightings assign the importance of each criterion in the evaluation of the level of autonomous control. For the determination of the criteria weights, a systematic method in form of a pairwise comparison is made (Eversheim and Schuh 1996) as illustrated in figure 2.4.

Pairwise Comparison

Legend

2 = A is more important than B

1 = A is equal to B

0 = A is less important than B

		A												
		Time behaviour of objective system	Organisational structure	Number of decision alternatives	Type of decision-making	Location of decision-making	System behaviour	Location of data storage	Location of data processing	Interaction ability	Resource flexibility	Identification ability	Measuring ability	Mobility
	Time behaviour of objective system		2	2	0	2	2	0	0	2	0	0	0	0
	Organisational structure	0		1	0	2	1	0	0	2	0	0	0	0
	Number of decision alternatives	0	1		0	2	1	0	0	2	0	0	0	0
	Type of decision-making	2	2	2		2	1	0	1	2	0	0	0	0
	Location of decision-making	0	0	0	0		0	0	0	1	0	0	0	0
	System behaviour	0	1	1	1	2		0	0	2	0	0	0	0
B	Location of data storage	2	2	2	2	2	2		2	2	1	2	2	1
	Location of data processing	2	2	2	1	2	2	0		1	0	1	1	0
	Interaction ability	0	0	0	0	1	0	0	1		0	0	0	0
	Resource flexibility	2	2	2	2	2	2	1	2	2		2	2	0
	Identification ability	2	2	2	2	2	2	0	1	2	0		2	0
	Measuring ability	2	2	2	2	2	2	0	1	2	0	0		0
	Mobility	2	2	2	2	2	2	1	2	2	2	2	2	
Total		14	18	18	12	23	17	2	10	22	3	7	9	1
Priority		6	3	3	7	1	5	12	8	2	11	10	9	13
Weighting		9	12	12	8	15	11	1	6	14	2	4	6	1

Fig. 2.4 Pairwise comparison

Using this evaluation method, every criterion is compared with each other regarding its importance to determine the level of autonomous control. Accordingly it is investigated if criterion K_n is more important, is equal or is less important than criterion K_{n+1}. The results of the pairwise comparison are compiled in a two-dimensional matrix. By computing the total values for each criterion, the priority and consequently the weighting of each criterion can be determined, which describes the importance of each criterion concerning the evaluation of the level of autonomous control. The weightings of this pairwise comparison are a first possible result, which allows an approximate rating of the importance of each criterion to describe autonomous control in logistics.

As a second step, the considered logistics system is evaluated concerning the fulfilment of each criterion by selecting the corresponding property (compare following chapter). Each property of a criterion contains a fulfilment value which is uniformly distributed in the range of 0 (absolutely conventionally controlled) and 3 (absolutely autonomously controlled) with an increasing level of autonomous control in their order from left to right in figure 2.3. After that, weighted evaluation values are calculated by multiplication of weight and fulfilment of respective criteria. Finally, the total evaluation, i.e. the level of autonomous control, can be calculated by summarizing the weighted evaluation values. As a consequence the level of autonomous control in an absolutely conventionally controlled logistics system is 0 because all fulfilment values are 0, whereas the level of autonomous control in an absolutely autonomously controlled logistics system comes to a total evaluation value of 468. In general, the level of autonomous control probably lies in between these extreme total evaluation values.

2.4.6 Application of the catalogue of criteria

In this chapter, criteria and properties as well as the methodical approach to determine the level of autonomous control of a considered logistics system are illustrated using a production logistics scenario. Figure 2.5 gives an overview of a scenario of two-stage job shop production. Each criterion characterises the behaviour of logistic objects and is assigned to the decision-making system layer, information system layer or execution system layer.

The first production stage entails the manufacturing of a part on two alternative machines (M_{ij}). The raw materials that are needed for production are provided by the source (So). In the second production stage, the assembly of the parts that were produced in the first stage is done alternatively on two machines (A_{ij}). The manufactured items leave the material flow net at the sink (Si). At a pre-determined time a disturbance occurs in the form of a breakdown of machine A_{21}. In conventionally controlled production systems a machine breakdown at night would cause at least a delay of many hours before the disturbance is recognised and the production plan is adjusted in the traditional way.

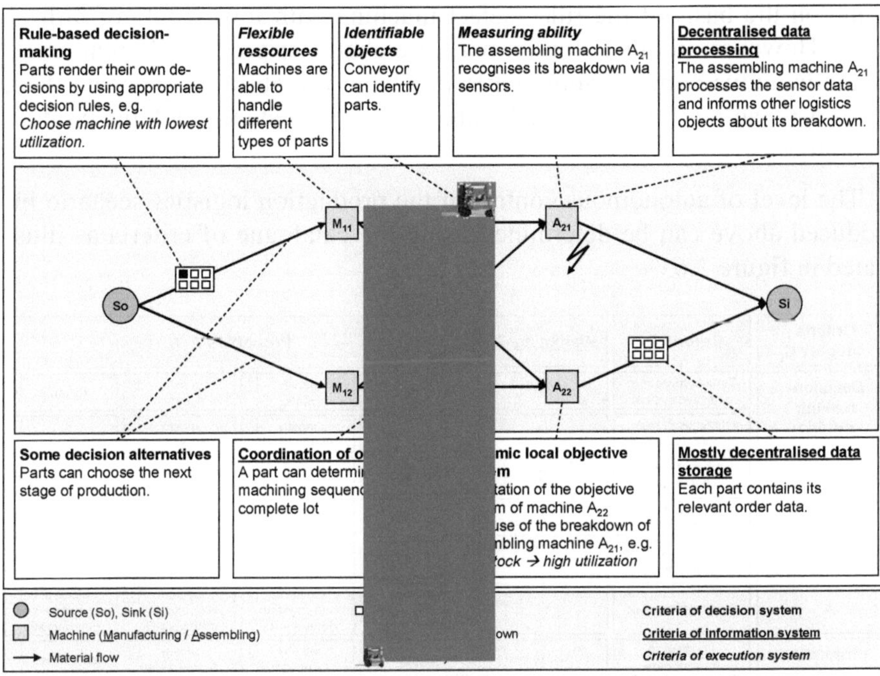

Fig. 2.5 Autonomously controlled production logistics scenario

The autonomous control of the machines provides the opportunity to react fast and flexibly to disturbances. Machine A_{21} autonomously recognises its breakdown by constant measuring and processing of the sensors data. Deviations of the sensors data are identified, analyzed and appropriate activities are initiated. In this scenario, the machine A_{21} immediately informs other logistic objects about its breakdown, especially machine A_{22}. Based on this information, machine A_{22} could adapt its dynamic local objective system by prioritizing the objective of high utilization instead of low stock to counteract the bottleneck of the assembly stage. Parts waiting in front of machine A_{21} are informed about the machine breakdown. Because of this information and their measuring ability, parts can define their position and initiate their own transport to the alternative machine A_{22}. Because of the identification ability, the conveyor is able to precisely identify the parts.

The existence of alternative manufacturing and assembling stages as well as the availability of local information allow parts to render decisions regarding their way through the production process. The decision-making process in this scenario is rule-based, i.e. logistic objects act according to defined rules. For example, a part could choose the manufacturing machine on the basis of the rule "select machine with lowest rate of utilization". However, in this scenario, parts are characterised by different levels of autonomous control. Some parts just have the ability to allocate data; other parts acting for the entire lot are able to navigate through the production process.

The level of autonomous control of the production logistics scenario introduced above can be determined using the catalogue of criteria as illustrated in figure 2.6.

Criteria category C_i	Criteria C_{ij}	Weighting G_{ij}	Properties P_{ij}			
Decision-making criteria	Time behaviour of objective system	9	static 0	mostly static 1	mostly dynamic 2	dynamic 3
	Organisational structure	12	hierarchical 0	mostly hierarchical 1	mostly heterarchical 2	heterarchical 3
	Number of decision alternatives	12	none 0	some 1	many 2	unlimited 3
	Type of decision making	8	static 0	rule-based 1,5		learning 3
	Location of decision making	15	system layer 0	subsystem layer 1,5		system-element layer 3
	System behaviour	11	elements and system deterministic 0	elements non-/ system deterministic 1	system non-/ elements deterministic 2	elements and system non-deterministic 3
Information processing criteria	Location of data storage	1	central 0	mostly central 1	mostly decentralised 2	decentralised 3
	Location of data processing	6	central 0	mostly central 1	mostly decentralised 2	decentralised 3
	Interaction ability	14	none 0	data allocation 1	communication 2	coordination 3
Decision execution criteria	Resource flexibility	2	inflexible 0	less flexible 1	flexible 2	highly flexible 3
	Identification ability	4	no elements identifiable 0	some elem. identifiable 1	many elem. identifiable 2	all elements identifiable 3
	Measuring ability	6	none 0	others 1	self 2	self and others 3
	Mobility	1	immobile 0	less mobile 1	mobile 2	highly mobile 3

Level of autonomous control

$$\sum_{i=0}^{n}\sum_{j=0}^{n} G_{ij} * p_{ij} = 220$$

C_i = Criteria category
C_{ij} = Criterion
G_{ij} = Weighting of criterion
P_{ij} = Property of criterion
p_{ij} = Fulfilment of criterion

Fig. 2.6 Application of catalogue of criteria

The properties of each criterion are ascertained on the basis of the description of the production logistics scenario. After that, weighted evaluation values are calculated by multiplication of criteria weighting as described in the preceding chapter and fulfilment of respective criterion. The total evaluation, which aggregates to 220, represents the total of all weighted evaluation values and is defined as level of autonomous control of the considered production system.

On the basis of this production logistics scenario it has been shown that each logistic system can be classified according to the level of autonomous control by means of the introduced catalogue of criteria. As a result the catalogue of criteria is an appropriate tool for comparing logistics systems regarding their level of autonomous control and therefore for evaluating fields of application of autonomous control, for example, by using simulation studies.

2.4.7 Conclusions and outlook

In this paper a catalogue of criteria was introduced to describe autonomous control in logistics systems. Based on the definition of autonomous control and its main characteristics in the context of logistics, the catalogue of criteria was developed. The catalogue of criteria represents an easy to use tool that affords an approximate analysis of a logistics system concerning its level of autonomous control. The catalogue of criteria allows both the characterisation of an existing as well as a future logistics system concerning its level of autonomous control by determination of the properties of each criterion. Furthermore, two different logistic systems can be compared regarding their level of autonomous control. The last mentioned point is of particular importance due to the fact that this comparison allows an evaluation of the fields of application of autonomy in logistics.

The application of autonomous control in logistics is based on the supposition that the allocation of planning and control tasks to autonomously controlled logistics objects results in a higher achievement of logistic objectives because of a better coping with high complexity in today's logistics systems. However, at a certain level of autonomous control, a decrease of the achievement of logistic objectives seems probable as a result of chaotic behaviour.

To verify this thesis an evaluation system for autonomously controlled logistics systems is necessary that meets the following demands:

- Determination of the level of autonomous control of the considered logistics system;
- Ascertaining of the level of complexity of the considered logistics system;
- Measuring of the logistic objective achievement of the considered logistics system.

Only if an evaluation system meets these demands, it is possible to make a statement on which level of complexity an autonomously controlled logistics system leads to a better achievement of logistic objectives compared to conventional control. Based on these demands an evaluation system of autonomously controlled logistics systems was developed, which is illustrated in figure 2.7.

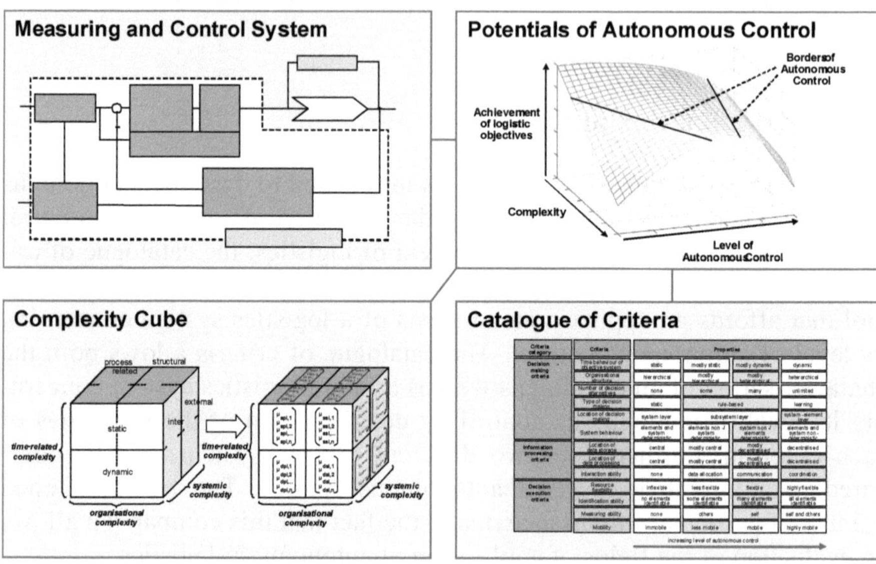

Fig. 2.7 Evaluation system of autonomously controlled logistics systems

Future research is directed to further development of the catalogue of criteria, especially detailing and completion of its criteria, as well as the advancement of the other components of the evaluation system pictured in figure 2.7. The complexity cube allows the description of the complexity of a considered logistics system regarding time-related, organisational and systemic aspects. By means of the measuring and control system, achievement of logistic objectives can be ascertained through comparison of target and actual logistic performance figures related to the objectives low work in process, high utilization, low throughput time and high due date punctuality. Through simulation studies using the developed evaluation system, it is anticipated that the borders of autonomous control can be found, specifying in which cases an increase of autonomous control does not lead to correspondingly higher performance of the logistics system.

References

Böse F, Lampe W (2005) Adoption of RFID in Logistics. In: Proceedings of 5th IBIMA International Business Information Management Association Conference, Cairo, pp 62-65

Eversheim W, Schuh G (1996) Hütte: Produktion und Management; Betriebshütte, Springer, Berlin

Flämig M (1998) Naturwissenschaftliche Weltbilder in Managementtheorien. Campus, Frankfurt New York

Frese E, Schmidt G, Hahn D, Horváth P (1996) Organisationsstrukturen und Management. In: Eversheim W, Schuh G (eds) Betriebshütte, Produktion und Management, Springer, Berlin New York

Goldammer E (2006) Heterarchie – Hierarchie: Zwei komplementäre Beschreibungskategorien. Download at 17.02.2006 from: http://www.vordenker.de/heterarchy/a_heterarchie.pdf

Krallmann H (2004) Systemanalyse in Unternehmen: partizipative Vorgehensmodelle, objekt- und prozessorientierte Analysen, flexible Organisationsarchitekturen. Oldenbourg, München Wien

Luczak H, Eversheim W, Schotten M (1998) Produktionsplanung und -steuerung: Grundlagen, Gestaltung und Konzepte. Springer, Berlin

Monostori L, Váncza J, Kumara SRT (2006) Agent-Based Systems for Manufacturing. In: Annals of the 56th CIRP General Assembly, Vol. 55/2

Philipp T, Böse F, Windt K (2006) Autonomously Controlled Processes - Characterisation of Complex Production Systems. In: Proceedings of 3rd CIRP Conference in Digital Enterprise Technology, Setubal, Portugal, forthcoming

Ropohl GJB (1979) Eine Systemtheorie der Technik – Grundlegung der Allgemeinen Theorie. Carl Hanser, München

Scherer E (1998) The Reality of Shop Floor Control – Approaches to Systems Innovation. In: Scherer E (eds) Shop Floor Control – A Systems Perspective. Springer, Berlin

Scholz-Reiter B, Windt K, Freitag M (2004) Autonomous Logistic Processes – New Demands and First Approaches. In: Proceedings of 37th CIRP International Seminar on Manufacturing Systems, Budapest, pp 357-362

Scholz-Reiter B, Windt K, Kolditz J, Böse F, Hildebrandt T, Philipp T, Höhns H (2006) New Concepts of Modelling and Evaluating Autonomous Logistic Processes. In: Chryssolouris G, Mourtzis D (eds) Manufacturing, Modelling, Management and Control, Elsevier, Oxford

Wiendahl H-P (2005) Betriebsorganisation für Ingenieure, Carl Hanser, München

Zangemeister C (1976) Nutzwertanalyse in der Systemtechnik: eine Methodik zur multidimensionalen Bewertung und Auswahl von Projektalternativen. Wittemann, München

Windt K, Hülsmann M (2007) Changing Paradigms in Logistics - Understanding the Shift from Conventional Control to Autonomous Co-operation and Control. In: Hülsmann M, Windt K (eds) Understanding Autonomous Cooperation and Control - The Impact of Autonomy on Management, Information, Communication, and Material Flow. Springer, Heidelberg

2.5 Business Process Modelling of Autonomously Controlled Production Systems

Felix Böse, Katja Windt

Department of Planning and Control of Production Systems, BIBA
University of Bremen, Germany

2.5.1 Introduction

Conventional production systems are characterised by central planning and control methods, which show a wide range of weaknesses regarding flexibility and adaptability of the production system to environmental influences. Centralised planning and control methods are based on simplified premises (predictable throughput times, fix processing times of production orders etc.), which lead to an inadequate and unrealistic description of the production system. The different centralised planning steps of the traditional ERP respectively MRP based PPC-Systems are executed sequentially, therefore adaptation to changing boundary conditions (e.g. planning data) is only possible within long time intervals. This means that changes to the job shop situation cannot be considered immediately, but during next planning run at the earliest. As a result, current planning is based on old data and the needed adaptation measures cannot be performed in time for a proper reaction of the discrepancy between the planned and the current situation (Scholz-Reiter et al. 2006). In case of disturbances or fluctuating demands, centralised planning and control methods are insufficient to deal with the complexity of the comprehensive planning tasks of centralised systems, which rises disproportionately to their size and heavily constrains fault tolerance and flexibility of the overall system (Kim and Duffie 2004; Prabhu and Duffie 1995).

These weaknesses of conventional logistic planning and control systems require a fundamental reorganisation. In recent scientific research, the concept of autonomously controlled logistic systems as an innovative approach of a decentralised planning and control system is investigated, which meets the increasing requirements of flexible and efficient order processing (Freitag et al. 2004; Pfohl and Wimmer 2006). To establish the logistic concept of autonomous control adequate modelling methods are needed which allow an exact description of autonomously controlled logistic processes.

In this paper a definition of the term autonomous control in logistics and a specification of its main criteria are introduced. Based on this, the ARIS (Architecture of Integrated Information Systems) concept as an integrated method for the modelling of processes and information systems is analysed regarding its suitability to describe autonomous control in production systems. Afterwards, changes in order processing are exemplarily illustrated in several views of a business process model using the ARIS concept. The paper ends with a short summary and an outlook in respect of further research activities.

This research is funded by the German Research Foundation (DFG) as part of the Collaborative Research Centre 637 "Autonomous Cooperating Logistic Processes: A Paradigm Shift and its Limitations" (SFB 637) at the University of Bremen.

2.5.2 Autonomous control in production systems

The idea of autonomously controlled logistic processes is to develop decentralised and heterarchical planning and control methods in contrast to existing central and hierarchical aligned planning and controlling approaches (Scholz-Reiter et al. 2006). According to the system theory, there is a shift of capabilities from the total system to its system elements (Krallmann 2004). Consequently, decision functions are transferred to autonomous logistic objects, which are defined as physical items (e.g. part, machine, conveyor) or logical items (e.g. production order) of a networked logistic system. Autonomous logistic objects have the ability to interact with other logistic objects of the considered system and are able to act independently according to their own objectives and navigate through the production network themselves (Windt et al. 2005). To achieve this, logistic objects are enabled to render decisions by themselves in a complex and dynamically changing environment by using new information and communication technologies as well as planning and control methods.

Based on the results of the work in the context of the above mentioned Collaborative Research Centre 637 "Autonomous Cooperating Logistic Processes - A Paradigm Shift and its Limitations" at the University of Bremen (Scholz-Reiter et al. 2004), autonomous control can be defined as follows: "Autonomous Control describes processes of decentralized decision-making in heterarchical structures. It presumes interacting elements in non-deterministic systems, which possess the capability and possibility to render decisions independently."(Windt and Hülsmann 2007). Based on this definition, the main constitutive criteria of autonomous control can be described as follows: heterarchical structures of the logistic system, decentralised decision-making by autonomous system elements with an own objective system, system element's ability of interaction as well as non-deterministic system behaviour (for a more comprehensive characterisation of the criteria compare Böse and Windt 2007; Windt 2006).

2.5.3 Business process modelling of autonomous control

To answer the question concerning the suitability of existing models to describe autonomously controlled logistic systems, several process studies using the ARIS concept are executed by means of existing reference models of logistic order processing (Loos 1992; Luczak et al. 1998; Scheer 1995; Schönsleben 2001). The ARIS concept as an integrated method for the modelling of processes and information systems provides several views on a system: the data view, the function view, the organisational view and the control view, which uses the EPC (Event-driven Process Chain) as modelling notation (figure 2.8).

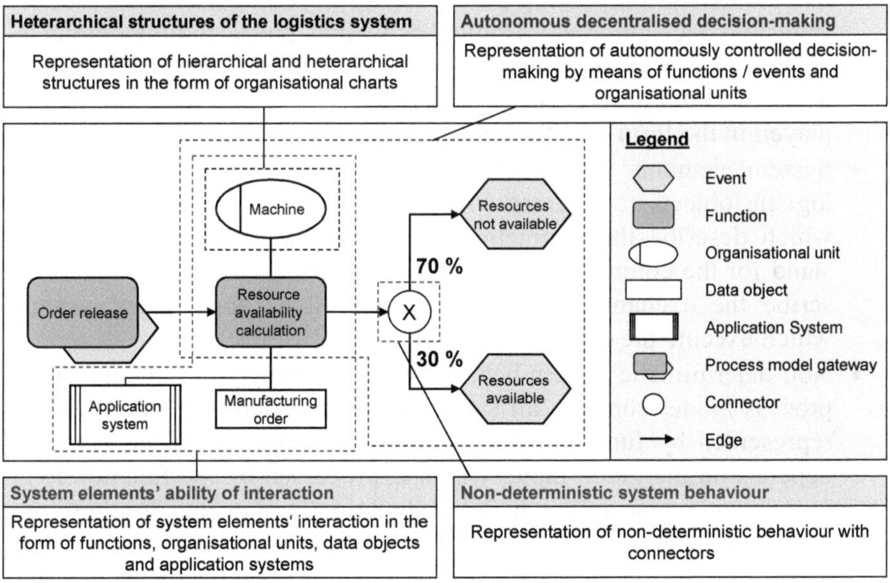

Fig. 2.8 Modelling criteria of autonomous control using EPC

A modelling method that is suited for modelling autonomous control must be able to represent the criteria of autonomous control described in the preceding chapter. Figure 2.8 illustrates the possibilities of EPC to model autonomously controlled logistic processes using as example the business process of the resource availability calculation. After order release a machine proves the availability of all needed resources to a given manufacturing order. Possible results of this function are the availability or unavailability of the necessary resources. In the given example the criteria of autonomous control can be represented as follows:

- Heterarchical structures of the logistic system: Both hierarchical and heterarchical organisational structures can be represented in the form of organisational charts. In some reference models of production systems, logistic objects (e.g. machines) are partly described as organisational units (Scheer 1995). Consequentially, in autonomously controlled logistic systems autonomous logistic objects acting as decision-making units are displayed in the form of organisational units;
- Autonomous decentralised decision-making: The criterion of decentralised and autonomously controlled decision-making executed by logistic objects can be described using several elements of EPC notation. The decision-making process is displayed by a function, the responsible decision-maker (logical as well as physical autonomous logistic objects) by an organisational unit and the possible results of decision-making by events. Various decision alternatives can be displayed in the form of different functions;
- System elements' ability of interaction: The ability of autonomous logistic objects to interact with others is represented by functions, which describe the interaction process, organisational units, which stand for the communicating logistic objects, data objects, which describe the exchanged information as well as application systems, which execute the data exchange on the software level;
- Non-deterministic system behaviour: A completely designed business process model contains all states of a considered system, which are represented by functions and events. The sequence of the functions depends on the given input, which is processed by the function to an output, and connectors, which link functions and events. Furthermore, connectors can present stochastic effects in the form of probabilities. Therefore, the specific sequence of functions and events results during run time, which leads to a non-deterministic behaviour of the considered system.

2.5.4 Changes in order processing by autonomous control

The definition of the term autonomous control in logistics and the description of its main criteria are the basis of a comprehensive investigation of the changes in order processing caused by establishing of autonomous control. Focus of interest is the question, to what extent existing models of logistic order processing are suited for modelling autonomously controlled logistic processes, respectively which modifications are necessary. The range of required modifications depends on the level of autonomous control of the considered production system. The definition of autonomous control explained in a preceding chapter describes the maximum level of imaginable autonomous control. But autonomously controlled logistic systems will probably contain both: conventionally managed and autonomously controlled elements and sub-systems.

In the following, essential modifications of existing reference models because of changes in logistic order processing due to establishing autonomous control are introduced. For this purpose the modifications in every single view of the ARIS concept are exemplarily illustrated (compare figure 2.9) and shortly described.

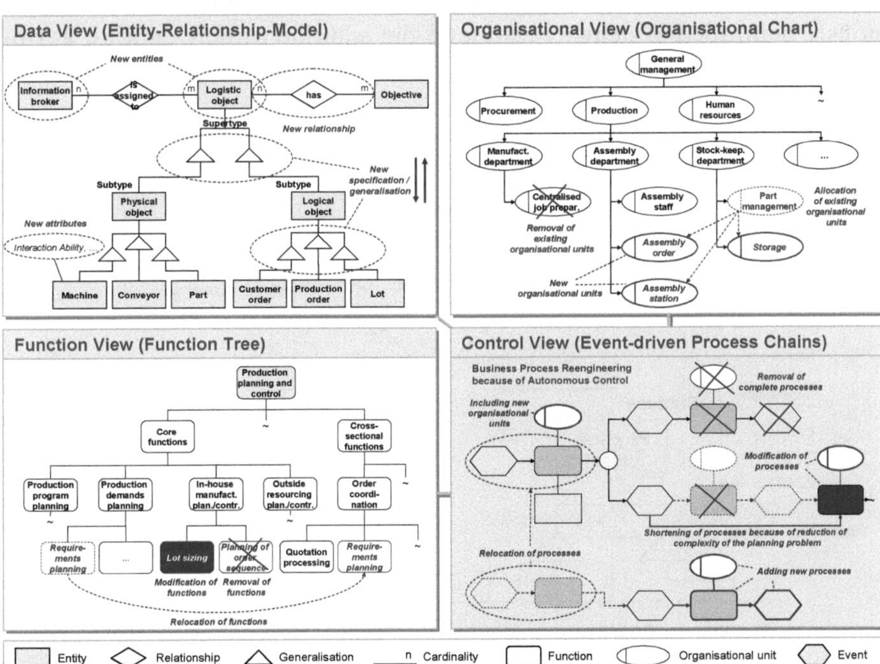

Fig. 2.9 Modelling views of autonomously controlled logistic processes

Data view

Existing data models, for example in the form of an ERM (Entity-Relationship-Model), are not sufficient to adequately describe entities in autonomously controlled logistic systems, but have to be extended by new entities, attributes, relationships as well as specification / generalisations (figure 2.10). As described in a preceding chapter, in autonomously controlled systems autonomous logistic objects, both physical and logical objects, have the ability to interact with other logistic objects of the considered system, to act independently according to their own objectives and navigate them through the production network. Considering these criteria of autonomously controlled logistic systems, the logistic object as well as the physical and logical object has to be added as new entities just as the belonging generalisation between these new entities. Furthermore, there is a new relationship from the entity objective to the new entity logistic object. A new entity information broker is introduced to represent special information broker objects. These objects are needed in autonomously controlled systems to register the logistic objects at certain process stations and provide communication links between them. Also, the interaction ability has to be added as an attribute of an entity, for example as attribute of the entity machine. Accordingly, the complexity of data models of autonomously controlled systems rises due to the addition of new entities, attributes, relationships as well as specifications / generalisations.

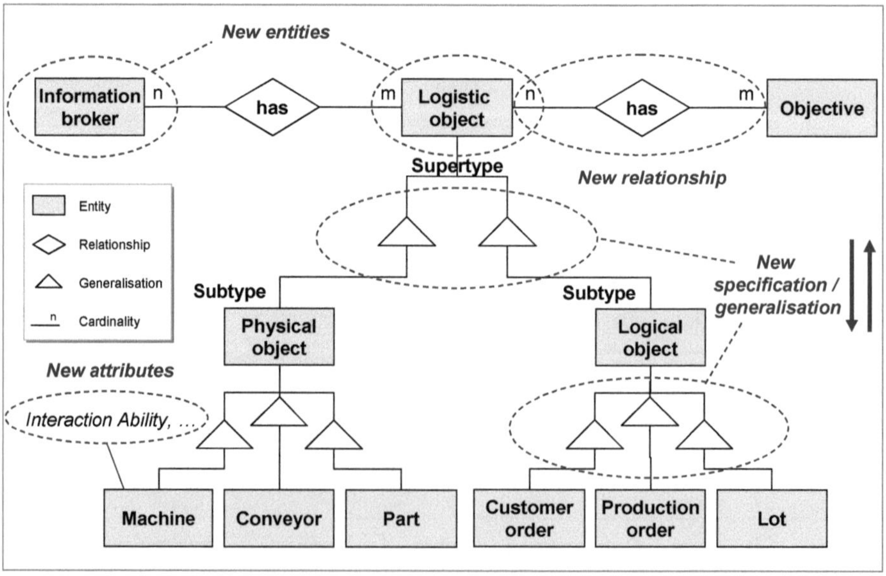

Fig. 2.10 Data view of an autonomously controlled logistic system

Function view

The modifications of the function models, pictured in the form of a function tree (figure 2.11), are determined by the decentralisation of the planning and control functions to the logistic objects, that requires a relocation respectively reorganisation of several functions. For example, in autonomously controlled logistic systems there is no longer a centralised requirements planning. Instead, this function is moved from the centralised system to the logistic object order and with it assigned to the superior function order coordination. Some functions of conventional planning and control systems, which are executed centralised for several logistic objects, are removed. For instance, the function planning of order sequence in the field of in-house manufacturing and control is no longer needed because the control of order sequence happens at run time by the machines themselves. Other functions still remain, but require an alteration. For example the activities within the function lot sizing is simplified due to the fact that based on the decentralisation of the planning and control functions there is no longer a centralised lot sizing, but a local lot sizing coordinated by a single machine.

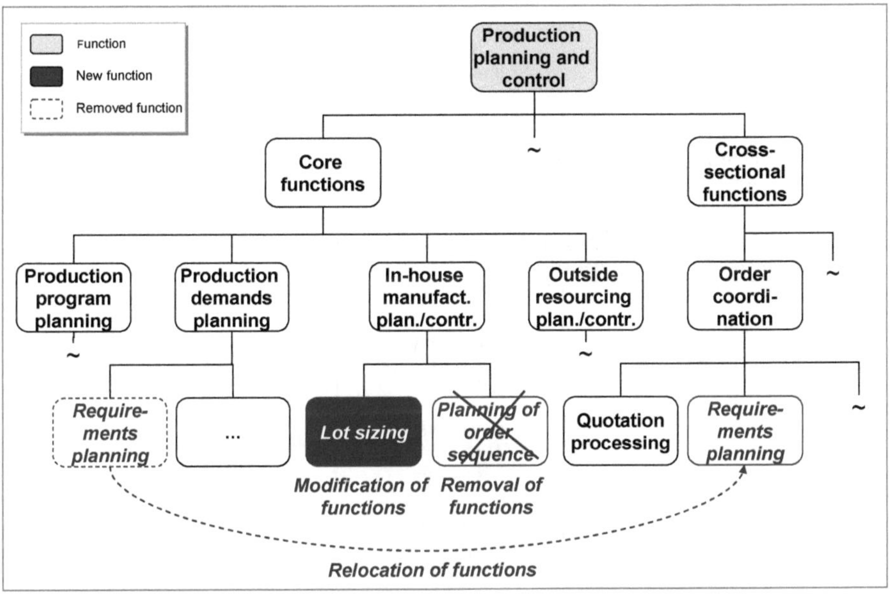

Fig. 2.11 Function view of an autonomously controlled logistic system

Organisational view

There are several changes concerning the organisational structure caused by establishing autonomous control in order processing due to the fact that in autonomously controlled systems logistic objects are able to initiate and execute functions (figure 2.12). Because of the relocation of functions to the logistic objects some centralized organisational units are no longer needed, for example the organisational unit centralised job preparation. Some organisational units are substituted by other, partly new organisational units. So the organisational unit part management is replaced by several logistic objects such as storage, assembly order and assembly station, which are added as new organisational units. Thus a logistic object can function both as an entity and organisational unit. Even though in autonomously controlled logistic systems logistic objects are able to initiate and execute functions within order processing, it is highly doubtful, whether they can take on a responsibility for the related functions or their results. On the contrary, it makes sense, that not the single logistic object but rather the superior "human" organisational unit is responsible for the results and (unintended) effects of the functions.

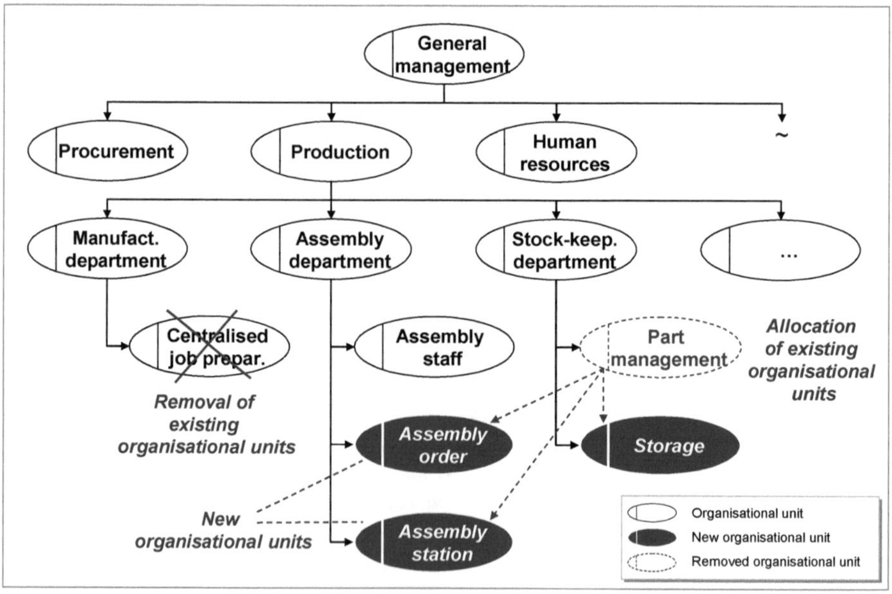

Fig. 2.12 Organisational view of an autonomously controlled logistic system

Control view

The modifications of the data, function and organisational view are reflected in the control view, which contains the adaptations of the business processes of the planning and control system caused by establishing autonomous control. The wide range of manifold modifications of the several views results in a corresponding number of modifications in the control view. In the context of this paper, only a common illustration of the changes is introduced (figure 2.13). As described above, new organisational units as well as new entities have to be included because of the existence of autonomous logistic objects. The decentralisation of planning and control functions to the logistic objects causes relocations of processes within the work flow, removals of complete processes as well as shortenings and modifications (modifications of functions, replacing organisational units, adding new entities etc.) of logistic processes. This results in two different effects on the complexity of the business process model. The decentralisation of the planning and control tasks reduces the need for long and complicated process chains of planning and control tasks. However, it also results in an increasing number of (redundant) processes and thus leads to a higher complexity of the business process model.

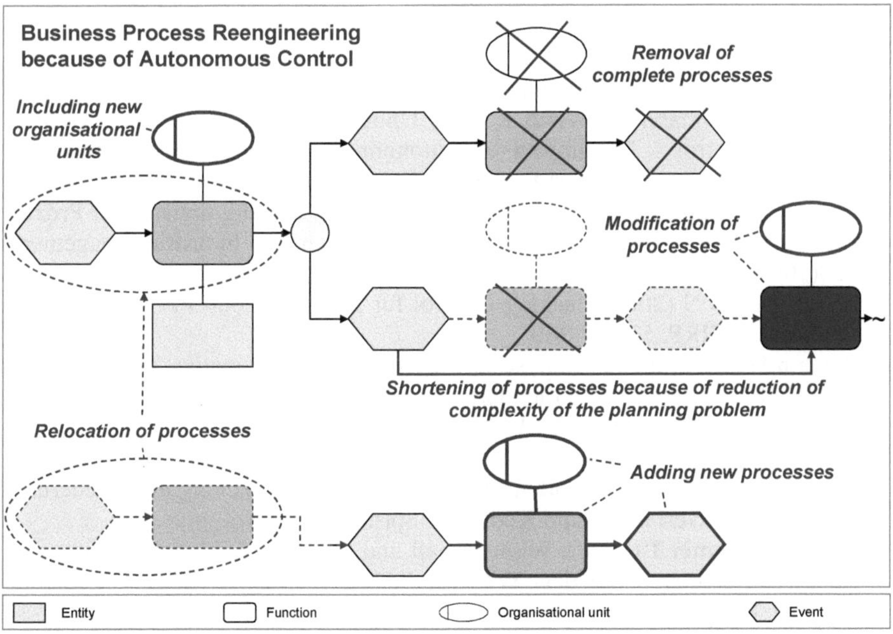

Fig. 2.13 Control view of an autonomously controlled logistic system

2.5.5 Conclusions

In the context of this paper the concept of autonomously controlled logistic systems was introduced as an innovative approach of a decentralised planning and control system, which meets the new requirements of flexible and efficient order processing. Based on the definition of the term autonomous control within the scope of logistics and the constitutive criteria, the ARIS concept was analysed regarding its suitability to describe autonomous control in production systems. Furthermore, it has been shown, that there are several changes in order processing of production systems caused by establishing autonomous control, which are not sufficiently considered in existing models of logistic order processing. Manifold modifications of existing models are necessary, which depend on the level of autonomous control of the considered production system. Using the ARIS concept, several modifications were introduced regarding data, function, organisational and control view. Future research is focused on the detailed investigation and modelling of changes in logistic order processing by establishing autonomous control. Main objective is the development of a reference model of the autonomously controlled logistic order processing using the example of a job shop manufacturing scenario.

References

Böse F, Windt K (2007) Catalogue of Criteria for Autonomous Control in Logistics. In: Hülsmann M, Windt K (eds) Understanding Autonomous Cooperation & Control - The Impact of Autonomy on Management, Information, Communication, and Material Flow. Springer, Heidelberg

Freitag M, Scholz-Reiter B, Herzog O (2004) Selbststeuerung logistischer Prozesse – Ein Paradigmenwechsel und seine Grenzen. In: Industriemanagement, 20(2004)1: 23-27

Kim J-H, Duffie N (2004) Backlog Control for a Closed Loop PPC System. Annals of the CIRP, 53:357-360

Krallmann H (2004) Systemanalyse in Unternehmen: partizipative Vorgehensmodelle, objekt- und prozessorientierte Analysen, flexible Organisationsarchitekturen. Oldenbourg, München Wien

Loos P (1992) Datenstrukturierung in der Fertigung, Oldenbourg, München

Luczak H, Eversheim W, Schotten M (1998) Produktionsplanung und -steuerung: Grundlagen, Gestaltung und Konzepte. Springer, Berlin

Pfohl H-C, Wimmer T (2006): Wissenschaft und Praxis im Dialog: Steuerung von Logistiksystemen – auf dem Weg zur Selbststeuerung. Deutscher Verkehrs-Verlag, Hamburg

Prabhu V, Duffie, N (1995) Modelling and Analysis of nonlinear Dynamics in Autonomous Heterarchical Manufacturing Systems Control. Annals of the CIRP, 44:425-428

Scheer A (1995) Wirtschaftsinformatik: Referenzmodelle für industrielle Geschäftsprozesse. Springer, Berlin

Schönsleben P (2001) Integrales Informationsmanagement: Informationssysteme für Geschäftsprozesse – Management, Modellierung, Lebenszyklus und Technologie, Springer, Berlin

Scholz-Reiter B, Windt K, Freitag M (2004) Autonomous Logistic Processes – New Demands and First Approaches. In: Proceedings of 37th CIRP International Seminar on Manufacturing Systems, Budapest, pp 357–362

Scholz-Reiter B, Windt K, Kolditz J, Böse F, Hildebrandt T, Philipp T, Höhns H (2006) New Concepts of Modelling and Evaluating Autonomous Logistic Processes. In: Chryssolouris G, Mourtzis D (eds) Manufacturing, Modelling, Management and Control, Elsevier, Oxford

Windt K, Böse F, Philipp T (2005) Criteria and Application of Autonomous Cooperating Logistic Processes. In: Gao J, Baxter D, Sackett P (eds): Proceedings of the 3rd International Conference on Manufacturing Research – Advances in Manufacturing Technology and Management, Cranfield

Windt K (2006): Selbststeuerung intelligenter Objekte in der Logistik. In: Vec M, Hütt M, Freund A (eds): Selbstorganisation – Ein Denksystem für Natur und Gesellschaft. Böhlau Verlag, Köln

Windt K, Böse F, Philipp T (2007) Autonomy in Logistics – Identification, Characterisation and Application. International Journal of Robotics and CIM, Pergamon Press Ltd, forthcoming

Windt K, Hülsmann M (2007) Changing Paradigms in Logistics - Understanding the Shift from Conventional Control to Autonomous Co-operation and Control. In: Hülsmann M, Windt K (eds) U Understanding Autonomous Cooperation and Control - The Impact of Autonomy on Management, Information, Communication, and Material Flow. Springer, Heidelberg

2.6 Strategic Decisions for Autonomous Logistics Systems

Lars Arndt, Georg Müller-Christ

Chair of Sustainable Management, Department of Economics and Business Studies, University of Bremen, Germany

2.6.1 Introduction

Logistics management is currently facing major challenges. The integration of the value chain and the growing importance of spatially and organisationally distributed production networks strongly increase the need for logistical coordination. Growing customer orientation requires product customization and increased responsiveness in order delivery, thereby raising flexibility and reactivity requirements within the whole supply chain. These developments contribute to the increase in structural and dynamic complexity of logistics systems, thus complicating central planning and control of logistics processes.

Research on autonomous cooperating logistics processes confronts these challenges by proposing to replace central planning and control with decentral, autonomous coordination. While former concepts of organisational decentralisation implied an increase in the autonomy of employees, autonomous cooperation in logistics is primarily based on the capability of logistics objects to decide and coordinate themselves. Scholz-Reiter et al. describe the scenario of autonomous cooperation in logistics as follows:

"Imagine decentralized distributed architectures of intelligent and communicating objects instead of today's centralized control of non-intelligent objects in hierarchical structures (...). The flow of goods is no longer controlled by a central instance. Instead, the package is finding its way through the transport network to the destination autonomously while constantly communicating with conveyances and nodes and considering demands, e.g. concerning delivery date and costs." (Scholz-Reiter et al.2004)

Autonomous cooperation in logistics promises higher efficiency as well as increased flexibility and robustness even in complex logistics systems.

While it is based on the application of several new technologies (cp. Scholz-Reiter et al. 2004), multi-agent technology plays the most prominent role in regard to the actual ability of local self-coordination. Although this technology is already applied on several layers of the supply chain, e.g. in industrial production (Van Dyke Parunak 2000) or in transport logistics (Graudina and Grundspenkis 2005; Davidsson et al. 2005), a comprehensive and integrative automation of decision making in the supply chain is still a vision for the future. Not only remaining technical restrictions but also organisational factors act as constraints on the application of multi-agent technology in practice. As Janssen notes, “the prospect of delegating routine supply chain decisions to software agents still makes many managers nervous” (Janssen 2005: 316).

While the question how to convince managers of the advantages of multi-agent technology has been addressed by several authors (Van Dyke Parunak 2000; Janssen 2005), the character of the underlying decision problem remains unclear. In this article, we deal with this decision problem by elaborating on its strategic nature, which has to be appropriately comprehended in order to understand the difficulties related to the decision about autonomous cooperation and possible ways to address them. For these purposes, this article

- describes autonomous cooperation in logistics as a particular form of delegation of decision making;
- attributes the strategic character of this delegation process to the necessity for organisations to open their boundaries;
- outlines a concept of boundary management in order to foster and regulate the boundary opening and thus to provide the appropriate organisational context for the decision to implement autonomous cooperation.

2.6.2 Autonomous cooperation in logistics as delegation of decision making

In this article, we suggest that it is not possible to capture the strategic relevance of autonomous cooperation by comprehending it as a mere technological innovation potentially providing a competitive advantage. Instead, we propose to focus on the issue of delegation of decision making, which shall be explained in the following.

It has already been indicated that multi-agent systems (MAS) play a crucial role in regard to the ability of logistics objects to coordinate and decide for themselves. MAS consist of interacting, intelligent agents, i.e.

of „autonomous, computational entities that can be viewed as perceiving their environment through sensors and acting upon their environment through effectors" (Weiss 1999: 2) and which are able to "pursue their goals and execute their tasks such that they optimize some given performance measures" (Weiss 1999: 2). Intelligent agents can fulfil different functions in logistics processes like representing individual logistics objects and the related objectives or mediating the coordination process between other agents. The possibility to represent distinct entities with potentially conflicting interests and the ability to act on the basis of local knowledge make MAS an attractive solution for the decentral coordination of logistics processes. Besides the agents' ability to learn, the particular problem solving capability of MAS is mainly based on the agents' cooperation, i.e. it emerges through their interactions (Chainbi et al. 2001; Odell 2002).

Considering the ability to learn and the emergence of the problem solving capability, the notion of technology reaches its limits in the context of MAS. Understanding technology ("Technik") as tight coupling of causal elements (Luhmann 2000), it is obvious that the notion of a technical system does not describe agent-based autonomous cooperation appropriately. Technology refers to the use of isolated causal relations in order to achieve some intended effects on the basis of defined preconditions (Baecker 2005). Autonomous cooperation, however, is supposed to enable problem solving in situations, where technology reaches its limits, i.e. where neither causal relations nor preconditions can be operationalised unambiguously and the intended effects are themselves dependent on the former.

From the perspective of the organisation, operations of MAS are characterized by a high degree of contingency[1] untypical for technology. Contingency refers to the large number of possible results these operations can achieve. Consequently, the organisation is confronted with uncertainty with regard to their outcomes and thus with a loss of control similar to the case of delegation of decision making to human agents (Laux and Liermann 2003). In order to substantiate this similarity we briefly address the question, whether agents' operations can be perceived as decision making[2]. In this article, we refer to the notion of decision brought forward by the so-

[1] The issue of contingency in the context of MAS is e.g. discussed in Dryer (1999) and Paetow and Schmitt (2002).

[2] The terms 'decision' and 'delegation' are sometimes referred to in the literature on MAS (Castelfranchi and Falcone 1998). However, we do not intend to review these discussions here. For our purpose, it is sufficient to understand how the related problems are perceived from an organisational perspective.

ciologist Niklas Luhmann (Luhmann 2000). According to Luhmann, decision making can be comprehended as a basic form of dealing with the contingencies organisations face in their everyday operations. Organisations use decisions to transform open contingency, i.e. the existence of several alternatives to act before the decision, into closed contigency after the decision, when one alternative has been chosen and the others remain in the background as excluded possibilities only (Luhmann 2000). Referring to this understanding, it can be argued that decisions process contingency. Technology as a causal simplification, in contrast, only works if these contingencies are suppressed. In order to successfully utilise technology, contingency has to be eliminated first. Yet, MAS function in a different manner; they actively and adaptively develop situation-aware methods to address contingency and uncertainty. This implies, however, that their actual behaviour cannot be easily predicted by an external observer. Paetow and Schmitt (2002) thus refer to MAS as technical systems with non-technical properties.

Consequently, from the point of view of the organisation, implementation of autonomous cooperation in logistics indeed can be viewed as a process of delegation of decision making, accompanied by a loss of control as a typical side effect, which is likely to be one of the main problems in the context of the decision about autonomous cooperation.

In the following, we use concepts from New Systems Theory (especially Luhmann's theory of social systems) to further analyse autonomous cooperation as delegation of decision making. We show that the strategic character of this delegation is based on the necessity for organisations to open their boundaries. In comparison with economic theories addressing the issue of delegation, like the agency theory, Luhmann's theory offers two advantages. Firstly, it relieves us from the necessity to deal with the applicability of restrictive theoretical assumptions (e.g. the agency theory's notion of bounded rational, opportunistic, self-interested agents) to MAS. Secondly, Luhmann understands organisations as recursive unities of decisions and connects the way these unities structure decision making processes to their ability to reproduce themselves. This understanding seems especially appropriate when dealing with the strategic nature of the delegation of decision making.

2.6.3 Delegation of decision making as a process of boundary opening and its strategic relevance

Speaking of boundary opening, we first have to address basic concepts of openness and closeness of organisations. The idea of organisations being open systems has a long tradition in organisation theory (cp. Scott 1998). It implies that organisations rely on a constant throughput of resources (flows of energy, material and information) to secure their reproduction. By particularly emphasising the issue of information and its processing within organisations, the open systems approach has itself laid the foundation for the notion of (informational) closure. This does not necessarily mean to give up the concept of openness. Remer (2002), for example, notes that organisations are able to sustain themselves only if they are materially open but closed with regard to 'ideal' matters like identity.

Considering it as the basic prerequisite for the organisation's self-reproduction, Luhmann (1984) offers the most consequent notion of informational closeness. He proposes to substitute the notion of self-referential closure for thc distinction between open and closed systems. The meaning of self-referential closure in the organisational context can only be grasped if organisations are understood as systems based on sense (Luhmann 2000). They emerge through sense-based selections referring to each other and thus stabilising as a condensed unity distinguishing itself from its environment through selectively reduced complexity. The boundary between an organisation and its environment thus marks a difference in complexity. On the inside of this boundary, the organisation can develop a specific identity, whereas the outside is perceived as environment. As the demarcation is the result of the organisation's internal activities, in a sense, the organisation constructs its own environment. As Seidl and Becker describe it, organisations "come into being by permanently constructing and reconstructing themselves by means of using distinctions, which mark what is part of their realm and what not" (Seidl and Becker2006: 9).

The sustainment of the organisation as a unity distinct from its environment is directly linked to the maintenance of its boundaries. Thus, the question of „boundaries is central, not peripheral to organisations" (Hernes 2004: 10). The same holds true for the issue of boundary maintenance which is not a function at the periphery of the organisation, but a core problem, which all operations refer to in one way or another.

In the context of sense systems, we can comprehend self-referential closure as simultaneity of closeness and openness. According to the New Systems Theory, openness is based on a double closure; double closure means that, first of all, systems are closed in regard to their basal self-reference (often termed 'autopoiesis'). We can speak of basal self-reference when systems reproduce their elements exclusively by means of already existing elements and their relations. Systems are considered doubly closed if they are able to refer to or reflect on themselves on the basis of this basal self-reference (Luhmann 1984). As we are dealing with sense systems, this can only be achieved by means of distinctions; the system refers to itself by internally operating on the distinction between system (self-reference) and the environment (external reference). Double closure thus, in a sense, enables openness towards the environment (Luhmann 1984). By openness, however, we mean a cognitive openness, which a self-referential social system uses to condition its own operations.

In order to fully comprehend the simultaneity of openness and closeness of organisations, we have to take a closer look at Luhmann's notion of organisation. According to Luhmann, organisations (re-)produce themselves as social orders by means of decisions about their practices and procedures. Thus, organisations have to be understood as recursive unities of decisions. They are self-referentially closed systems as one decision has to connect to another decision to secure their continued existence. They are cognitively open systems, however, because their decisions permanently refer to their environment. Decisions represent organisations' specific form of operations, by which they conduct sense-based selections and thus distinguish what belongs to their 'realm' and what belongs to the environment. They are means to transform the uncertainty related to contingency ("What is the right choice?") into a temporary, self-produced relative certainty to which further decisions can refer.

We have already indicated that autonomous cooperation can be perceived as a process of boundary opening. Yet, if organisations are permanently characterised by simultaneity of openness and closeness, which meaning has the notion of boundary opening?

According to the above remarks, boundary opening refers to an organisation's cognitive openness and implies an expansion of the part of the world which has been made accessible by the organisation. On the basis of such an enhanced view of the world, the organisation is potentially able to modify its operations. This, however, can only be realised if external references are successfully connected to the own operations on the basis of reflexive closure. Therefore, we can argue that opening and closure condition themselves reciprocally. They are two different sides of the same process, namely the positioning of the system within its environment and thus the permanent operational confirmation or modification of the system's boundaries. Luhmann (2000) notes that systems oscillate between external references and self-reference. Organisational boundaries are the result of this oscillation process and as such in permanent motion. At every point in time they represent the organisation's only temporarily valid understanding of itself and its environment. As a result of previous operations they contain knowledge of successful or failed strategies of the past and thus offer hints for the future development; at the same time, however, they restrict the possibilities of organisations to change. Hernes (2003) correspondingly speaks of the "enabling and constraining properties of organisational boundaries".

Oscillation between opening and closure, i.e. the permanent operational confirmation or modification of organisational boundaries, enables the organisation to stabilize in its environment. When this process is interrupted, for example by rigidly clinging to given boundaries, the viability of the organisation is endangered as its fit with the environment is at risk. Earlier, we have emphasised that boundary maintenance is an internal achievement of the organisation and that thereby the organisation in a sense constructs its own environment. Yet, this does not imply that the world does not provide surprises. Organisational boundaries do not cut through causal relations and – when neglected – these causal relations transform the world to a source of permanent, potential threats to the organisation (cp. Schreyögg and Steinmann 2005). Especially in dynamic and systemically differentiated environments, strong and complex interdependencies require a constant adaptation of boundaries and thus a permanent reconfiguration of the relation between opening and closure.

We emphasise again that this process of reconfiguration is not a peripheral function. Rather, all operations of the organisation in some way refer to the duality of opening and closure. The same holds true for common criteria of differentiation applied to organisations. Table 2.1 gives some examples and relates them to openness and closeness respectively.

Table 2.1 Organisational criteria of differentiation related to openness and close ness

Openness	Closeness
Increase in complexity	Reduction of complexity
Variety	Redundancy
Flexibility	Inflexibility
Viability	Optimisation
Loose coupling	Tight coupling
Resource slack	Leanness

The notions in the same columns can be considered correlative concepts. The properties they refer to occur together, yet cannot be arranged in a strict causal hierarchy. They point to the same problem in regard to the self-reproduction of the organisation but from different perspectives. These different perspectives can be used to strengthen the understanding of the notions of openness and closeness. Increasing complexity, variety, flexibility, viability, loose coupling and resource slack stand for organisational openness, whereas reduced complexity, redundancy, inflexibility, optimisation, tight coupling and leanness refer to its closeness. We stress again that the mentioned concepts do not represent antipodes but condition each other reciprocally. Correspondingly, each organisation is characterised by a specific relation between openness and closeness. Otherwise its reproduction would be impossible.

Finally, we propose to relate the duality of autonomous cooperation and external control to the duality of openness and closeness. In order to substantiate this suggestion, we have to develop an understanding of autonomous cooperation which fits the theoretical context outlined in this section. For this purpose, we comprehend autonomous cooperation as a problem of the internal structuration of the organisation as decision system. We have already argued that autonomous cooperation can be understood as a form of decentral, heterarchical decision making in contrast to external control as central, hierarchical decision making. Thus, we can clarify the meaning of autonomous cooperation on the basis of this difference.

According to Baecker (2005), the function of hierarchy with regard to the structuration of the organisation is to ensure the connectivity of decisions in two ways; firstly, hierarchy supports organisations in referring to decisions as their own operations. Everything that is confirmed by means of hierarchy can be expected to be valid and thus binding in an organisation. On the one hand, autonomous cooperation reduces this effect and thus the probability of successful connections. Yet, this negative influence on the organisation's self-reproduction is compensated by an increase in the variety of decisions on the other hand; while hierarchical control strongly predetermines decision making processes, thereby excluding many options and serving as a cognitive constraint, autonomous cooperation allows to process a high number of external references. It literally helps organisations to broaden their horizons and to find "proper reductions" (Baecker 2005) instead of reducing complexity at any price. Yet, while autonomous cooperation increases the variety of options the organisation is potentially able to realise, it complicates the realisation of each particular option as the organisation gives up the reference points for decision making provided by a hierarchical decision making structure. In the language of New Systems Theory, we can say that it becomes more difficult to ensure the connectivity of decisions.

The second function of hierarchy is related to this problem and refers to the solution of possible conflicts between different decisions. Autonomous cooperation makes it more difficult to deal with this problem and organisations have to find functional alternatives to hierarchy (Ehnert et al. in press; Dembski and Timm 2005).

On the basis of the outlined systems theoretical understanding of organisations, it is possible to appropriately frame the strategic meaning of autonomous cooperation. Here, it should be explicitly emphasised that our notion of strategy refers to the long-term viability of an organisation in relation to its environment. Despite the current dominance of the resource-based view, the idea that strategy is related to an organisation's performance in its environment is still widely prevalent in the strategic management literature (Sydow and Windeler 2001). As Mintzberg and Lampel note, strategic management is generally "concerned with how organisations use degrees of freedom to manoeuvre through their environments" (1999: 25). Our notion of strategy, however, is distinct from conventional concepts as it directs the attention to the organisation's viability in its environment and thus to Luhmann's concept of systems rationality (Luhmann 1970, 1984) instead of simple means/ends-relations and purposive reasoning (Schreyögg 1984).

If we substantiate the notion of strategy with Luhmann's understanding of systems rationality (Schreyögg 1984) we can easily grasp its strategic meaning. Autonomous cooperation provides the organisation with more options to operate and thus potentially enhances its problem solving capability. The rising number of external references that can be processed increases the organisation's sensibility towards the environment, thereby raising the probability to find ways to evolve in accordance with it. Hence, opening on the level of decision making structures increases the probability that the organisation finds viable solutions. Yet, this opening comes at a price. With the growing number of options, it becomes more difficult to realise particular ones and an excess of external references endangers the connectivity of decisions. From the perspective of the organisation, this dilemma appears as an increase in contingency and uncertainty.

Recalling that organisations are permanently striving to reduce contingency and uncertainty, the difficulties with regard to the decision about autonomous cooperation become obvious and it is comprehensible why the delegation of decision making to a technical system with non-technical properties might face resistance from within the organisation.

2.6.4 Boundary management as an enabling tool for the implementation of autonomous cooperation

Understanding autonomous cooperation as a process of boundary opening, we finally have to address ways to regulate this process and thus to provide a context in which managers can decide in favour of the implementation of autonomous cooperation. In the following, we outline a concept of boundary management for these purposes.

Reflecting our understanding of boundaries, boundary management is not conceptualised as a particular management function at the periphery of the organisation, but rather a necessary, managerial process of reflection focusing on the viability of the organisation in its environment. Thus, in contrast to other concepts of boundary management[3], we understand it as a kind of meta-management with a strategic, reflexive character.

---

[3] There are two main perspectives on boundaries and their management in management theory. The first perspective constitutes what has been termed the boundary school of strategic management (cp. Foss 2001). It deals with the strategic importance of the boundaries of the firm and is strongly influenced by the transaction cost approach. The boundary school reflects the growing tendency for hybrid, interorganisational arrangements. Thereby, it is related to the second, more design-oriented perspective, which perceives boundary management as part of the management of interorganisational relations (cp. Windeler 2001). In some cases, this concept of boundary management is also applied to intergroup relations. In our view, both perspectives rely on an insufficient understanding of organisational boundaries; boundaries are neither condensed results of efficiency

What are the issues that have to be addressed by this process of reflection? Following the previous considerations it seems appropriate to direct the attention to Luhmann's notion of decision premises. Decision premises – Luhmann explicitly refers to decision programmes, communication channels, persons and organisational culture (Luhmann 2000) – condition and structure the organisation as a recursive unity of decisions able to reproduce itself; to put it in Luhmann's words, decision premises "articulate" the interior of organisational boundaries (Luhmann 2000: 239) and thus regulate the internal processes of their maintenance.

While management is generally concerned with deciding about decision premises, the particular contribution of boundary management is to reflect on and modify these decision premises with reference to the viability of the organisation in its environment. Hence, boundary management influences the decision about autonomous cooperation in two different ways. First of all, it can directly address the decision premise communication channels, i.e. the way the organisation structures its decision making processes. Facing increasing internal and environmental complexity the organisation might indeed consider reconfiguring its communication channels, e.g. by granting more autonomy to local decision makers. Existing approaches of decentralisation in practice confirm this. Yet, arguing that reflecting on the need for autonomous cooperation is sufficient to solve the related decision problem seems unsatisfactory if we recall the argument laid out in this article. Thus, we should direct the attention to the second way boundary management influences the decision about autonomous cooperation. For this purpose, it is important to note that the mentioned decision premises are not independent from each other but condition each other reciprocally. Hence, it is possible to influence the decision about autonomous cooperation, i.e. opening with regard to the decision premise communication channels, by modifications on the level of decision programmes, persons and organisational culture. Table 2.2 shows some aspects which should be addressed in the context of these decision premises. These aspects represent exemplary design problems that can be derived from our concept of boundary management.

deliberations nor well-defined design problems at the periphery of the organisation. Referring to their central meaning for the organisation's self-reproduction, our understanding of boundary management connects to the problem of systems rationality (cp. Tacke 1999). Yet, we do not suggest that other concepts are meaningless. Especially the design-orientied perspective offers important contributions to be integrated with our perspective.

Table 2.2 Selected aspects to be addressed by boundary management

Decision programmes	Persons	Organisational culture
Collective strategies	Boundary roles	Reflection of contingency of culture
Resource and cost allocation	Qualification of boundary spanners	Culture development
Profit sharing	Personal identification	Management of cultural artefacts

Decision programmes are "what would usually be called procedures or plans – they specify how decisions should be made, (…) or what goals should be pursued" (Mingers 2002: 110). They are adopted "to provide guidelines for evaluating the correctness of decisions" (Luhmann 2002: 45). The reflection and modification of decision programmes in regard to viability is an important aspect of boundary management. Issues especially relevant in regard to fostering autonomous cooperation in logistics are collective strategies, agreements about resource and cost allocation as well as profit sharing.

Persons within an organisation function as decision premises as well. As such, they potentially play an important role in mediating processes of boundary opening. The comprehensive amount of literature on 'boundary spanners' (Adams 1976; Aldrich and Herker 1977; Kiessling et al. 2004) indicates that management theory is aware of the relevance of persons with regards to managing boundaries. Boundary spanners are defined as "persons who operate as exchange or linking agents at the periphery or boundary of the organisation with elements outside it and who link two or more systems whose goals and expectations are likely partially conflicting" (Halley, 1997: 153). Important aspects of boundary management with regard to persons are reflecting and establishing boundary roles as well as qualifying boundary spanners for their task. Measures supporting personal identification can contribute to the closure of the organisation on the level of persons.

Here, we put a special emphasis on organisational culture as one aspect of boundary management. Organisational culture is usually defined as "pattern of basic assumptions that a given group has invented, discovered, or developed in learning to cope with its problems of external adaptation and internal integration" (Schein, 1984: 3). As such, it is implicit in all actions of the organisation's members. Luhmann (2000) argues that organisational culture is largely based on values, i.e. existing preferences which function as reference points for decisions without being explicitly referred to. The history of an organisation manifests in these values. Rather than a "knowledge repository" (Lemon and Sahota 2004) organisational culture thus functions as a pool of preferences, which – of course – are strongly related to the knowledge the organisation has acquired throughout its history. Processes of boundary opening can only be successful if the corresponding changes can be communicated as important innovations rather than unwished deviations from organisational culture. Utilizing organisational culture for managing boundaries presupposes the development of a corresponding managerial sensitiveness in this respect.

While it is readily comprehended that organisational culture has an important influence on the configuration of organisational boundaries, it is less obvious how organisational culture can be developed to support the management of boundaries. As Czarniawska-Joerges notes, "in order to control through culture, one had to be able to control culture first." (1992: 174). Yet, as Luhmann argues, organisational culture is the only decision premise which cannot be decided upon (Luhmann 2000).

How can this dilemma be solved, i.e. how can organisational culture be regulated in order to mediate the process of boundary opening? The first and maybe the most important aspect is once more reflection; even if we assume that changes in organisational culture have to be understood as evolutionary processes, a fundamental condition to influence these processes is a proper reflection of culture and its impact on the organisation's operations. This reflection, optimally taking place at all levels of the organisation, induces processes of boundary opening by revealing the contingency of traditional patterns within the organisation. Understanding this contingency implies the insight that things could be handled in a different way; it is a first step towards organisational change. Fundamental convictions like the refusal to cooperate with competitors can suddenly be questioned. Revealing the contingency of culture can be considered a prerequisite of a directed process of culture development. Probst and Büchel (1994) bring forward a concept of culture development that strongly emphasises the development of corporate visions. These, however, represent only a single aspect of organisational culture. A more comprehensive notion of influencing the development of organisational culture is provided by authors focusing on cultural artefacts. Shrivastava (1985) identifies several "cultural products" being the result of organisational culture (like myths and sagas, language systems and metaphors, symbolism, ceremony and rituals as well as value systems and behaviour norms) and relates them to strategic change. Higgins and McAllaster (2004) underscore this proposition by bringing forward a case study to emphasise the possibility of managing the aforementioned "cultural artefacts" to support strategic change.

2.6.5 Conclusions

Having to decide about the implementation of autonomous cooperation in logistics, managers are confronted with a difficult task. In this article, we have argued that the strategic meaning of this decision and the related difficulties do not stem from the implementation of autonomous cooperation as a new technology as such, but from its particular 'non-technical' character. From an organisational perspective, autonomous cooperation can be perceived as delegation of decision making confronting organisations with the necessity to open their boundaries. Drawing on Luhmann's theory of social systems, we analysed the strategic nature of this process of boundary opening. Finally, we proposed a concept of boundary management that supports building the context for decisions in favour of autonomous cooperation and thus functions as an enabling tool.

We have argued that due to the importance and the central character of boundary maintenance in organisations, it is not indicated to conceptualise boundary management as a set of predefined managerial measures. Successful management of boundaries rather starts with a process of reflection of decision premises as a basis for subsequent changes. This process of reflexion is the first step in building a context for decisions in favour of autonomous cooperation.

References

Adams JS (1976) The Structure and Dynamics of Behaviour in Organisational Boundary Roles. In: Dunnette MD (ed) Handbook of Industrial and Organisational Psychology. Rand McNally, Chicago, pp 1175-1190

Aldrich HE, Herker D (1977) Boundary Spanning Roles and Organisation Structure. Academy of Management Review 2(2): 217-230

Baecker D (2005) Organisation als System. Suhrkamp, Frankfurt am Main

Castelfranchi C, Falcone R (1998) Towards a theory of delegation for agent-based systems. Robotics and Autonomous Systems 24: 141-157

Chainbi W, Ben-Hamadou A, Jmaiel M (2001) A Belief-Goal-Role Theory for Multiagent Systems. International Journal of Pattern Recognition and Artificial Intelligence 15(3): 435-451

Czarniawska-Joerges B (1992) Exploring complex organisations: a cultural perspective. Sage, Newbury Park, Calif

Davidsson P, Henesey L, Ramstedt L, Törnquist J, Wernstedt F (2005) An analysis of agent-based approaches to transport logistics. Transportation research Part C 13: 255-271

Dembski N, Timm IJ (2005) Contradictions between Strategic Management and Operational Decision-Making - Impacts of Autonomous Processes to Decision-Making in Logistics. In: Pawar KS, Lalwani CS, Crespo de Carvalho J, Muffatto M (eds) Innovations in Global Supply Chain Networks. Proceedings of the 10th International Symposium on Logistics, Lisbon, Portugal, pp 650-656

Dryer C (1999) Getting Personal With Computers: How to Design Personalities For Agents. Applied Artificial Intelligence 13:273-295

Ehnert I, Arndt L, Müller Christ G (in press) A Sustainable Management Framework for Dilemmas and Boundaries in Autonomous Cooperating Transport Logistics Processes. International Journal of Environment and Sustainable Development

Foss NJ (2001) The Boundary School. In: Volberda HW, Elfring T (eds) Rethinking Strategy. Sage, London

Graudina V, Grundspenkis J (2005) Technologies and Multi-Agent System Architectures for Transportation and Logistics Support: An Overview. In: Rochev B, Smrikarov A (eds) Proceedings of the International Conference on Computer Systems and Technologies – CompSysTech'05, Varna, Bulgaria, pp IIIA.6.-1–IIIA.6.-6

Halley AA (1997) Applications of Boundary Theory to the Concept of Service Integration in the Human Services. Administration in Social Work 21(3-4): 145-168

Hernes T (2003) Enabling and Constraining Properties of Organisational Boundaries. In: Paulsen N, Hernes T (eds) Managing Boundaries in Organisations: Multiple Perspectives. Palgrave, New York, pp 35-54

Hernes T (2004) Studying composite boundaries: A framework of analysis. Human Relations 57(1): 9-29

Higgins JM, McAllaster C (2004) If You Want Strategic Change, Don't Forget to Change Your Cultural Artifacts. Journal of Change Management 4(1): 63-73

Janssen M (2005) The architecture and business value of a semi-cooperative, agent-based supply chain management system. Electronic Commerce Research and Applications 4: 315-328

Kiessling T, Harvey M, Garrison G (2004) The Importance of Boundary-Spanners in Global Supply Chains and Logistics Management in the 21st Century. Journal of Global Marketing 17(4): 93-116

Laux H, Liermann F (2003) Grundlagen der Organisation. Springer, Berlin Heidelberg New York

Lemon M, Sahota PS (2004) Organisational culture as a knowledge repository for increased innovative capacity. Technovation 24(6): 483–498

Luhmann N (1970) Soziologische Aufklärung. In: Luhmann N (ed) Soziologische Aufklärung 1. Aufsätze zur Theorie sozialer Systeme. Westdt Verl, Opladen

Luhmann N (1984) Soziale Systeme. Grundriss einer allgemeinen Theorie. Suhrkamp, Frankfurt am Main

Luhmann N (2000) Organisation und Entscheidung. Westdeutscher Verlag, Opladen Wiesbaden

Luhmann N (2002) Organisation. In: Bakken T, Hernes T (eds) Autopoietic Organisation Theory: Drawing on Niklas Luhmann's Social Systems Perspective, Abstrakt, Oslo, pp 31–52

Mingers J (2002) Observing Organisations: An Evaluation of Luhmann's Organisation Theory. In: Bakken T, Hernes T (eds) Autopoietic Organisation Theory: Drawing on Niklas Luhmann's Social Systems Perspective, Abstrakt, Oslo, pp 103-122

Mintzberg H, Lampel J (1999) Reflecting on the Strategy Process. Sloan Management Review 40(3): 21-30

Odell J (2002) Agents and Complex Systems. Journal of Object Technology 1(2): 35–45

Paetow K, Schmitt M (2002) Das Multiagentensystem als Organisation im Medium der Technik. Zur intelligenten Selbststeuerung künstlicher Entscheidungssysteme. In: Kron T (ed) Luhmann modelliert: Sozionische Ansätze zur Simulation von Kommunikationssystemen. Leske + Budrich, Opladen

Probst GJB, Büchel BST (1994) Organisationales Lernen: Wettbewerbsvorteil der Zukunft. Gabler, Wiesbaden

Remer A (2002) Management: System und Konzepte. REA, Bayreuth

Schein E (1984) Coming to a New Awareness of Organisational Culture. Sloan Management Review 25(2): 3-16

Scholz-Reiter B, Windt K, Freitag M (2004) Autonomous Logistic Processes: New Demands and First Approaches. In: Monostri L (ed) Proceedings of the 37th CIRP International Seminar on Manufacturing Systems, Budapest, Hungaria, pp 357-362

Schreyögg G (1984) Unternehmensstrategie: Grundfragen einer Theorie strategischer Unternehmensführung. De Gruyter, Berlin

Schreyögg G, Steinmann H (2005) Management: Grundlagen der Unternehmensführung. Konzepte, Funktionen, Fallstudien. Gabler, Wiesbaden

Scott WR (1998) Organisations: rational, natural, and open systems. Prentice Hall, Upper Saddle River, NJ

Seidl D, Becker KH (2006) Organisations as Distinction Generating and Processing Systems: Niklas Luhmann's Contribution to Organisation Studies. Organisation 13(1): 9-35

Shrivastava, P (1985) Integrating Strategy Formulation with Organisational Culture. The Journal of Business Strategy 5(3): 103-111

Sydow J, Windeler A (2001) Strategisches Management von Unternehmungsnetzwerken – Komplexität und Reflexivität. In: Ortmann G, Sydow J (eds) Strategie und Strukturation. Strategisches Management von Unternehmen, Netzwerken und Konzernen. Gabler, Wiesbaden, pp 129-142

Tacke V (1997) Systemrationalisierung an ihren Grenzen – Organisationsgrenzen und Funktionen von Grenzstellen in Wirtschaftsorganisationen. In: Schreyögg G, Sydow J (eds) Managementforschung 7: Gestaltung von Organisationsgrenzen. De Gruyter, Berlin New York, pp 1-44

Van Dyke Parunak H (2000) Agents in Overalls: Experiences and Issues in the Development and Deployment of Industrial Agent-Based Systems. International Journal of Cooperative Information Systems 9(3): 209-227

Weiss G (1999) Prologue. In: Weiss G (ed) Multiagent Systems. A Modern Approach to Distributed Artificial Intelligence. MIT Press, Cambridge, MA London, pp 1-23

Windeler A (2001) Unternehmungsnetzwerke: Konstitution und Strukturation. Westdt Verl, Wiesbaden

2.7 Autonomous Units: Basic Concepts and Semantic Foundation

Karsten Hölscher[1], Renate Klempien-Hinrichs[2], Peter Knirsch[1], Hans-Jörg Kreowski[1], Sabine Kuske[1]

[1] Faculty for Mathematics and Computer Science, University of Bremen, Bremen, Germany

[2] Faculty for Production Engineering, University of Bremen, Bremen, Germany

2.7.1 Introduction

Today, most data processing systems and most logistic systems comprise various, possibly distributed, components. These components typically act autonomously, but they may also communicate and interact with each o-ther, spontaneously linking up to form a network. These components do not necessarily need to be stationary. Sometimes they even move or are carried around. Although the components act autonomously, the task to be solved is handled by their interaction and the system as a whole. In this paper the concept of autonomous units for modeling such systems is proposed. Autonomous units form a community with a common environment, in which they act and which they transform. Autonomous units are based on rules, the applications of which yield changes in the environment. They are also equipped with an individual goal, which they try to accomplish by applying their rules. A control condition enables autonomous units to select at any time and in any situation the rule that should actually be applied from the set of all applicable rules.

The motivation for introducing autonomous units as a modeling concept arises from the Collaborative Research Centre 637 Autonomous Cooperating Logistic Processes. This interdisciplinary collaboration focuses on the question whether and under which circumstances autonomous control may be more advantageous than classical control, especially regarding time, costs and robustness. The guiding principle of autonomous units is the possibility to integrate autonomous control into the model of the processes. This provides a framework for a semantically sound investigation and comparison of different mechanisms of autonomous control. In more detail the aims are the following:

- Algorithmic and particularly logistic processes shall be described in a very general and uniform way, based on a well-founded semantics;
- The range of applications and included methods should comprise methods starting from classical process chain models like the one by Kuhn (Kuhn 2002) or Scheer (Scheer 2002) and the well-known Petri nets (Reisig 1998) leading to multiagent systems Weiss 1999) and swarm intelligence (Kennedy and Eberhart 2001);
- The fact that autonomous units are based on rules provides the foundation for the dynamics of the processes. The application of these rules causes local changes in the common environment, yielding the steps of the processes, transformations, and computations. Archetypes for this behavior are grammatical systems of all kinds (Rozenberg and Salomaa 1997) and term rewriting systems (Baader and Nipkow 1998) as well as the domain of graph transformation (Rozenberg 1997; Ehrig et al. 1999a; Ehrig et al. 1999b) and DNA computing (Păun et al. 1998). The rule-based approach is meant to ensure the possibility of executing the semantics as well as to lay the foundation for formal verification;
- The autonomous control should become apparent on two levels. On the one hand a system comprises a community of autonomous units in an underlying environment. On this level all the units are considered equal in the sense that they may act independently of other units (provided that the state of the environment is suitable for the application of the desired rules). Since no further control exists, the units act autonomously. On the other hand transformation units as rule-based systems are typically nondeterministic, since at any time several rules may be applicable, or the same rule may even be applicable at different positions. In this case the autonomous control facilitates the selection of the different possibilities.

The following section introduces autonomous units. In Sects. 3 to 5 the semantics of a community of autonomous units is defined in three stages. First of all a simple sequential semantics is introduced. This semantics is merely suitable for systems that allow only one action at a time. This covers not only many algorithms and sequential processes, but also card and board games. The sequential semantics of autonomous units is also presented and investigated in (Hölscher et al. 2006b). On the second stage a parallel semantics is defined. Here a number of actions take place in parallel at the same time. This allows for an adequate description of parallel derivations in L-systems (Rozenberg and Salomaa 1998), the firing of Petri nets, and parallel algorithms and processes. While the parallel actions in this semantics occur sequentially, the third stage defines a concurrent semantics with no chronological relations between the acting units. Here the autonomous units may act independently, unless a causal relationship demands a certain order of actions.

The concept of autonomous units is illustrated employing two examples. On the one hand place-transition systems are modeled so that every transition corresponds to one autonomous unit. On the other hand a transport network with packages and trucks is described as a system of autonomous units. Here every package as well as every truck is modeled as an autonomous unit. The paper ends with a short conclusion.

It should be pointed out that autonomous units generalize the concept of transformation units, which has been investigated in e.g. (Janssens et al. 2005; Kuske 2000; Kreowski and Kuske 1999; Kreowski et al. 1997). Here the derivation process is controlled by a main transformation unit and no changes of the environment can occur outside of this control. First steps towards distributed transformation units can be found in (Knirsch and Kuske 2002).

2.7.2 Autonomous units

In this section, the concept of autonomous units is introduced as a modeling approach for data processing systems with autonomous components. Autonomous units form a community with a common environment, which they may transform.

For the sake of simplicity we represent the environments as graphs. But graphs are used in a quite generic sense, including all sorts of diagrams. They may be directed, undirected, labeled or attributed. Since graphs may comprise different subgraphs and different connected components it is also possible to use sets, multisets, and lists of graphs or even arbitrarily structured graphs as environments.

Every autonomous unit is equipped with a goal, rules and a control condition, which autonomously manages the application of the rules in order to accomplish the given goal. Rules transform the environment through their application, thus defining a binary relation of environments as their semantics. Since the control condition determines which rules may be applied to the current environment, its semantics is also defined as a binary relation of environments. Goals are formulated as class expressions, the semantics of which is a class of environments in which the goal is accomplished.

All available environments, rules, control conditions and class expressions form a transformation approach. Its rules, control conditions and class expressions provide the syntactical ingredients of autonomous units. Additionally class expressions are used to define the initial environments and the overall goals of system models.

A transformation approach $\mathcal{A}=(\mathcal{G}, \mathcal{R}, \mathcal{X}, \mathcal{C})$ consists of a class of graphs $\mathcal{G}$, called environments, a class of rules $\mathcal{R}$, a class of environment class expressions $\mathcal{X}$ and a class of control conditions $\mathcal{C}$. Every rule $r \in \mathcal{R}$ specifies a binary semantic relation $SEM(r) \in \mathcal{G} \times \mathcal{G}$. Every pair $(G,H) \in SEM(r)$ is a rule application of r, which is also called a direct derivation and denoted as $G \Rightarrow_r H$. The semantics of a class expression $X \in \mathcal{X}$ is specified as a set $SEM(X) \subseteq \mathcal{G}$ of environments. A control condition defines a binary relation $SEM(C) \subseteq \mathcal{G} \times \mathcal{G}$ on environments as semantics.

A community of autonomous units COM = (Goal,Init, Aut) consists of an environment class expression Goal, defining the terminal environments and thus the overall goal, an environment class expression Init, specifying the initial environments, and a set Aut of autonomous units.

An autonomous unit is a tuple aut=(goal,rules,control) with goal $\in \mathcal{X}$ being the individual goal, rules $\subseteq \mathcal{R}$ being the set of rules, and control $\in \mathcal{C}$ being the control.

Example place/transition systems

Place/Transition (*P/T*) systems are a frequently used kind of Petri nets that can be modeled as a system of autonomous units. The *P/T* net with its marking is regarded as the environment. Transitions are modeled as rules. The firing of a transition defines a rule application that changes the marking in the usual way. Class expressions may be single markings, which define themselves as semantics. A further class expression *all* is also needed, meaning that all environments are permitted as goals. The control condition consists solely of the standard condition *free*, which defines all pairs of environments and imposes no restrictions on the application of rules.

If every transition t is considered as an autonomous unit $aut(t)=(all, \{t\}, free)$ a *P/T* net N with the set of transitions T and initial marking m_0 is modeled as the community of autonomous units $COM(N,m_0)=(all,m_0,\{aut(t) \mid t \in T\})$.

Example transport net

As a further illustration, a simplified example from the domain of transport logistics is sketched.

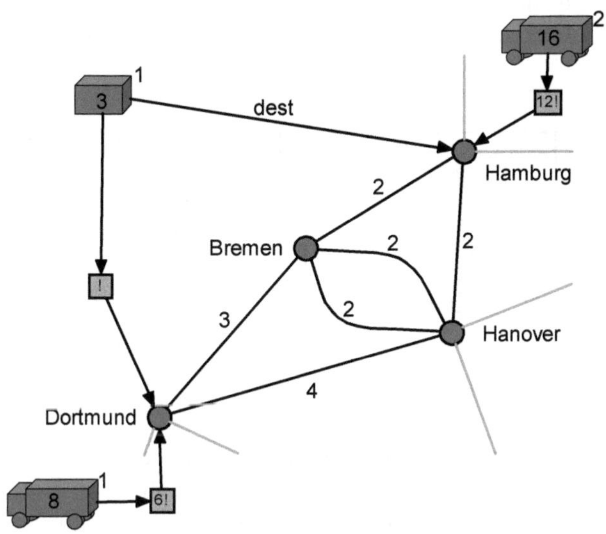

Fig. 2.14 A transport net represented by a graph

A transport net is a graph in which nodes represent locations, e.g. depots, where packages may be picked up and to which packages may be delivered. The edges represent the connections between the depots. Every edge is labeled with the time that is needed to travel along the connection that is represented by the edge. Fig. 2.14 shows a small excerpt of a transport net containing depots in the cities Dortmund, Bremen, Hamburg and Hanover. Trucks and packages are modeled as autonomous units, which use the transport net as underlying environment. Instances of these autonomous units are represented as special nodes with unique identifiers. The transport net contains two trucks (1, 2) and one package (1). The truck nodes are labeled with a number, which represents the amount of time the truck may be moving around. In the given example truck 1 is permitted to move around eight hours, while truck 2 may move around 16 hours (because it may be equipped with two drivers). Both truck nodes are connected to a rectangular tour node which is labeled with a number and an exclamation mark. The number defines the payload capacity of the truck, in our example specified in tons. Truck 1 has a payload of 6 tons, and truck 2 may load up to 12 tons of cargo. The exclamation mark indicates the current tour node. A package node is labeled with a number which specifies its weight. It is also connected to a rectangular tour node, which in turn is connected to the depot that currently holds the package. Analogously to the truck tour node the exclamation mark indicates the current package tour node. An edge labeled "dest" connects the package node with its destination depot, i.e. the depot to which the package has to be delivered.

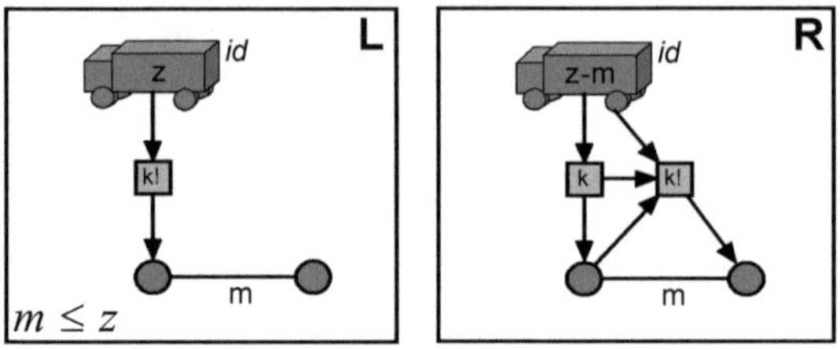

Fig. 2.15 Arranging a truck tour

The transformation unit *truck* contains a rule for planning a tour. This rule is depicted in Fig. 2.15. The application of this rule extends the current truck tour. This is done by adding a tour section leading from the current depot to an adjacent depot. Here the remaining travel time z of the truck must be at least as great as the travel time m between the depots, denoted by the application condition $m \leq z$. Such an application condition has to be evaluated to true, otherwise the rule may not be applied. The application of the rule defines the newly added tour section (represented by the added tour node) as current, and reduces the travel time of the truck by the time that is needed to drive to the adjoining depot.

A package unit has a tour planning rule that is similar to the rule of the truck units. It is depicted in Fig. 2.16.

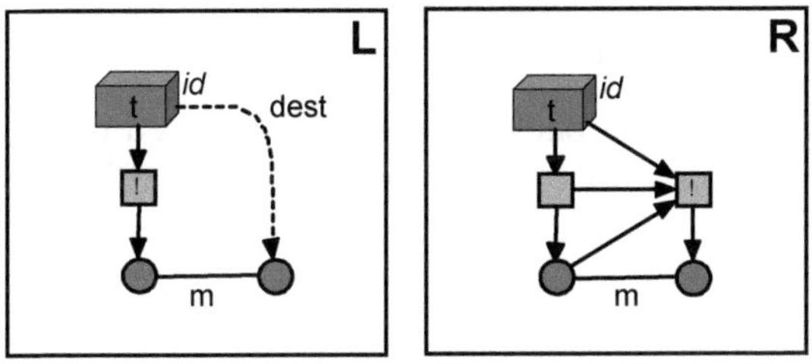

Fig. 2.16 Planning a package tour

The application of this rule extends the package tour by adding a new package tour node and connecting it to an adjacent depot. Analogously to the truck rule the newly added package tour node becomes the current one. This rule should only be applicable if the package is not planning its final tour section. This is modeled in the left-hand side of the rule by the dashed edge connecting the package node with a depot. This edge is labeled with "dest", indicating that the depot is the place to which the package should be delivered. The dashed edge is called a negative application condition (*NAC*) (Habel et al. 1996). If a situation as specified in the NAC is present in the transport net, the rule cannot be applied. Hence, the rule must not be applied if the adjacent depot is already the target depot of the package.

If this is the case the second tour planning rule of the package unit is needed. It is depicted in Fig. 2.17.

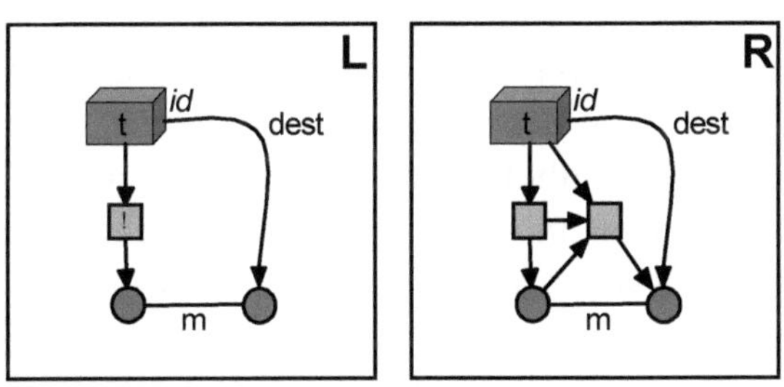

Fig. 2.17 The final part of the package tour

Here the adjacent depot must be the destination depot of the package, as indicated by the edge labeled with "dest". Basically the application of the rule yields the same changes as the first tour planning rule of the package unit. The only difference is that the newly added package tour node is not labeled with an exclamation mark. This is due to the fact, that no current tour node is needed anymore, because the package has finished its tour planning. Given these tour planning rules, truck units as well as package units may independently plan their tours.

After planning its tour a package should be picked up by a truck. Therefore, a package unit contains a rule that makes an offer to a passing truck. This rule is depicted in Fig. 2.18. The rule may be applied if a tour section of a truck coincides with a tour section of the package and the payload capacity k of the truck for this tour section is sufficient for the transport of the package (as indicated by the application condition $t \leq k$). The application of the rule inserts a new edge into the environment, connecting the package tour node to the truck tour node. It is labeled with the actual offer n and a question mark, indicating that an offer for transportation has been made for the amount of n currency units.

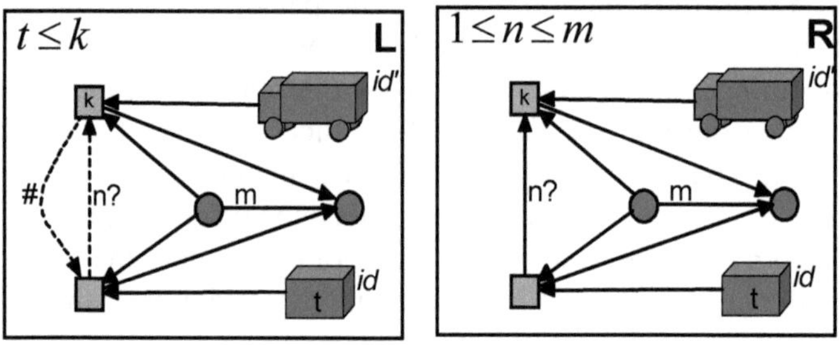

Fig. 2.18 A package offer

The dashed edges in the left-hand side L of the rule again represent negative application conditions. They guarantee that no previous offer was made by the package to the truck if one of the dashed edges with the given labels connect the tour nodes in the specified way. No offer can be made if either the package unit has already made an offer with some amount n, or if the truck unit has finally rejected the offer (indicated by the label "#"). The right-hand side of the rule also contains the post-condition $1 \leq n \leq m$. Such a post-condition has to hold after the rule is applied. In this case, the post-condition guarantees that the package will always offer an amount that is proportionally related to the distance.

The truck unit contains two rules which handle package offers. The first rule is depicted in Fig. 2.19.

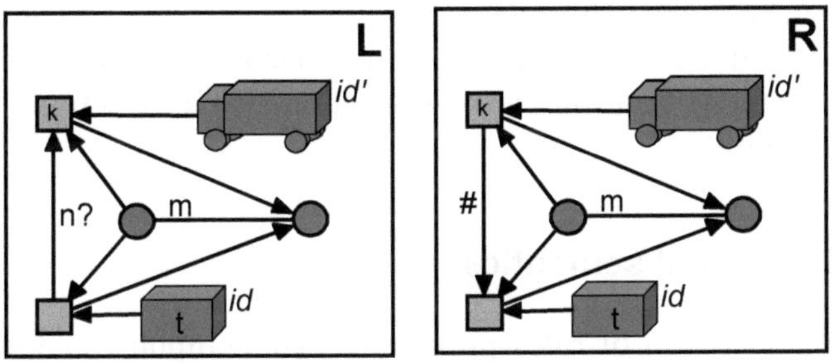

Fig. 2.19 Refusal of an offer

It specifies the rejection of a package offer by deleting the edge representing the offer and inserting a reversely directed edge labeled "#". In this case a package unit cannot make another offer, because the NAC of the offer rule prohibits the existence of such an edge. The second rule is depicted in Fig. 2.20 and specifies the acceptance of an offer.

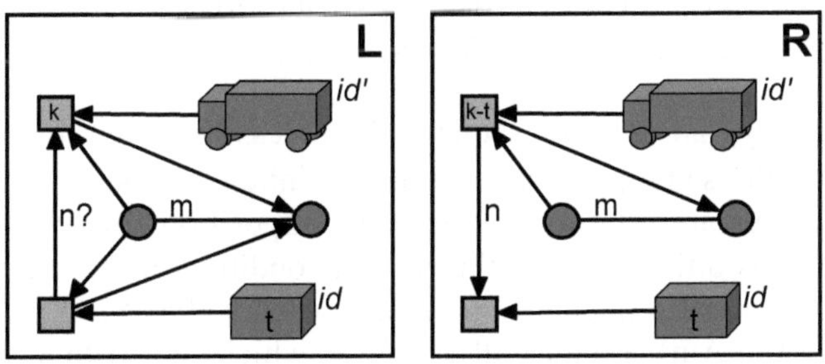

Fig. 2.20 Accepting a package

Similar to the first rule, the edge representing the offer is removed and a reversely directed edge is inserted. But in this case the edge is labeled with n, indicating that the truck transports the package in this section of its tour for a payment of n. Additionally the weight t of the package is subtracted from the payload capacity k of the truck for the corresponding tour section. The connections of the package tour node to the depots are removed, since the actual route of this tour section is described by the tour node of the truck. This removal also ensures that the package does not make any further offers for this tour section.

In the following sections the semantics of communities of autonomous units is defined in three variants. The simplest one is the sequential semantics, which is merely suitable for systems that allow only one action at a time. The parallel semantics allows for activities to take place in parallel, i.e. in a synchronized way. The third variant covers true concurrency. Only causally related activities (e.g. one action needs something that is created by another action) occur in chronological order. Other activities may happen at any time.

2.7.3 Sequential semantics

Since the application of rules provides single computational steps, a first simple semantics for communities of autonomous units is obtained by sequential composition of these steps. This yields finite as well as infinite sequential processes.

Let COM = (Goal, Init, Aut) be a community of autonomous units. A finite sequential process, also called derivation or computation is then defined by $(G_i)_{i\in[n]}$ with $[n]=\{0,\ldots,n\}$ for $n \in \mathbb{N}$, where the following holds for all $i=1,\ldots,n$:

An autonomous unit $aut_i = (goal_i, rules_i, control_i)$ and a rule $r_i \in rules_i$ exist such that $G_{i-1} \Rightarrow_{r_i} G_i$ and $(G_{i-1}, G_i) \in SEM(control_i)$.

Analogously, an infinite sequential process is given by a sequence $(G_i)_{i\in\mathbb{N}}$ with the same properties as in the finite case, but for $i \in \mathbb{N}$. In this sense processes are arbitrary sequential compositions of rule applications by autonomous units, obeying the control condition of the currently active unit. The set of all sequential processes is denoted as SEQ(Aut). Accordingly, SEQ(Init,Aut) contains all processes which start with an initial environment, and SEQ(Goal,Init,Aut) = SEQ(COM) contains all finite processes which additionally terminate in an environment that meets the goal.

In the latter case the semantics can also be defined by an input-output relation, which describes the computation without intermediate steps: we have $(G,H) \in REL_{SEQ}(COM)$ if $(G_i)_{i\in[n]} \in SEQ(COM)$ exists such that $G=G_0$ and $H = G_n$. Even for arbitrary processes the goal specification makes sense, since it can be determined whether Goal has been reached for processes $(G_i)_{i\in\mathbb{N}}$ in intermediate steps: $G_{i_0} \in SEM(Goal)$ for some $i_0 \in \mathbb{N}$?

Analogously, a sequential semantics for a single autonomous unit aut = (goal,rules,control) can be defined taking into account that besides the rule application of the considered unit other units may also change the environment.

Let $CHANGE \subseteq \mathcal{G} \times \mathcal{G}$ be a binary relation on environments, which describes all changes that are not performed by aut. Let further

$N=[n]=\{0,\ldots,n\}$ for an $n \in \mathbb{N}$ or $N=\mathbb{N}$. Then a sequential process of aut is a sequence $(G_i)_{i \in N}$ such that for all $i \geq 1$ either $(G_{i-1},G_i) \in$CHANGE or for an $r \in$ rules $G_{i-1} \Rightarrow_r G_i$ and $(G_{i-1},G_i) \in$SEM(control). The set of sequential aut processes is denoted as $SEQ_{CHANGE}(aut)$.

The sequential processes SEQ(Aut) of a set Aut of autonomous units and the sequential processes of one of its members are strongly connected:

$$SEQ(Aut)=SEQ_{SEQ(Aut-\{aut\})}(aut). \quad (2.1)$$

So every sequential process is an aut process for every autonomous unit in Aut and vice versa, provided that the changes in the environment are precisely the sequential processes of the other autonomous units.

Example place/transition systems

Let COM(N,m0) be the system of autonomous units that corresponds to a P/T system. Then the application of a rule yields the same effect as the firing of a transition. In this way sequential processes correspond to the firing sequences of the P/T system.

Example transport net

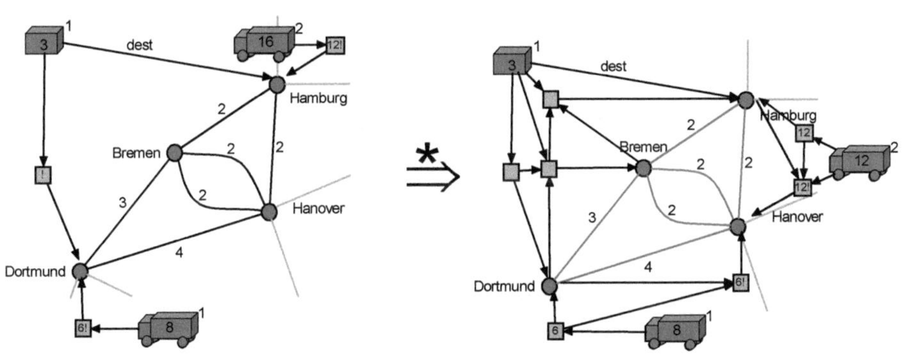

Fig. 2.21 Derivation

The planning of package and truck tours can be regarded as sequential processes in a transport net. Fig. 2.21 depicts a process for the tour planning of package 1, which intends to be transported from Dortmund to Hamburg via Bremen, while truck 1 and truck 2 each planned a tour to Hanover, with truck 1 originating in Dortmund and truck 2 starting in Hamburg.

2.7.4 Parallel semantics

In many cases it is rather unrealistic to consider a system of autonomous units that transform the shared environment in a sequential way. The actual processes in most data processing systems are more suitably modeled by allowing more than one activity on the environment at the same time. This includes in particular the fact that different events which do not influence each other, can happen in parallel.

In order to obtain a suitable formal definition of parallel processes it is necessary to extend the assumptions on the given graph transformation approach. So far we have considered situations where only one rule is applied at a time. For the parallel semantics definition let us now consider situations where a multiset of rules may be applied to the environment. This means that a number of different rules may be applied or even a single rule may be applied multiple times. For this purpose let $\mathcal{A}=(\mathcal{G}, \mathcal{R}, \mathcal{X}, \mathcal{C})$ be a parallel transformation approach, meaning that a binary semantic relation SEM(R) $\subseteq \mathcal{G}\times\mathcal{G}$ exists for every multiset R of $\mathcal{R}$. Instead of (G,H) $\in$SEM(R) this may also be denoted as $G \Rightarrow_R H$ and may also be called direct parallel derivation.

Parallel processes are then defined analogously to the sequential case. First the occurrence of a single rule application has to be replaced by the application of a multiset of rules. Secondly the definition for obeying the control condition has to be changed. The corresponding sets of parallel processes PAR(Aut), PAR(Init,Aut), and PAR(Goal,Init,Aut) = PAR(COM) as well as an input-output relation REL_{PAR}(COM) are then obtained analogously to the sequential case.

Example place/transition systems

A parallel transformation approach is obtained by defining parallel firing of a multiset of transitions in a P/T system in the usual way. For the system COM(N,m_0) the parallel processes correspond exactly to the firing sequences of multisets of transitions.

Example transport net

The tour planning of package and truck units can also be regarded as parallel processes in a transport net. The process depicted in Fig. 2.21 can be modeled as a parallel procedure of one package tour planning step and both truck tour planning steps, followed (or preceeded) by the remaining package tour planning step.

In general, sequential and parallel processes may produce very different results. Consider for instance cellular automata, where a transition step of all linked finite automata depends on the state of their neighbors. Here a parallel computational step of some automata would change the context of the other automata such that later steps yield different configurations. In other approaches, like e.g. Petri nets, term replacement, or most approaches to graph transformation, parallel changes do not affect the final output, but yield a reduced number of transformation steps. This is due to the fact that the parallel actions may also occur sequentially in an arbitrary order without affecting the final result. This phenomenon is called true concurrency. In order to obtain true concurrency in the context of parallel transformation approaches the following has to hold:

Let $\mathrm{R} = \mathrm{R}' + \mathrm{R}''$ be the sum of two multisets of rules and $G \Rightarrow_R X$ be a parallel derivation step. Then parallel derivation steps $G \Rightarrow_{R'} H$ and $H \Rightarrow_{R''} X$ exist for a suitable environment H.

Remember that every multiset is the commutative sum of its single elements. For this reason true concurrency implies that every parallel step could also be executed as an arbitrarily ordered sequence of the corresponding single rule applications, yielding the same result. Parallel processes and their sequentialization are called equivalent in the context of concurrency. Consider an equivalence class of a parallel process, i.e. all processes that are equivalent to each other. Then the chronological order of two rule applications can only be determined if the one causally depends on the other. Otherwise they can be applied in parallel or in an arbitrarily ordered sequential way.

Since every sequential transformation step is a special case of a parallel transformation step, the sequential semantics of autonomous units is contained in the parallel semantics, i.e. $SEQ(A) \subseteq PAR(A)$ is true for the processes of a set of autonomous units A. Furthermore an equivalent process $\bar{s} \in SEQ(A)$ can be found for every process $s \in PAR(A)$. For a system of autonomous units S this implies in particular

$$\mathrm{REL}_{\mathrm{SEQ}}(\mathrm{S}) = \mathrm{REL}_{\mathrm{PAR}}(\mathrm{S}). \tag{2.2}$$

2.7.5 Concurrent semantics

Like the sequential process semantics, the parallel process semantics may not be suitable for every application situation. This is due to the fact that components which act autonomously and independently, do not necessarily start and finish their activities simultaneously, as is the case with parallel steps. If such components act far away from each other, or work on completely different tasks without influencing each other it may even not be possible to determine simultaneity. Anyway, demanding or enforcing simultaneity would not make any sense in this case. A chronological order of concurrent and distributed processes is only given in the case that one activity needs something that another activity provides. Such causal relationships can be expressed by directed edges between these activities. In the case of concurrent processes this results in an acyclic graph of activities. Such a graph yields a concept for concurrent processes in communities of autonomous units. This is basically the same idea as in the notion of processes of Petri nets.

Let COM = (Goal,Init,Aut) be a system of autonomous units over a parallel transformation approach $\mathcal{A} = (\mathcal{G},\mathcal{R},\mathcal{X},\mathcal{C})$. Then a concurrent process consists of an initial environment G_0 and an acyclic, directed graph run=(V,E,lab), with a set of nodes V and a set of edges $E \subseteq V \times V$. The nodes are marked with lab:$V \rightarrow \mathcal{R}$, which maps every node to a rule. The following must also hold for G_0 and run:

1. Every node in run must be reachable via a path originating in an initial node, i.e. a node without incoming edges;

2. Every complete beginning part of run, i.e. every subgraph which contains all initial nodes and with every node also all paths from the initial nodes to that node, is either finite or contains an infinite path;

3. For every complete beginning part a parallel process $(G_i)_{i \in N}$ together with a bijection between the nodes of the subgraph and the applied rules can be found for $N = [n]$, $n \in \mathbb{N}$ or $N = \mathbb{N}$. These rules conform to the markings of the nodes. This bijective relation keeps the causal dependency. This means that a rule which marks the source of an edge in the subgraph is always applied in an earlier step than the rule which marks the target of this edge.

The first condition enforces that run does not contain infinite paths without start. Otherwise there would be a path with no corresponding process. The second condition implies that only finitely many nodes are causally independent of each other. The third condition guarantees that concurrent processes are actually executable.

Example place/transition systems

With the notion of occurrence nets at least the special case of Condition/Event(C/E) nets has a similar process concept. If every path of length 2 that runs along a condition is replaced with a directed edge in such an occurrence net, then a concurrent process in the aforementioned sense is obtained.

Example transport net

In the transport net example the tours of trucks as well as packages can be planned concurrently. The negotiation for transport of different truck-package pairs may also occur in a concurrent way.

An elaborated description of the relation between parallel and concurrent processes goes beyond the scope of this paper and thus has to be deferred to further work. But it is noteworthy to mention that in the case of true concurrency a strong relation between concurrent processes and canonical derivations exists. This has been investigated in (Kreowski 1978) in the context of graph transformation employing the double-pushout approach. Such canonical derivations represent equivalence classes of parallel derivations in a unique way by enforcing maximum parallelity and an application as soon as possible.

2.7.6 Conclusions

In this chapter we have introduced the new concept of autonomous units. This rule-based concept is meant to model data processing systems comprising different distributed components and processes. These components may act autonomously but they may also communicate and interact with each other. The operational semantics of such systems has been defined in three stages. Sequential and parallel processes establish a chronological order of the activities in such a system. In the context of concurrent processes only the chronological order of causally related activities is fixed. The approach employs graphs and graph transformation rules allowing visual models, as illustrated by the example of transport nets. At the same time the concept is flexible by allowing to embed different modelling approaches, which provides the opportunity of semantical comparisons between different modeling methods. The example of Petri nets gives a hint in this direction, which has to be substantiated by further research in future work. Anyhow a number of aspects have not been addressed so far and some questions have been left unanswered in this introductory work. This includes, among others, the following:

- The sequential and parallel processes are composed of application of either single rules or multisets of rules. This is closely related to the semantics of labeled transition systems, which are frequently used for the semantic foundation of communication and distributed systems. This relation demands further investigation;
- As indicated at the end of Sect. 5, a strong relation exists between concurrent processes and canonical derivations, the latter being special kinds of parallel processes. The detailed investigation of this relationship would be interesting;
- A remarkable aspect of the classical transformation units is the structuring principle. This is achieved by the import feature of transformation units, which allows them to import other transformation units and utilize them to solve subtasks. So far only autonomous units with sequential semantics have been defined with an additional structuring principle (cf. (Hölscher et al. 2006b)). But it would generally make sense for autonomous units to modularize the solution of a task or to let subtasks be handled by other autonomous units. For this reason, future work should also concentrate on structured autonomous units in the parallel and concurrent cases;
- The main task for further investigation of autonomous units will be to investigate the means of control. On the one hand specific control mechanisms allowing for autonomy have to be investigated. This will comprise in particular concepts for the evaluation of the environment and for a more goal oriented control. On the other hand, the control, which is currently only defined for single steps, has to be enhanced to also cover extended processes, as this is already the case with classical transformation units and sequential autonomous units;
- The significance and suitability of autonomous units as a modeling approach will be proved by a number of case studies. These will comprise studies reaching from games over agent systems and artificial ant colonies to the conventional approaches of process modeling. A first example can be found in (Hölscher et al. 2006b), where the board game ludo is modelled with autonomous units;
- A further theoretical investigation of autonomous units together with existing theoretical results would be useful for the practical application of autonomous units. This includes for example decidability re-

sults of control conditions (Hölscher et al. 2006a) or class expressions as well as (automated) correctness proofs.

References

Baader F, Nipkow T (1998) Term Rewriting and All That. Cambridge University Press, Cambridge

Ehrig H, Engels G, Kreowski H-J, Rozenberg G (eds) (1999) Handbook of Graph Grammars and Computing by Graph Transformation, vol 2: Applications, Languages and Tools. World Scientific, Singapore

Ehrig H, Kreowski H-J, Montanari U, Rozenberg G (eds) (1999) Handbook of Graph Grammars and Computing by Graph Transformation, vol 3: Concurrency, Parallelism, and Distribution. World Scientific, Singapore

Janssens D, Kreowski H-J, Rozenberg G (2005) Main Concepts of Networks of Transformation Units with Interlinking Semantics. In: Kreowski H-J, Montanari U, Orejas F, Rozenberg G, Taentzer G (eds) Formal Methods in Software and System Modeling, Lecture Notes in Computer Science vol 3393. Springer, Berlin Heidelberg New York, pp 325-342

Habel A, Heckel R, Taentzer G (1996) Graph Grammars with Negative Application Conditions. Fundamenta Informaticae, 26:287-313

Hölscher K, Klempien-Hinrichs R, Knirsch P (2006a) Undecidable Control Conditions in Graph Transformation Units. In: Moreira Martins A, Ribeiro L (eds) Brazilian Symposium on Formal Methods (SBMF 2006), pp 121-135

Hölscher K, Kreowski H-J, Kuske S (2006b) Autonomous Units and their Semantics – the Sequential Case. In: Corradini, A, Ehrig H, Montanari U, Ribeiro L, Rozenberg G (eds) Proc. 3rd International Conference on Graph Transformations (ICGT 2006), Lecture Notes in Computer Science vol 4178, Springer, Berlin Heidelberg New York, pp 245-259

Kennedy J, Eberhart RC (2001) Swarm Intelligence. Morgan Kaufmann Publishers, San Francisco

Kreowski H-J, Kuske S (1999) Graph Transformation Units with Interleaving Semantics. Formal Aspects of Computing 11(6):690-723

Knirsch P, Kuske S (2002) Distributed Graph Transformation Units. In: Corradini A, Ehrig H, Kreowski H-J, Rozenberg G (eds) Proc. First International Conference on Graph Transformation (ICGT), Lecture Notes in Computer Science vol 2505, Springer, Berlin Heidelberg New York, pp 207-222

Kreowski H-J, Kuske S, Schürr A (1997) Nested Graph Transformation Units. International Journal on Software Engineering and Knowledge Engineering, 7(4): 479-502

Kreowski H-J (1978) Manipulationen von Graphmanipulationen. Ph.D. thesis, Berlin

Kuhn A (2002) Prozessketten - ein Modell für die Logistik. In: Wiendahl H-P (ed) Erfolgsfaktor Logistikqualität. Springer, Berlin Heidelberg New York, pp 58-72

Kuske S (2000) Transformation Units-A Structuring Principle for Graph Transformation Systems. Ph.D. thesis, Bremen

Păun G, Rozenberg G, Salomaa A (1998) DNA Computing-New Computing Paradigms. Springer, Berlin Heidelberg New York

Reisig W (1998) Elements of Distributed Algorithms-Modeling and Analysis with Petri Nets. Springer, Berlin Heidelberg New York

Rozenberg G (1997) Handbook of Graph Grammars and Computing by Graph Transformation, vol 1: Foundations. World Scientific, Singapore

Rozenberg G, Salomaa A (1997) Handbook of Formal Languages, vol 1-3. Springer, Berlin Heidelberg New York

Rozenberg G, Salomaa A (1998) Lindenmayer Systems. Springer, Berlin Heidelberg New York

Scheer AW (2002) Vom Geschäftsprozeß zum Anwendungssystem. Springer, Berlin Heidelberg New York

Weiss G (1999) Multiagent Systems-A Modern Approach to Distributed Artificial Intelligence. The MIT Press, Cambridge, Massachusetts

2.8 Mathematical Models of Autonomous Logistic Processes

Bernd Scholz-Reiter[1], Fabian Wirth[1], Michael Freitag[1], Sergey Dashkovskiy[2], Thomas Jagalski[1], Christoph de Beer[1], Björn Rüffer[2]

[1] Department of Planning and Control of Production Systems, BIBA, University of Bremen, Germany

[2] Center for Industrial Mathematics, University of Bremen, Germany

There exist various approaches to the mathematical modelling of dynamic processes occurring in shop floor logistics. These include methods from queuing theory or use dynamical systems given by ordinary or partial differential equations (fluid models). If the number of elements within the process is large it can become prohibitively complex to analyse and optimize a given logistic process or the corresponding mathematical model using global strategies. A new approach is to provide for an autonomy of various smaller entities within the logistic network, i.e. for the possibility of certain elements to make their own decisions. This necessitates changes in the appropriate mathematical models and opens the question of stability of the systems that are designed. In this paper we discuss the fundamental concepts of autonomy within a logistic network and mathematical tools that can be used to model this property. Some remarks concerning the stability properties of the models are made.

2.8.1 Introduction

In a production network (e.g. on shop floor level), the flow of parts is usually pre-planned by a central supervisory or control system. This approach fails for large scale networks in the presence of highly fluctuating demand or unexpected disturbances (Kim and Duffie 2004). One of the reasons for this phenomenon is that in practice the complexity of centralized control architectures tends to grow rapidly with the size of the network, resulting in rapid deterioration of fault tolerance, adaptability and flexibility (Prabhu and Duffie 1995).

An advantageous alternative is the management of the dynamic behaviour according to the requirements of production logistics. In this sense the development of decentralised and autonomous control strategies is a promising research field (Scholz-Reiter et al. 2004). Here autonomous control describes a decentralised coordination of intelligent logistic objects (parts, machines etc.) and the allocation of jobs to machines by the intelligent parts themselves. Therefore, there are no standard policies for production logistics that may be readily applied. Instead, strategic policies have to be derived that enable the parts to decide autonomously, instantaneously and using locally available information only to choose between different alternatives. The application of autonomous control in production networks leads to a coalescence of material flow and information flow and enables every part or product to manage and control its manufacturing process autonomously (Bonabeau et al. 1999). The dynamics of such a system depends on the local decision-making processes and produces a system's global behaviour that has new emerging characteristics (Helbing 2001).

In the literature several attempts may be found to explain the emergent behaviour of large scale structures that arise from autonomous control policies. First intuitive approaches suggest to set up a policy like 'go to the machine with the shortest processing time' or 'go to the machine with the lowest buffer level' (Scholz-Reiter et al. 2005a, 2005b) etc. More sophisticated autonomous control strategies can be found in biological systems. Camazine et al. (Camazine et al. 2001) give a good overview and some case studies of self-organized behaviour in biological systems. Their case studies comprise social insects, slime moulds, bacteria, bark beetles, fireflies and fish. According to the authors biological self-organisation can be found in group-level behaviour that arises in most cases from local individual actions that are influenced by the actions of neighbours or predecessors and in structures that are build conjointly by individuals. They identify positive feedback as a "key ingredient" of self-organisation. Positive feedback is a method that enables and endorses change in a system. In ant colonies for example, a scout ant that has found food lays down a pheromone trail as it returns to the nest. By changing the environment, succeeding ants may simply follow the trail and find the food, which in turn reinforce the trail with their pheromone (Parunak 1997).

Ant colony optimization (ACO, see e.g. Bonabeau et al. 1999; Dorigo and Stützle 2004) uses positive feedback with the help of artificial pheromones and is used to solve discrete optimization problems like the travelling salesman problem and the quadratic assignment problem. Logistics applications of the ACO concept can be found for example in Gambardella et al. (Gambardella et al. 1999), where the authors find solutions to vehicle routing problems with time windows and in Bautista et al. (Bautista and Pereira 2002), where ACO is applied to an assembly line balancing problem for a bike factory. Applications of the pheromone concept for manufacturing control can be found in Peeters et al. (Peeters et al. 1999) and Armbruster et al. (Armbruster et al. 2006a) where pheromones are used to find a control system for a flexible shop floor.

Brückner et al. (Brückner et al. 1998; Brückner 1999) suggested implementing the pheromone concept to organize production systems as multi-agent systems. The authors call the approach a "synthetic ecosystems" and present a formal software infrastructure as well as a real-world example. In their "guided manufacturing control system" they combine distributed and reactive control in their control subsystem with a global advisory subsystem.

A concept that uses the interaction between nearest neighbours but does not rely on pheromones is the idea of a bucket brigade, which was introduced by Bartholdi et al. (Bartholdi and Eisenstein 1996). A bucket brigade is a production line setup, where workers independently follow simple rules that determine what to do next. The rules are: a) Process your work until you meet a downstream worker. If so, give him your work. b) If you do not have work, go upstream until you meet another worker and continue with his job. c) If you are the first worker and you do not have work, then start a new job. d) If you are the last worker, then finish the job and follow rule b). The authors show that such a bucket brigade is self-balancing and results in a global optimum if the workers are sequenced from slowest to fastest. The concept has been extended to bucket brigades with worker learning by Armbruster et al. (Armbruster et al. 2007).

In order to develop and analyse autonomous control strategies dynamic models are required. For production systems several model classes have been investigated. These can be divided in discrete and continuous models.

Discrete models are based on the consideration of individual parts in a network of machines. Queuing networks (e.g. with re-entrant lines) can be used to model complex manufacturing systems such as wafer fabrication facilities. The advantage of such models is the possibility to assign decision rules to machines and parts. Stability of such networks is defined probabilistically in terms of Harris recurrence and is often hard to check. For single class networks, which are also called generalized Jackson networks, with work-conserving disciplines such as the FIFO priority discipline or the processor sharing discipline, stability is guaranteed by the usual traffic condition, which requires that the load is less than the capacity at each machine.

However, this condition is not sufficient for multiclass open queuing networks (Chase et al. 1993). Nonetheless, there are fluid limits models that allow the investigation of the stability question for such networks (Bramson 1994; Chase et al. 1993). These are continuous models obtained with help of the functional strong law of large numbers.

A further model class can be derived within the framework of dynamical systems. By time averaging over a representative time period, it is possible to obtain a system of differential equations describing the behaviour of a queuing process as a continuous approximation, see e.g. (Dashkovskiy et al. 2004). The advantage of this approach is that methods from the theory of dynamical systems can be used. E.g., stability criteria for a class of such systems were recently developed in (Dashkovskiy et al. 2004, 2006a, 2006b). Continuous models and some stability conditions will be presented later on. Here the term continuous denotes the continuous material flow. In the literature continuous flow models of production systems are often called hybrid models (Armbruster et al. 2004; Chase et al. 1993; Peters et al. 2004), meaning that the material flow is modelled as a continuous flow that is controlled by discrete actions. This discrete control is typical for production systems.

2.8.2 Logistic processes

Within this paper, we focus on logistic processes on shop floors. Production logistics in this sense encompasses planning, control and monitoring of manufacturing processes. Enterprises face the problem of reacting to dynamically changing market competition in order to deploy and establish high quality products with a reasonable price possibly in a very short time. Thus, production logistics covers the interdisciplinary task between production planning and control, engineering and strategic management. It takes care of the operational control of material and information flows to guarantee efficient and flexible production processes (Chase et al. 1993).

The main goal of production logistics is to design and organise production processes according to high utilisation, low inventory and work-in-process, short throughput times and high adherence to delivery dates. The first two aims are at operational level, whereas the two latter aims are customer driven. It is obvious that these four aims are mutually contradictory; an enterprise has to find a trade-off between these goals and to position itself according to its own interpretation of their importance.

The main tasks of production logistics can be derived from the main goals. The allocation of orders or jobs to resources comprises of getting (i) the right products or services (ii) at the right time (iii) in the right amount (iv) to the right place. In this section we will discuss how autonomous control can meet these demands in presence of high dynamics.

Autonomy in logistic processes

By autonomy of a logistic process we understand the capability of the process to determine how to react to given changes in the environment, be they fluctuations in demand or in required production rate, failures in some components or changes in the function required of the process. Mathematically speaking we model an autonomous process as an input-output system that is regulated by its own feedback loop with a possibly dynamic feedback, i.e., a feedback capable of using the memory of the system to calculate the control input, see figure 2.22.

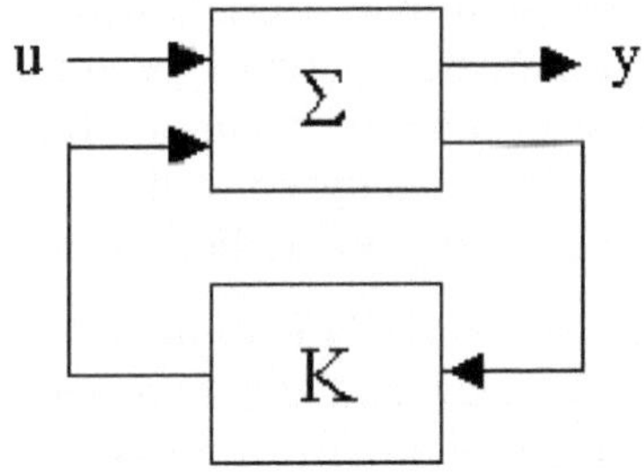

Fig. 2.22 A feedback loop

From an abstract point of view it may seem difficult to call a system with inputs autonomous, since in general an input can be used to regulate a system from the outside. The distinction arises through the classification of inputs into inputs directly aimed at low-level control and others. We will call those systems autonomous that receive only inputs in terms of material and information, that needs to be processed, as well as high level demands. The decision on how these high level demands are met using the available resources rests with the control loop of the system. Clearly, the concepts we are using here are not defined in mathematical terms but would depend on the interpretation of different objects within a concrete scenario[4].

[4] We note that the usage of the word 'autonomy' in this paper does not correspond to terminology that is widely used within mathematical systems theory. Here a system is either called autonomous if the laws governing the evolution of the system do not explicitly depend on time (Sontag 1998), or within the framework of behavioural systems, a system is called autonomous, if the behaviours of the system are not parameterised by inputs (Polderman and Willems 1997).

As an example, consider a two machine two buffer system: Assume that due to customer demand a certain part has to be processed within the system. In the conventional approach a central controlling entity decides based on global information on which buffer-machine system the part is processed. In contrast autonomous control would enable the part to choose the buffer-machine system autonomously based on local information the part actually has access to.

2.8.3 Mathematical modelling of logistic processes

There are fundamental discrepancies in the interpretation of what constitutes a model depending on different fields of research. In this paper we will take a modest mathematical point of view. We wish to understand the dynamics of logistic processes, that is, the laws by which certain logistic objects or quantities evolve in time. Here logistic objects may be parts in a factory, containers in a transport network or similar things. A model will therefore mostly consist of a set of equations for the time behaviour of a process. These models can be analysed to derive certain global properties of the system or simulated to obtain predictions for specific cases.

The aim of deriving such models is to be able to analyse the behaviour from a qualitative point of view and also to provide predictive models, that is models that are accurate enough to provide good estimates of what is happening in the real process. Based on such a model, control or optimisation strategies may be derived.

Due to the discrete nature of many logistic processes, the earliest models of such processes were in terms of discrete systems with an emphasis on the stochastic nature of the processes, arrival processes and other factors. We describe such models in the ensuing Section 3.1. In this approach processes are modelled by a number of servers with a processing rate. Each server has one or several queues to which possibly different types of customers arrive. The customers wait in these queues until they are served and after completion of the particular task they go on to the next server or leave the network. Concrete examples where such a modelling approach can be used are job shops where individual machines are interpreted as servers and customers are the parts that have to be processed. In the later sections we present continuous models in which parts and also production stage are not modelled as discrete variables.

Discrete models and fluid approximations

Let J be the number of single machines denoted by index $i=1,...,J$. There are K classes of parts being processed. Each class $k=1,...,K$ has its own exogenous arrival process with interarrival times $t_k(n)$, $n=1,2,...$ with $t_k(n)=\infty$ for all n for some class k meaning that there are no external arrivals for this class.

Parts of class k require service at machine $s(k)$ and their service times are $T_k(n)$, $n=1,2,...$. After being processed at station $s(k)$ a class k part becomes a part of class l with probability P_{kl} or exits the network with probability $1-\Sigma_l P_{kl}$, independent of all previous history, where $P=(P_{kl})$ is a sub-stochastic matrix which is called routing matrix. Such a network is called an open multiclass queuing network, or briefly multiclass network. In case there is only one class with exogenous arrivals and the entries of the routing matrix satisfy $P_{k,k+1}=1$, for $k=1,...K-1$ and zero otherwise, then the multiclass network is called a re-entrant line, see figure 2.23.

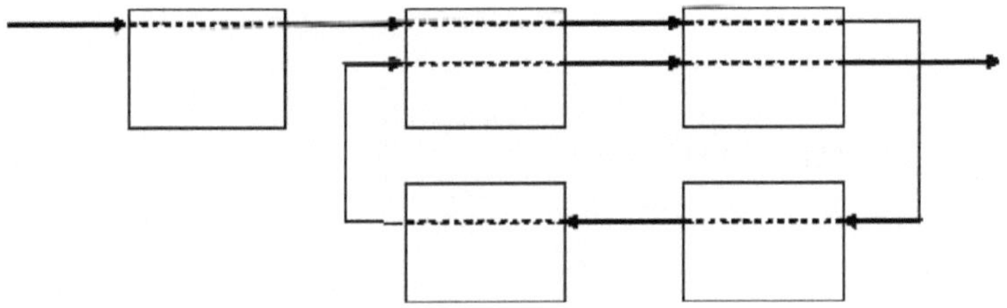

Fig. 2.23 A seven buffer five machine re-entrant line

Such models have been considered by many authors, see e.g. (Dai 1995). The fluid limit models for multiclass networks and re-entrant lines were considered by, e.g., (Dai 1995; Dai and Weiss 1996), where the stability question is discussed and stability criteria via fluid models are obtained.

Within this modelling framework autonomous control can be introduced as follows. If the transition probabilities P_{kl} are dependent on the current buffer level of classes, this dependence can reflect the ability of parts to decide where to go to. Furthermore, the distribution of T_k can also depend on the state of the queues; this reflects the ability of machines to change their own processing rate. Finally, servers may be able to decide in which order to process the waiting parts on the base of their buffer levels, i.e., the serving discipline is changing with time. Stability investigation and fluid models have yet to be developed for such re-entrant lines with autonomous control.

Continuous models: partial differential equations

We now describe a modelling approach based on partial differential equations. We introduce the variable x taking values in $[0,1]$ which signifies the completion stage within a certain production process (Armbruster et al. 2004). So material at the stage $x=0$ stands for raw material, while the material has reached stage $x=1$ when production process is completed. In this approach we are interested in the density function $\rho(x,t)$ which denotes the amount of material that has reached completion stage x at time t. The approach is now to write down a partial differential equation for ρ. The first of the following equations represents conservation of mass, while the second is an equation for the local velocity within the production system, cf. (Armbruster et al. 2006b).

$$\begin{aligned} &\frac{\partial \rho}{\partial t} + \frac{\partial(\rho v(\rho))}{\partial x} = 0, \quad x \in (0,1), \\ &v(\rho(x,t)) = v_0 \left(1 - \frac{\rho(x,t)}{\rho_m}\right). \end{aligned} \tag{2.3}$$

The advantages of this modelling approach lie in the relative ease with which model based simulations can be performed. For logistic processes with a large number of production stages it is also plausible to justify the transition from a finite number of production stages to a continuum. However, the approach does not lend itself easily to the modelling of autonomy because it is not obvious how to incorporate the behaviour of autonomous parts in the PDE. For instance one of the problems occurring is that for autonomous parts there may not be an ordered set of stages that has to be completed, so that it does not really seem appropriate to model completion by a variable taking values in $[0,1]$. While this does not mean that the approach is not suitable for modelling autonomous processes, the derivation of such models is an open problem.

Continuous models: ordinary differential Equations

In this section we first consider a single autonomous machine that can be modelled in a continuous modelling framework. Then we will show how such machines can be combined in a logistic network.

A single machine

Let $x = (x^1, \ldots, x^n)$ be the vector representing the state of a machine at time t and let $u = (u^1, \ldots, u^k)$ be the vector of inputs representing both external disturbances and inputs from other machines, see figure 2.24. The evolution of the state x with time t is described by a differential equation

$$\frac{dx}{dt} = f(x,u) \tag{2.4}$$

with initial condition

$$x(0) = x^0 . \tag{2.5}$$

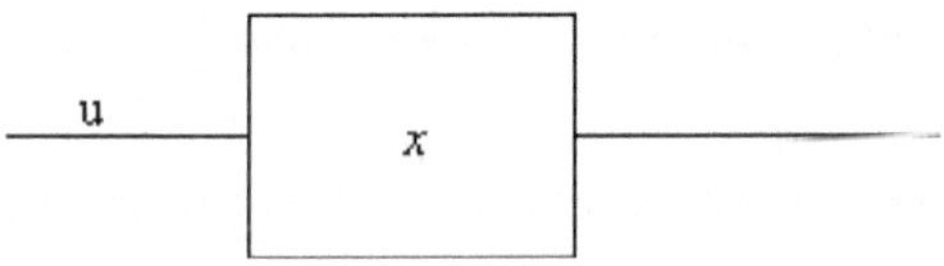

Fig. 2.24 A single machine

The decision rules of the machine are included in the function f. The input u accounts also for the decisions of the processed parts. Stability properties of such a nonlinear system can be described in terms of input-to-state stability (ISS, Sontag 1989).

A production network

Consider a shop floor with several, that is m machines. To each of these we associate its state vector denoted by $x_i = (x^1, \ldots, x^n) \in R^n$, $i = 1 \ldots, m$, and denote the total state of the network by $x = (x_1, \ldots, x_m) \in R^{nm}$. Let us combine these machines in a network, see figure 2.25. This network may be represented as a directed graph, where the nodes are individual machines and edges describe an influence of the state of one machine on the state of another machine.

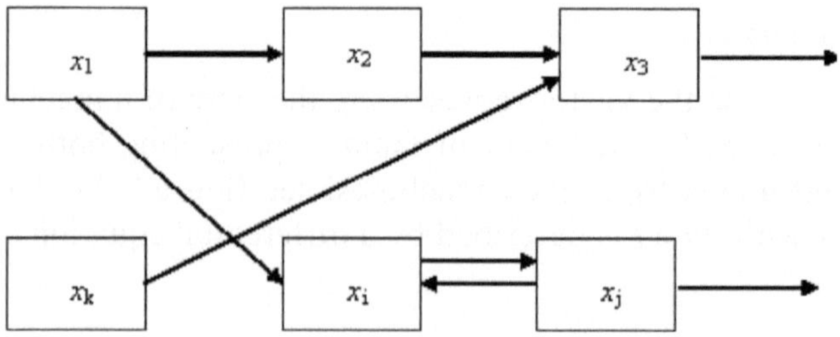

Fig. 2.25 A network of machines with mutual influences represented as a directed graph

The notion of ISS incorporates a measure of influence of the magnitude of the input to the magnitude of state, called nonlinear gain. A nonlinear gain γ_{ij} from machine x_i to machine x_j is a strictly increasing continuous function with $\gamma_{ij}(0)=0$ (Sontag 1989). These gains can be gathered into a matrix, setting $\gamma_{ii} \equiv 0$, which is a weighted adjacency matrix of the graph representation of the production network. Based on this a stability condition can be derived.

The dynamical behaviour of this network is given by a system of differential equations

$$\begin{aligned} \frac{dx_1}{dt} &= f_1(x_1,\ldots,x_m,u) \\ &\vdots \\ \frac{dx_m}{dt} &= f_m(x_1,\ldots,x_m,u) \end{aligned} \tag{2.6}$$

with initial conditions

$$x_1(0) = x_1^{\,0},\ldots,x_m(0) = x_m^{\,0}. \tag{2.7}$$

Modelling autonomy in logistic processes

As we have seen in the brief discussions of the previous sections it is not obvious how to include the concept of autonomy in the mathematical models, depending on the modelling approach. In general existing models aim for a global understanding of the system and are suitable for the derivation of global control strategies. The implementation of such strategies may be unfeasible due to the size of the network, problems in making information available globally within a network and the like. This is the intrinsic motivation for studying autonomous control processes. Autonomy of processes suggests to model each process individually and to derive a model for the overall systems by coupling the autonomous components. Such an approach has been studied in the area of decentralised control, which we will now briefly discuss.

In the field of control theory decentralised control has been actively investigated starting in the early 80s of the last century, see (Siljak 1990; Trave et al. 1989) for an account and an introduction to the available results. The basic paradigm of decentralised control is that in contrast to the situation depicted in figure 2.22, a system is to be controlled by several controllers each of which only has access to a subset of the measured variables and to the control inputs to perform its task. This raises the question under which conditions a global control goal can be reached via the implementation of several local controllers. Especially for linear systems several results have been obtained that characterise stabilisability and optimisation of systems in which only an approach using decentralised strategies is possible (Siljak 1990; Trave et al. 1989). For nonlinear systems however, many basic questions remain unsolved.

From a certain point of view the problem of designing logistic processes with several autonomous components can be viewed as a variant of the problems treated in the field of decentralised control. Also in the logistic context the goal is to achieve certain tasks by the actions of several independent processes, each of which has limited access to the information. One of the fundamental difficulties in this approach is that very often logistic processes are governed by nonlinear laws. In other cases, one wishes to introduce nonlinearities to achieve certain control goals. In this area many mathematical problems are still unsolved.

2.8.4 Autonomous control and its effects on the dynamics of logistic processes

Here we give some examples, how autonomous control can be introduced into the models discussed above and we consider how it affects the solutions of these models.

First consider the re-entrant line discussed above. As we have noted there, the possibility to choose where to go to be served for the parts can be described in terms of the transition probabilities P_{ij}, making them dependent on the current situation, e.g., on the queue lengths. From the other side, if the machines are able to increase their processing rate when their queues are long or to decrease it once the queues become short, the service times $T_k(n)$ become also functions of the queue lengths. Appropriately chosen rules of the autonomous control may improve the dynamics of the production line in the sense that it becomes more efficient and robust. The resources of idling machines can be utilised. The parts automatically go to an idling machine, i.e., one with an empty queue, if the others are busy, i.e., have longer queues. In case of failure of a machine the parts route themselves to other machines. The ability to change service rates may help to avoid bottlenecks. These are potential advantages of an autonomous control. However the rules of an autonomous control should be chosen carefully. There are examples (Bramson 1994) of networks satisfying the usual traffic condition that the nominal load of the whole network is less then one, but that are nonetheless unstable, i.e., the queues grow unboundedly.

2.8.5 An illustrative example

Let us consider a couple of simple deterministic scenarios to demonstrate what a continuous model looks like in case of autonomous control. We consider a two machine production network. In this network there are two types of parts arriving at rates a_i, $i=1,2$, to receive service at the two different machines. The first machine is designed to process the first type of parts at rate b_{11}, however, it is able to process parts of the second type at a reduced rate $b_{12}<b_{11}$. Similarly, for the second machine we have the two processing rates $b_{22}>b_{21}$, for serving the second and the first type, respectively. If there is no control of the particle routing, parts of each type are always served at the machine designed for their type, i.e., a part of type i goes always to the i-th machine. This situation we will call Scenario 1.

In the second scenario the parts are able to decide by themselves at what machine they want to be serviced. They use certain decision rules that form the autonomous control and that have to be defined in advance. For example, a part might choose the machine with the shortest queue. Here we will use the following decision rule: A part of type i is routed to the machine $j \neq i$ only if the queue in front of machine j is empty and at the same time the queue in front of machine i is positive. Otherwise, it chooses the machine i. In case of $a_i>b_{ii}$, $i=1,2$, both queues eventually become positive and each part of type i goes to the i-th machine. This case is not interesting for us. The situation is similar if $a_i<b_{ii}$, $i=1,2$. An interesting setup is $a_1<b_{11}$, and $a_2>b_{22}$. In this case the first machine, which would idle periodically in the first scenario, every now and then receives parts of the second type. Hence the total throughput should be not less than in the first scenario.

We consider also the following Scenario 3, but with a different autonomous control. The parts first arrive at a common buffer. Then, when the *i*-th machine completes service, it orders a part of type *i* from the buffer. If there are no parts of type *i*, it orders a part of the other type. One can say that in this scenario the machines are autonomously controlled. The machines decide which type of part to process next. One way to compare these three scenarios is via discrete event simulation, which we do before we turn to continuous models.

Discrete-event simulation

It is clear that the interesting case is $a_1<b_{11}$ and $a_2>b_{22}$. To perform the simulation we normalise the maximum arrival rate of the parts of the second type to be one and set $a_1=1/24$, $b_{11}=b_{22}=1/16$, $b_{12}=b_{21}=1/20$. The arrival rate of the second type is varied between $1/16<a_2<1$. The simulation result of a time period of 500 time units is presented in figure 2.26, where the total amount of parts processed by both servers is plotted. Dashed, solid, and dotted lines correspond to Scenarios 1, 2, and 3, respectively. In the first scenario there are no decision rules, and hence the total throughput depends only on the processing rates, but not on the arrival rates. The second scenario is more efficient than the first one for most choices of arrival rate a_2. As expected, the third scenario has an even higher throughput than the first two. For longer interarrival times $1/a_2$ of parts of the second type all three graphs coincide. This is clear, since in this case the second machine can serve all arriving parts of the second type.

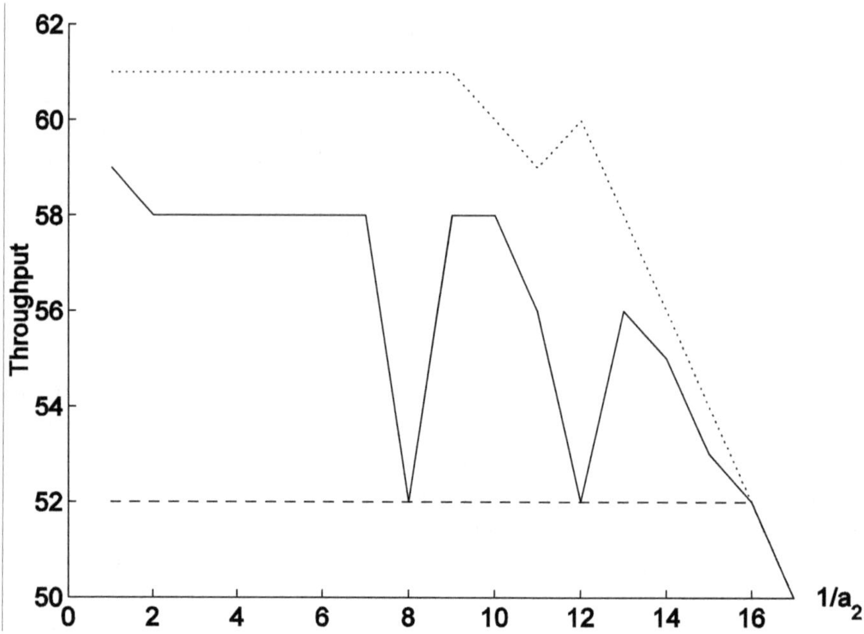

Fig. 2.26 Total throughput depending on the arrival rate of parts of second type

The continuous model

Let $x_i(t)$, $y_i(t)$ denote the number of parts of the first and second type, respectively, waiting in buffer i. Denote by $0 \le p_i(t) \le 1$ the fraction of parts of type i that are routed to machine 1 at time t. The evolution of these state variables can be described by ordinary differential equations as

$$\begin{aligned}
\dot{x}_1(t) &= p_1(t)a_1(t) - b_1(t)\frac{x_1(t)}{x_1(t)+y_1(t)} \\
\dot{y}_1(t) &= p_2(t)a_2(t) - b_1(t)\frac{x_1(t)}{x_1(t)+y_1(t)} \\
\dot{x}_2(t) &= (1-p_2(t))a_2(t) - b_2(t)\frac{x_2(t)}{x_2(t)+y_2(t)} \\
\dot{y}_2(t) &= (1-p_1(t))a_1(t) - b_2(t)\frac{x_2(t)}{x_2(t)+y_2(t)}
\end{aligned} \tag{2.8}$$

see (Dashkovskiy et al. 2004). The processing times of the machines are not constant but depend on the mixture of served parts, i.e., their fractions, which may change over time due to autonomous control of the parts. Moreover the processing rates are discontinuous functions of time and their expressions depend on the situation at the queues. If both queues are nonempty then

$$b_i(t) = \frac{x_i(t) + y_i(t)}{\frac{x_i(t)}{b_{ii}} + \frac{y_i(t)}{b_{ij}}}, \qquad i = 1,2, \tag{2.9}$$

see (Dashkovskiy et al. 2004) for details. If the first buffer is empty, $x_1(t)+y_1(t)=0$, i.e., $x_1(t)=y_1(t)=0$, it holds that

$$b_1(t) = \min\left(p_1(t)a_1(t) + p_2(t)a_2(t), \frac{p_1(t)a_1(t) + p_2(t)a_2(t)}{\frac{p_1(t)a_1(t)}{b_{11}} + \frac{p_2(t)a_2(t)}{b_{12}}} \right),$$

$$b_2(t) = \min\left((1-p_1(t))a_1(t) + (1-p_2(t))a_2(t), \frac{(1-p_1(t))a_1(t) + (1-p_2(t))a_2(t)}{\frac{(1-p_1(t))a_1(t)}{b_{11}} + \frac{(1-p_2(t))a_2(t)}{b_{12}}} \right), \tag{2.10}$$

The rules of autonomous control are encoded in the functions p_1 and p_2, which are in general functions of t, x_1, x_1, y_1, y_2 and, vice versa, given the rules of an autonomous control, the fractions p_1, p_2 can be calculated. For the Scenarios 2 and 3 the corresponding expressions can be found in (Dashkovskiy et al. 2004). Like the processing rates, so are their expressions different for different situations at the queues. Obviously, if both queues are non-empty at time t, then

$$p_1(t) = 1 \quad \text{and} \quad p_2(t) = 0 \tag{2.11}$$

hold. For $x_1(t)+y_1(t)=0$, $x_2(t)+y_2(t)>0$ one can derive

$$p_1(t) = 1, \quad p_2(t) = \min\left(1, \frac{(b_{11} - a_1(t))_+ b_{12}}{b_{11} a_2(t)}\right). \tag{2.12}$$

The corresponding expressions in the other cases, also for the third scenario, can be found in (Dashkovskiy et al. 2004). We note that the autonomous control rules in these scenarios are assigned to the parts. One can also allow the machines to decide in which order to process the parts or how fast to process them. In the latter case the processing rates b_{ij} become functions of t, x_1, x_1, y_1, y_2.

These simple examples illustrate how autonomous control can be defined, how it enters the equations and how it affects the dynamic behaviour of a logistic network.

2.8.6 Conclusions

We have classified possible models for autonomous logistic processes and discussed how an autonomous control enters these models and what its effects on the dynamics and stability of the processes are. An example illustrates the answers to these questions. We discussed the advantages of autonomous control and pointed out the related stability problem.

References

Armbruster D, de Beer C, Freitag M, Jagalski T, Ringhofer C (2006a) Autonomous Control of Production Networks Using a Pheromone Approach. Physica A 363: 104-114

Armbruster D, Gel ES, Murakami J (2007) Bucket brigades with worker learning. In: European Journal of Operations Research 176(1): 264-274, in press

Armbruster D, Marthaler D, Ringhofer C, Kempf K, Jo TC (2006b) A Continuum Model for a Re-entrant Factory. Operations Research, to appear

Armbruster D, Marthaler D, Ringhofer C (2004) Kinetic and Fluid Model Hierarchies for Supply Chains. SIAM J. on Mutiscale Modeling and Simulation 2: 43-61

Bartholdi JJ, Eisenstein DD (1996) A production line that balances itself. Operations Research 44(1): 21-34

Bautista J, Pereira J (2002) Ant algorithms for assembly line balancing. In: Proceedings of ANTS 2002 – Third International Workshop vol 2463: 65-75

Bonabeau E, Dorigo M, Theraulaz G (1999) Swarm Intelligence – From Natural to Artificial Systems. Oxford University Press, Oxford

Bramson M (1994) Instability of FIFO queueing networks. The Annals of Applied Probability 4: 414-431

Brückner S (1999) Return from the Ant: Synthetic Ecosystems for Manufacturing Control. In: The European Network of Excellence for Agent-Based Computing (AgentLink) Newsletter vol 4: 12-14

Brückner S, Wyns J, Peeters P, Kollingbaum M (1998) Designing Agents for Manufacturing Control. In: Proceedings of the 2nd AI and Manufacturing Research Planning Workshop, Albuquerque

Camazine S, Deneubourg JL, Franks NR, Sneyd J, Theraulaz G, Bonabeau E (2001) Self-Organisation in Biological Systems, Princeton University Press, Princeton

Chase C, Serrano J, Ramadge P (1993) Periodicity and chaos from switched flow systems. In: Proc. IEEE Transactions on Automatic Control: 70-83

Dai JG, Weiss G (1996) Stability and instability of fluid models for re-entrant lines. Mathemetics of operations research 21: 115-135

Dai JG (1995) On positive Harris Resurrence of multiclass queueing networcs: A unified approach via fluid limit models. Annals of Applied Probability 5: 49-77

Dachkovski S, Wirth F, Jagalski T (2004) Autonomous control in Shop Floor Logistics: Analytic models. Manufacturing, Modelling, Management and Control 2004, Edited 2006 by Chryssolouris G, Mourtzis D. Elsevier, Amsterdam

Dashkovskiy S, Rüffer B, Wirth F (2006) Discrete time monotone systems: Criteria for global asymptotic stability and applications. In: Proc. of the 17th International Symposium on Mathematical Theory of Networks and Systems, July 24-28, Kyoto, to appear

Dashkovskiy S, Rüffer B, Wirth F (2006) An ISS Lyapunov function for networks of ISS systems. In: Proc. of the 17th International Symposium on Mathematical Theory of Networks and Systems, July 24-28, Kyoto, to appear

Dorigo M, Stützle T (2004) Ant Colony Optimization, MIT Press, Cambridge

Gambardella LM, Taillard ED, Agazzi G (1999) MACS-VRPTW: A multiple ant colony system for vehicle routing problems with time windows. In: New Ideas in Optimization. McGraw Hill, London, pp. 63–76

Helbing D (2001) Traffic and related self-driven many particle systems. Reviews of modern physics 73: 1067-1141

Kim JH, Duffie NA (2004) Backlog Control for a Closed Loop PPC System. Annals of the CIRP 53: 357-360

Parunak H (1997) Go to the Ant: Engineering Principles from Natural Multi Agent Systems. Annals of Operations Reserch 75: 69-101

Peeters P, v Brussel H, Valckenaers P, Wyns J, Bongaerts L, Heikkilä T, Kollingbaum M (1999) Pheromone Based Emergent Shop Floor Control System for Flexible Flow Shops. In: Proceedings of the International Workshop on Emergent Synthesis, pp 173-182

Peters K, Worbs J, Parlitz U, Wiendahl HP (2004) Manufacturing systems with restricted buffer sizes, Nonlinear Dynamics of Production Systems. Wiley, New York

Prabhu VV, Duffie NA (1995) Modelling and Analysis of nonlinear Dynamics in Autonomous Heterarchical Manufacturing Systems Control. Annals of the CIRP 44: 425-428

Polderman JW, Willems JC (1997) Introduction to mathematical systems theory. A behavioral approach. Springer, Berlin Heidelberg New York

Scholz-Reiter B, Freitag M, Windt K (2004) Autonomous logistic processes. In: Proc. 37th CIRP International Seminar on Manufacturing Systems: 357-362

Scholz-Reiter B, Freitag M, de Beer C, Jagalski T (2005a) Modelling dynamics of autonomous logistic processes: Discrete-event versus continuous approaches. Annals of the CIRP 55: 413-416

Scholz-Reiter B, Freitag M, de Beer C, Jagalski T (2005b) Modelling and analysis of autonomous shop floor control. In: Proc. 38th CIRP International Seminar on Manufacturing Systems

Siljak DD (1990) Decentralized control of complex systems, Mathematics in Science and Engineering 184. Academic Press, Boston

Sontag ED (1989) Smooth stabilization implies coprime factorization.In: Proc. IEEE Trans. Automat. Control 34(4): 435-443

Sontag ED (1998) Mathematical control theory. Deterministic finite dimensional systems. Springer, Berlin Heidelberg New York

Trave L, Titli A, Tarras AM (1989) Large scale systems: decentralization, structure constraints and fixed modes. In: Lecture Notes in Control and Information Sciences 120. Springer, Berlin Heidelberg New York

2.9 Autonomous Decision Model Adaptation and the Vehicle Routing Problem with Time Windows and Uncertain Demand

Jörn Schönberger, Herbert Kopfer

Chair of Logistics, University of Bremen, Germany

2.9.1 Introduction

The instruction of resources in logistic systems in order to ensure an effective as well as efficient usage is a very sophisticated task. At lot of data and requirements have to be considered simultaneously. For this reason computerized decision support (Makowski 1994) is strongly recommended (Bramel and Simchi-Levi 1997; Crainic and Laporte 1998).

A prerequisite for the application of computerized methods to tackle logistics decision problems is the representation of the current decision situation in a formalized fashion, a so-called decision model, normally described in terms of mathematical expressions (Williams 1999). Such a model, often of optimisation type, is than tackled by, typically heuristic, algorithms (Ibaraki et al. 2005; Michalewicz and Fogel 2004) in order to derive one (best possible) solution that is the instruction predicting the future activities in the logistic process execution system.

If the decision problem in the real world changes, the existing problem model becomes void and a re-modelling is required. Additional knowledge about the current system state and performance enters the model in order to propagate the problem changes to the used decision support system. However, this topic has received only minor attention so far in the scientific literature although it is of very high practical relevance and importance.

In this contribution, we investigate generic procedures and rules for an automatic feedback controlled adaptation of decision models for a variant of the well-known Vehicle Routing Problem with Time Windows. The considered problem differs from the generic problem because the customer sites, which require a visit, emerge successively over time so that a plan revision becomes necessary. In Subsection 2.9.2 we present the considered decision problem in more detail. Subsection 2.9.3 introduces the algorithmic framework for an autonomous adaptation of the decision model and in Subsection 2.9.4 we prove the framework's general applicability within numerical simulation experiments.

2.9.2 The vehicle routing problem with time windows and uncertain demand

This section is about the investigated decision problem. The problem is non-stochastic, e.g. requests are released consecutively but we do not know anything about their arrival times. In Subsection 1 we survey the scientific literature related to the problem considered here. Subsection 2 outlines the problem informally. The life cycle model of a request is presented in Subsection 3 and the decision problem that requires a solving whenever at least one additional request arrives is stated in 4. The construction of artificial test cases developed for a numerical simulation of selected problem instances is subject of Subsection 5.

Gendreau and Potvin (1998) survey vehicle routing and scheduling problems with incomplete planning data. Psaraftis (1988) and Psaraftis (1995) discuss the differences between vehicle routing and scheduling problems with deterministic and with probabilistic or incomplete planning data.

Jensen (2001) understands robust planning as the generation of plans that maintain their high or even optimal quality after subsequent modifications. He defines flexible planning as the generation of plans whose quality does not significantly decrease after the execution of algorithmic rescheduling and alterations of the so far used plans.

Jaillet (1998), Jaillet and Odoni (1988) as well as Bianchi et al. (2005) propose a robust transport scheduling approach. They construct optimal a-priori-routes. Such a route has a minimal expected length among all possible routes through the potential customer sites. However, this approach assumes that probability distributions of the actual demand at the customer sites are known. As soon as a vehicle has visited a customer site and the corresponding demand becomes sure it has to be decided whether a replenishment visit at a depot has to be executed before the next customer (again with uncertain demand) is met.

Flexible planning approaches do not require any knowledge about future events. An existing plan is updated consecutively and reactively. Sequenced planning problem instances P_i are solved one after another. Such a sequence of decision problems P_1, P_2,... is called an online planning problem according to Fiat and Woeginger (1998). A survey of online vehicle routing and scheduling problems is provided by Krumke (2001). Special cases are addressed by Ausiello et al. (2001). Theoretical results for online repairmen dispatching strategies are found in Bertsimas and van Ryzin (1989) as well as Irani et al. (2004).

Slater (2002) as well as Gayialis and Tatsiopoulos (2004) propose dispatching systems for transport planning tasks. Ghiani et al. (2003), Gendreau et al. (1999), Fleischmann et al. (2004), Séguin et al. (1997) and Gutenschwager et al. (2004) investigate dispatching systems in which decisions have to be derived in real time without any delay.

Gendreau and Potvin (1998) give a survey of applications for vehicle routing type problems requiring a re-planning. Brotcorne et al. (2003) report about an application of operations research methods to a relocation problem in medical rescue service. Chen and Xu (2006) as well as Savelsbergh and Sol (1999) investigate sophisticated algorithms for the repeated plan update in real world transport applications.

Informal problem description

Similar to the vehicle routing problem with time windows (VRPTW), we are looking for a decision support system that generates automatically a set of route for the available vehicles so that they fulfil customer orders and then travels back to a depot. Time windows restrict the intervals in which a customer order can be served. The problem we are investigating in this contribution comes along with three generalisations compared with the generic VRPTW version:

- (SOFT TIME WINDOWS) A customer site is allowed to be visited after the corresponding time window has been closed. However, lateness will produce additional penalty costs to be paid to the customer. The amount pen(δ) to be paid increases linearly with the temporal distance δ from the arrival time to the latest allowed arrival time but is limited to a certain amount PENMAX;
- (SUBCONTRACTION) Each customer request can be subcontracted. In such a situation the considered company, that maintains the fleet to be routed, orders another logistics service provider (LSP) to fulfil a particular request. The LSP receives a certain amount for this service

but ensures that the request is fulfilled within the specified time window;

- (UNCERTAIN DEMAND) Only a subset of all requests to be fulfilled is known to the planning authority at the time when the subcontraction is decided and the routes for the own vehicles are generated. Whenever one or more additional requests become known, it has to be decided whether these requests are subcontracted or not. In the latter case, the additional requests are integrated into the so far existing routes. For some of the requests, so far expected to be served by an own vehicle, subcontraction can become more attractive now. This is the result of a postponement of these requests in order to serve a recently released request in time. Attention has to be paid that a once subcontracted request cannot be reintegrated into the route of an own vehicle because the contract with the ordered LSP is binding.

We refer to this decision problem as the vehicle routing problem with time windows and uncertain demand (VRPTWUD). The goal is to find a transportation plan (Crainic and Laporte 1997) that describes which requests are served internally or externally by LSPs and how the requests are served by the own vehicles.

The SOFT TIME WINDOWS property allows a more flexible route generation because minor window violations are penalized only slightly. Due to the SUBCONTRACTION property, requests not fitting with the remaining portfolio do not have to be considered in the route generation so that a more advantageous request consolidation is achieved. However, the UNCERTAIN DEMAND issue requires a transport plan adaptation every time when additional requests are released.

Although each particular request in allowed to be late, there is a general guideline that predicts a global punctuality. More detailed, for a given transportation plan update time t, the percentage of the requests served in time has to be larger than p^{target}. We consider only those requests completed within the last t^- time units and whose completion is scheduled within the next t^+ time units. This means, only recent service quality information are used because the relevant consideration time window $[t^- - t; t+t^+]$ moves with ongoing time.

The goal of the planning support to be developed is to establish a planning system that allows the generation and repeated update as well as adaptation of flexible transportation plans for the field teams including decisions about externalization of selected requests. The flexibility is important because the customer requests are received successively and their arrival times cannot be predicted or forecasted so that only a reactive transport plan revision is realizable. Furthermore, in order to maintain the flexibility of the transport plans even in situations with an extreme workload, it is allowed to violate the agreed time windows but the corresponding customers are paid compensation.

Online request state update

In order to consider the successively arriving additional requests, we propose to update the existing transportation plan reactively after the additional requests become known (Schönberger and Kopfer 2007).

Let t_i denote the i-th time when additional requests become available and let $R^+(t_i)$ represent the set of additional requests, released at t_i. After the last transportation plan update at time t_{i-1}, several requests have been completed. These requests are stored in the set $R^C(t_{i-1},t_i)$. Then the request stock $R(t_i)$ at time t_i is determined by $R(t_i) := R(t_{i-1}) + R^+(t_i) - R^C(t_{i-1},t_i)$.

The life of a single request r consists of a sequence of states to which r belongs. Initially, when r enters the transportation system it is known but not yet scheduled (F). If r is assigned to an own vehicle for execution it is labelled by (I) or by (E) in case that r is assigned to an external service partner. A request whose completion work at the corresponding customer site has been started but not yet finished is labelled as (S). The final stage (C) of r indicates that r is completed.

Every time a transportation plan update becomes necessary, the current states of known requests from $R(t_i)$ are updated. The state (F) is assigned to all new requests from $R^+(t_i)$. For all requests contained in $R^C(t_{i-1},t_i)$, their state is updated from (I) or (E) to (C) and requests whose on-site execution have been started but not yet completed receive the new state (S) that replaces their former state (I) or (E). Now, the scheduling algorithm is started that carries out the necessary transportation plan updates. From the updated transportation plan the information about the intended type of request execution of all requests labelled as (F) or (I) is taken. The state of an (I)-labelled requests is updated to (E) if it has been decided to out source this request. Otherwise, the state of this request remains unchanged. Finally, all (F)-labels of externalized requests are replaced by (E)-labels for subcontracted requests and (I)-labels replace the (F) labels for the remaining requests from $R^+(t_i)$.

Statement of the scheduling problem

The decision whether a request should be assigned to an own team or given to an external partner cannot be solved uniquely for each request. A complex decision problem must be solved every time the currently valid transportation plan has to be updated, considering simultaneously all assignable requests, which are labelled by (I) or (F). It has to be decided for all these requests whether they are definitively subcontracted and given to a service partner for execution or if they should be assigned for the first time to one of the available own vehicles represented by the elements of set V(t). In order to find the minimal cost assignment, we propose the following optimisation model.

Let $\Omega(t)$ denote the set of all possible request sequences $p=(p_1,\ldots,p_{n(p)})$ representing the order in which the contained customer requests, selected from R(t), are visited. The vehicle v selected for request r in the last transportation plan is denoted as $\Psi(r)$. If r is labelled as (I) then $\Psi(r)\in V(t)$, otherwise $\Psi(r)=\{\}$.

We assume that each $p\in\Omega(t)$ holds for the following two properties:

- The final entry $p_{n(p)}$ of p refers to the depot to which all vehicles return;
- If the first entry p_1 refers to a request labelled currently as (S) then the departing time from p_1 cannot precede the finishing time of this request.

The following two binary decision variable sets are used to code the necessary decisions. The variable u_p is set to 1 if and only if sequence $p\in\Omega(t)$ is selected for vehicle $v\in V(t)$. Furthermore, y_r is set to 1 if and only if request $r\in R(t)$ is subcontracted.

We are looking then for instantiations of the above decision variables that minimizes the costs $C(\{x_{pv}\},\{y_r\})$ but considering that

- Each vehicle is assigned to exactly one (maybe an empty) path from $\Omega(t)$;
- Each request is contained in at most one of the selected paths.
- A request r labelled by (S) cannot be assigned to another vehicle as $\Psi(r)$;
- If request r is labelled by (E) then $y_r=1$.
- If vehicle v is assigned to p then p_1 must correspond to the current location of vehicle v.

We desist from giving the formal mathematical statement of the above five constraints since we do not need them in the remaining presentation.

The objective function $C(\{x_{pv}\},\{y_r\})$ calculates the costs associated to the instantiations of the two decision variable sets. It is the sum of the travel costs for the own deployed vehicles plus the sum for subcontraction fees and penalties to be paid for late arrivals at customer sites. Therefore, it denotes the costs for the associated transportation plan.

Artificial test cases

In order to evaluate different dispatching approaches and to control the severeness of the observed scenario, we have derived a set of artificial test instances. Each instance is defined by a special instantiation of a set of parameters. The adjustment of these parameters models different scenarios.

Two different kinds of routing scenarios with successively arriving requests are reported in the scientific literature. In the first scenario type, the number of demands that are released during a specific time interval remains unchanged. It is possible to adapt the available resources in such a situation so that all additional demands can be served in time. For this reason, such a scenario is called a *balanced scenario*. Examples can be found in Pankratz (2002), Lackner (2004) and Mitrović-Minić et al. (2004). In case that the number of additionally released demands during a specific time interval varies, the scenario is denoted as a *peak scenario*. Here, it is hardly possible to adapt the available resources in advance. Gutenschwager et al. (2004), Sandvoss (2004) as well as Hiller et al. (2005) deal with real world examples that do not allow a parameterization and classification for scientific analysis purposes.

In order to determine a competitive and comparable tariff for calculating the fare to be paid to an LSP for subcontraction, we compare the travelled and the demanded distances in the best-known VRPTW solutions for the Solomon benchmark Problem (Solomon, 1987) as described in detail in Schönberger (2005). For each request, the amount of the subcontraction costs is calculated and assigned to the particular request.

To simulate peak scenarios we first generate a balanced stream of incoming customer demands over the complete observation time period. A second stream is generated for a part of the observation period. Both streams are than overlaid so that during the period in which the second stream is alive, the balanced stream is interrupted and a higher number of requests must be scheduled.

The balanced stream of incoming demands for the observation period $[0,T_{max}]$ is generated by successively drawing requests from the Solomon instance P. At time $t^{rel}=0$, n_0 demands are drawn randomly from P. Then, the release time is updated by $t^{rel}:= t^{rel} +\Delta_t$. For this new release time, n demands are drawn from P at random. For each selected demand r, its release time is set to t^{rel}. The original service time window $[e_r,l_r]$ of r is replaced by $[t^{rel}+e_r, t^{rel}+l_r]$. Additional demands are generated as long as $t^{rel} \leq T_{max}$.

The second stream of demands is released to simulate a peak of demands. For the first generated release times $0, \Delta_t, 2\Delta_t, \ldots, n_1\Delta_t$ no demands are released. For the next n_2 release times $(n_1+1)\Delta_t, \ldots, (n_1+n_2)\cdot\Delta_t$ Δ_m demands are specified as described above. For the remaining release times, no additional demands are given.

All vehicles specified in P can be used.

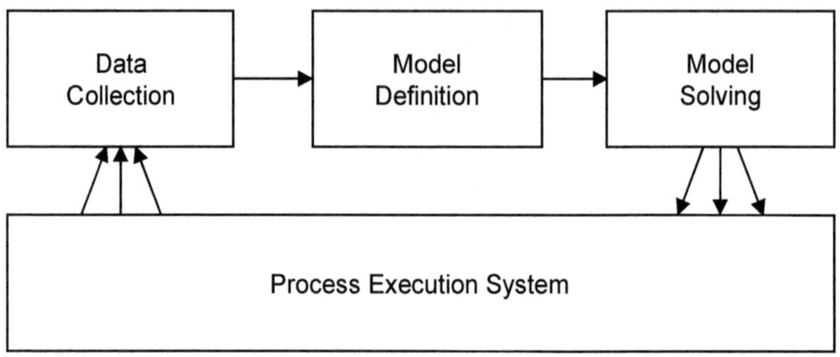

Fig. 2.27 Model-based replanning

Consequently, each scenario is described by the triple (P, d^{peak}, Δ_m). In this investigation, we use the four Solomon cases R103, R104, R107 and R108 to generate request sets. Furthermore, it is $n_0=n=50$ and $\Delta t=100$ time units. The peak duration has been set to $d^{peak}=200$ time units and the peak high is fixed at $\Delta_m=100$ additional request.

2.9.3 Model-based planning in dynamic environments

This section is dedicated to the presentation of an algorithmic framework realizing an automatic, autonomously controlled adaptation of a generic decision model to a particular situation. In Subsection 0, we compile the open issues that contradict the realization of an automatic adaptation of a decision model. Subsection 0 describes generic approaches for modifying an optimisation model. Subject of Subsection 0 is the proposal of an iterative procedure that controls the adaptation of a logistic process exploiting a feedback-triggered adaptation and in 0 explicit adaptation rules are presented for the VRPTWUD.

Model-based replanning

The re-active planning of logistic processes in a changing environment for adapting a process to a new situation requires the repeated execution of the three basic steps data collection, model statement and model solution. Figure 2.27 represents the steps to be executed. As soon as relevant changes are detected that affect the so far executed process, the available data are collected and prepared. From these data, a new decision model is set up and, next, this model is solved. Finally, the new process (the solution of the recently solved decision problem) is broadcasted back into the logistic system for execution.

The re-start of the planning cycle is a response to a modification in the underlying real world decision problem. Clearly, these modifications have to be considered in the compilation and solving of the new decision model. Up to now, some technical as well as conceptual challenges have to be overcome before the autonomous and appropriate redefinition and solving of a decision model can be exploited to the largest possible extend. The collection of the required problem data and the re-setup of the decision model as well as its solving are faced with some deficiencies that have not been solved satisfactorily yet.

Data collection. The technical availability of the data is high but a lot of effort and intelligence has to be spent in order to get helpful, consistent and reliable as well as complete data for the next instance of a decision model in an online decision problem. In a deterministic environment, there is enough time to discuss the adequacy of the data and to look for missing data but if the changing environment predicts a rapid and reliable reaction on environment changes and requires the revision of former decisions then a strong automatic data pre-processing support is unconditional necessary. Contributions from artificial intelligence and/or ideas methods borrowed from Data Mining should be incorporated to calibrate, to complete the collected data, and to prepare the setup of the next decision model.

Automatic adaptation of the planning goal. The objective belonging to the model of the next optimisation problem instance requires an automatic, feedback-triggered adaptation with respect to the congruence with superior objectives that predict the development of the logistic system over a longer time horizon. Consequently, an analysis of the currently collected data with respect to the current system performance is necessary in order to decide what the next goal to be followed will be and which data are required for the definition of the objective function.

Vectors of data. Beside the consideration of current problem data, it is necessary to consider both its development direction and the velocity of change as well. Therefore, additional data are necessary in order to assess and describe the system evolution adequately.

Data extraction and data interpolation. Data provided by information technology-based services have to undergo a substantial pre-processing before these data can be used for the setup of the next model. Redundant data have to be eliminated and missing data must be integrated.

These three items describe special requirements for the collection and preparation of the modelling of a decision situation in an online scenario.

Model building. In order to allow the automatic re-definition of an adequate decision model, the so far existing straightforward techniques require some extensions.

Flexible representation of decision alternatives. The setup of a particular decision model is currently compromised by inflexible representations of the decision alternatives. Here, future research efforts should be spent to the development of more flexible and adaptable representation methods so that a modified decision problem can be coded easily.

Automatic adaptation of the decision alternative evaluation. It is necessary to re-think the worthiness of a certain decision alternative after the problem under consideration has changed. Often, the usefulness of a certain alternative is given only if some assumptions are met (enough resources, enough time, …). If these assumptions become void, the worth of a particular decision alternative runs into danger to alter.

Automatic feedback-controlled adaptation of the search space. The search space of an optimisation problem instance represents the decision alternatives currently available. In order to identify adequate solutions that support the strive for the fulfilment of longer term planning goals, it is necessary to prune some alternatives that are currently feasible but, on the long run, to not lead to the intended system development. Furthermore, additional solution alternatives should be allowed if the existing solution alternatives do not comply with the current situation.

Model solving. The derivation of a solution (proposal) is left to automatic software procedures (algorithms). They have been applied successfully to problems in static environments for several decades. In order to use the observed findings in scenarios with varying system environments, the software procedures must undergo some specific modifications in order to apply them successfully to process adaptations.

Autonomous re-parameterization and re-configuration. In order to enable the software to deal with quantitative as well as with qualitative differences in problem instances it is unconditional necessary to equip the software with a problem interpreter to analyse the current decision problem. Furthermore, depending of the results of the problem analyse, the software has to decide autonomously about the adjustment of their search parameters as well as their hardware usage.

Simultaneous addressing of feasibility recovery and update improvement. Decision software for process updating in systems with uncertain problem data has to consider simultaneously the recovery from event-based infeasibility as well as the improvement or even optimisation of the updated processes in order to achieve highest process quality and reliability.

Limitation of decision times. In order to provide the logistic system with an adapted process proposal after a process disruption event the update time has to be kept as short as possible. In case that the pre-specified update answer time is very short (or even close to zero), it is tried to provide a feasible update first and to improve it afterwards if time is still available (Gutenschwager et al. 2004).

The application of software decision support and planning algorithms is unconditional necessary to cope with the complexity of recent decision problems. The availability of an adequate final decision model is a prerequisite for the successful application of automatic software procedures. An enlargement of the quality of the provided data is currently subject of different research disciplines, e.g. data mining (Clifton et al. 2002). Furthermore, certain researchers (Holzer 2003) also address the automatic re-configuration and the speed-up of decision algorithms. However, the exploitation of feedback-information from the underlying process execution system for the adaptation of formal decision (e.g. optimisation) models to the currently observed system state and to the currently waiting decision problem is not yet subject of any scientific work. In the remainder of this contribution, we propose some generic ideas to target this topic.

Generic approaches for optimisation model adaptation

Any formal optimisation problem consists of a description of the search space that includes all feasible solutions and of an objective function that assigns a numerical or vector value to each element of the search space. One or both of these components can be subject of modifications in order to adjust and adapt an existing generic optimisation model according to externally given rules.

Modification of the objective function. The main reason for defining an objective function is to evaluate each solution alternative in order to distinguish different solutions as well as to rank them. An automatic solving procedure exploits the objective function and uses the objective function value(s) as a feedback when

- Comparing different branches for further exploration of the search space and selecting one branch to be searched next or
- pruning some branches from a further exploration due to a reliable estimation about an unsatisfying solution quality to be found in this branch.

For this reason, a modification of the objective function can be interpreted as an adaptation of the search direction of the applied search algorithm. The main goals of this adaptation kind are (i) guiding the search process away from solutions that are currently unattractive and (ii) allowing the search algorithm to find adequate solutions quicker.

Adaptation of the search space. The search space can be modified by excluding (pruning) solution alternatives from the search space defined in a given model or adding additional solutions to the proposed set of solution alternatives. The pruning of solutions can be achieved by strengthening existing restrictions or adding new restrictions and the search space can be enlarged by relaxing or skipping so far valid restrictions. Pruning of solutions aims at prohibiting the selection of certain solutions. This technique is often used if the evaluation scheme cannot be used effectively to prevent the selection of low quality solutions. The promotion of additional solution alternatives is a response if no adequate solution can be identified in the so far maintained search space.

Basic algorithm

The adaptation of the decision model of the current problem instance should be based on the current system performance measured in term of the instantiation of one or more key indicators. Therefore, the current system performance is determined and the observed values are then compared with some major guidelines predicted by a superior authority (SA) that has the right to instruct the planning authority (PA). This concept is shown in a formalized fashion in figure. 2.28 Initially, SA receives feedback information, e.g. performance information, from the process execution system (1). It compares the observed performance values with the values predicted by SA. In case that a discrepancy is detected, it instructs PA to adapt its decision rules in a fashion that supports the achievement of the SA guidelines (2). A confirmation of the adaptation is submitted from PA to SA (3). As soon as a replanning becomes necessary, PA pulls the required planning data from the process execution system (4), derives a new process using the currently valid planning rules and delivers the process information to the execution system (5). The interactions (2) and (3) form the planner adaptation cycle and the interactions (4) and (5) are the process control cycle. Both cycles are concatenated by the feedback interaction (1). The planner adaptation cycle enables the adaptation of the process control as soon as the system performance requires an adaptation. The overall planning system (consisting of the two mentioned interacting cycles) is therefore able to update the processes in the process execution system autonomously without any intervention of human assistance even if the underlying problem changes significantly.

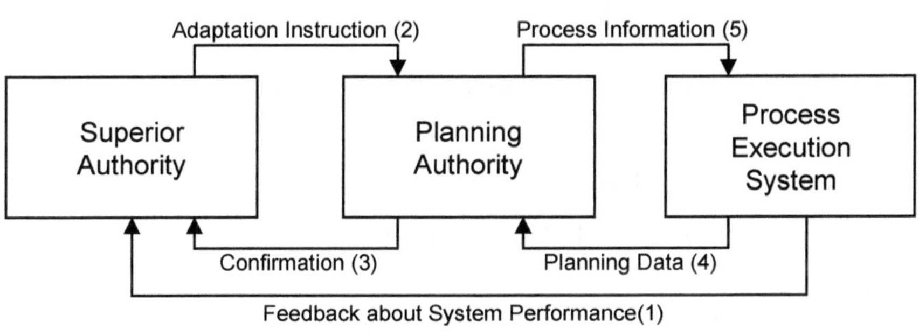

Fig. 2.28 Interaction of superior authority and planning authority

A generic process management algorithm is proposed in Fig. 2.29. It exploits an externally specified adaptation rule to adjust a generic decision model to the current performance. Such a generic decision model represents the underlying decision problem but does not exploit any feedback information about the current system performance or its environment.

Initially, the so far followed process InitialSolution is submitted together with the AdaptationRule to be followed (1). Then, the valid CurrentSolution is set (2). Now the procedure waits until the CurrentSolution is completed or additional requests collected in R become known (3). In the first case, the procedure terminates because nothing is to do anymore (4). In the latter case, it is check, whether R corrupts CurrentSolution (5). If this is not true then the procedure waits again otherwise it starts updating CurrentSolution (5). Therefore, the current time is fetched (6) and a generic decision model dm is derived that includes the additional requests (7). Next, the current performance of the logistic system is calculated (8). Now, the generic decision model is adapted with respect to the currently observed performance following the predicted AdaptationRule (9). The adapted decision model is solved and a new CurrentSolution is generated from the adapted decision model dm (10). This new solution is broadcasted to the process execution system (11). The update iteration is terminated (12) and the procedure waits again.

```
(1)  PROCEDURE process_management(InitialSolution,rule);
(2)  CurrentSolution := InitialSolution;
(3)  wait until (CurrentSolution is completed)
          or (ExternalEvents R are released);
(4)  if (CurrentSolution is completed) then
          goto (13);
(5)  if not (SolutionCorrupted(CurrentSolution,R)) then
          goto (12);
(6)  time:= GET_CURRENT_TIME();
(7)  dm:= GENERIC_MODEL(time,CurrentSolution,R);
(8)  perform-
   ance:=SYSTEM_PERFORMANCE(time,CurrentSolution);
(9)  dm:=ADAPT_MODEL(dm,performance,rule);
(10) CurrentSolution := SOLVE_MODEL(dm);
(11) BROADCAST(CurrentSolution);
(12) goto (3)

END PROCEDURE;
```

Fig. 2.29 Process management algorithm

This algorithmic describes the framework for the reactive management of a logistic process in response to external events that are detected over time.

In the VRPTWUD context, a generic decision model consists of a search space in which all feasible transportation plans are coded and the standard objective function C, that represents the costs associated with each single transportation plan (cf. 0).

To solve the adapted model, we apply the Memetic Algorithm framework introduced in Schönberger (2005). We desist of a detailed description of the algorithm components and configuration but refer to the previously given literature.

Adaptation rules

An adaptation rule maps a constellation of key indicator values to an instruction that describes the modification of the current generic model in order to implement additional knowledge about the current process performance into the model.

In this contribution, we apply three different adaptation rules in the process management algorithm for solving the VRPTWUD. All four rules read the currently observed punctuality p_t and derive some model modification instructions from this value. The proposed modifications are applied immediately.

For purposes of comparison, we define the rule **NONE**, that do not apply any adaptation. Consequently, the generic decision model is solved in each iteration. No performance feedback is exploited.

The adaptation of the search direction is targeted in the experiment with the rule **SDAD** (**S**earch **D**irection **AD**aptation). The generic approach consists in the modification of the objective function of the generic decision model. In particular, it is aimed at adjusting the costs caused by a too late arrival at a customer location. The idea is to give less weight to a time window violation if the system load is very high so that time window violations cannot be prevented at all. Instead, decisions about subcontraction or self-fulfilment should be favoured. Furthermore, in case that the system load is low and time window violations can be prevented by subcontraction, such a time window violation should be penalized very hard. In order to realize the adaptation of the objective function, we replace the proposed cost function C in the memetic algorithm solver by an extended cost function that adjusts the weighting of the cost drivers to the current search state. The cost function C is replaced by

$$\widetilde{C}(s) := \left(\frac{e^{1+\chi(s,k)}}{e^{1}} \right)^{k} \cdot C(s) \tag{2.13}$$

where s denotes the current solution alternative to be evaluated, k refers to the search iteration of the applied Memetic Algorithm solver. The parameter $\chi(s,k)$ represents the fraction of the time-window-constraint-violations within s compared to all time window constraint violations observed in the k-th population generated by the Memetic Algorithm. $\widetilde{C}$ enlarges the costs of s compared to the other maintained solution proposals if s contains an above-average number of too-late-arrivals. On the other hand, it awards s if it comes along with a below-average number of too-late-arrivals at customer sites.

The second proposed rule aims at adjusting the constraint set of the given model with respect to the currently detected performance. **CSAD** (**C**onstraint **S**et **AD**aptation) shrinks decision variable domains of selected indicator variables determining the subcontraction of a request. More concretely, if the current punctuality falls below p^{target}, then the subcontraction of a certain subset of requests is enforced by shrinking the set of possible values for the corresponding binary decision variables y_r from $\{0,1\}$ down to $\{1\}$. If the least expected punctuality p^{target} is re-achieved, then 0 is added again to the domain of the affected y_r. A detailed description of the control of CSAD can be found in Schönberger and Kopfer (2007).

The third proposed adaptation rule **SDCS** (**S**earch **D**irection and **C**onstraint **S**et Adaptation) combines the features of the SDAD and CSAD rule. Both the search for appropriate solutions as well as the predetermination of subcontraction decisions is addressed for adapting the decision model.

2.9.4 Numerical experiments

In this subsection, we report about the executed numerical experiments in which the proposed framework is assessed in combination with the proposed adaptation rules. The setup of the experiments is described in 0 and the achieved results are presented and discussed in 0.

Experimental setup

We have simulated the four scenarios within three independent runs for each combination of rule and scenario and for each of the four adaptation rules NONE, SDAD, CSAD and SDCS. Overall, 3×4×4 = 48 experiments have been performed.

In each of the experiments, we have set the target punctuality p^{target} to 0.8. The configuration of the CSAD rule is as follows. A first constraint set modification (intervention) is applied as soon as the currently observed punctuality falls below 0.85 percent and the maximal intervention intensity takes place as soon as p_t falls below 0.75. If the punctuality p_t falls below 0.85 the intensification of the application of the specified rule increases proportionally until it reaches the maximal possible intensity 1 for $p_t \leq 0.75$.

At the transportation plan update time t only requests contained in the interval [t-500;t+500] are considered for the calculation of p_t. A delay of less than 10 time units at a customer site does not cause any penalization costs. If a vehicle arrives more than 100 time units after the associated time window has been closed, a penalty amount of PENMAX=100 money units has to be paid for this out of time window arrival to the affected customer. With increasing delays larger than 10 time units, the corresponding penalties increase proportionally up to the maximal penalty amount of 100 money units.

Since we want to demonstrate the ability of the model adaptation to overrule the short-term cost minimization objective, we enlarge the subcontraction costs by the prohibitive factor 20.

Results

We have recorded the observed punctualities within the moving time window specified above. The value p_t^{RULE} denotes the averagely observed punctuality at time t within the experiments where the adaptation is carried out according to RULE.

In a reference experiment with the application of rule NONE, p_t^{NONE} reduces by 60.4% after the demand peak has occurred (compared to p_{500}^{NONE}). Furthermore, it can be seen from figure 2.30 that in 89% of the observation interval, the observed punctuality lies below p^{target}. Immediately after the demand peak occurs at time t=1500, p_t^{NONE} falls below p^{target} and does not recover throughout the ongoing simulation experiment.

The logistic system performs better if the rule SDAD is applied in the process management procedure. In this case, a decrease by 21.9% of p_t^{SDAD} is observed after t=1500 and in only 21% of the length of the observation interval, the target punctuality is not reached. The duration for the recovery of p_t^{SDAD} is 1000 time units (from t=2100 where the target punctuality is not achieved for the first time until t=3100 when it is re-achieved). From the presented results, we state that the adaptation of the search direction boosts the system performance with respect to the current punctuality of the system.

A further increase in the system performance is observed for the application of the CSAD rule. Here, the maximal loss of p_t^{CSAD} after the demand peak is limited to 17% and in 20% of the observation interval, the least desired punctuality p^{target} is not achieved. Furthermore, the decrease of p_t^{CSAD} starts immediately at t=1700 units but the target punctuality has been re-achieved after 900 time units at time t=2600. From this time on, the punctuality does not fall again below 80%.

The simultaneous adaptation of the search direction as well as of the constraint set as realized in the SDCS rule outperforms the other two adaptation rules and produces very convincing results. After the demand peak has been started, the punctuality p_t^{SDCS} does not reduce by more than 8% with respect to p_{500}^{SDCS}. Throughout the overall observation period, the punctuality does not fall below 0.80.

To conclude the presentation of the observed results we state that the adaptation of the decision model to be solved after additional problem knowledge was known is necessary in a scenario with strongly varying system load. Instead of an overall quality reduction by 62,6% (comparing the maximal and the minimal observed punctuality values) in the reference experiment without any adaptation, the adjustment of the search direction results in a significant performance increase. The maximal reduction of p_t^{SDAD} is 22,4%. However, this value is further reduced down to 13,4% if the constraint set is adapted but the most convincing results are observed for the simultaneous adaptation of both the search direction and the constraint set (8% variation). Therefore, the main result we learn from this experiment is that the adaptation of the decision model is able to reduce the impacts of system load peaks and helps to keep the performance on a nearly unchanged level.

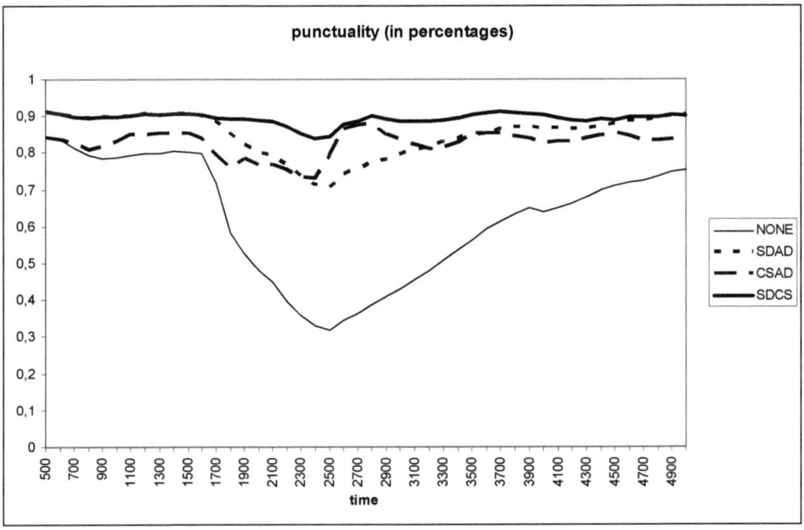

Fig. 2.30 Observed punctualities p_t^{RULE}

It is obvious that the stabilization of the punctuality is not achieved "free of charge" because the application of SDAD, CSAD or SDCS overrules the repeated cost minimization. For a couple of requests not the cheaper self service but the much more expensive subcontraction fulfilment mode has been selected to keep the punctuality on a sufficiently high level. In order to compare the impacts of the application of the different adaptation rules, we have computed the relative increase cc_t^{RULE} in the cumulated costs observed up to a time t with respect to the cumulated costs CC_t^{NONE}.

$$cc_t^{RULE} := \frac{CC_t^{RULE}}{CC_t^{NONE}}, \text{ for RULE} = \text{SDAD, CSAD, SDCS,} \qquad (2.14)$$

It is where CC_t^{RULE} denotes the averagely observed cumulated costs up to time t observed in the experiment with RULE. The observed values for cc_t^{RULE} are summarized and presented in figure 2.31.

Closely after the demand peak has been occurred, the additional costs explode. However, for later observation times, the relative costs reduce. For SDAD and SDCS it seems to converge asymptotically towards the value 2. However, CSAD seems to produce costs that will be four time larger than in the NONE experiment but it should be stated that CSAD seems to be worried in early times (t<1500) producing an overreaction.

2.9.5 Conclusions

Within this contribution, we have introduced a generic framework for the reactive adaptation of logistic processes to unpredictable change in its environment. This framework has been tested successfully for artificial benchmark instances representing the VRPTWUD. Different adaptation rules have been assessed.

From the observed results we deduce the general applicability of the proposed framework as well as of the proposed adaptation rules for adjusting a generic optimisation problem to the currently observed system performance. However, the additional costs produced by a deviation from the pure cost minimization objective are significant larger.

The adaptation rule concept allows an autonomous self-adjustment of the two-cycle planning system to varying planning assumptions. Feedback from the process execution system (the real world) is exploited explicitly.

Future research will include the investigation of more complex adaptation rules. Furthermore, it should be investigated how the gap between the costs for the different available fulfilment modes (self-fulfilment and subcontraction) influences the applicability of the model adaptation.

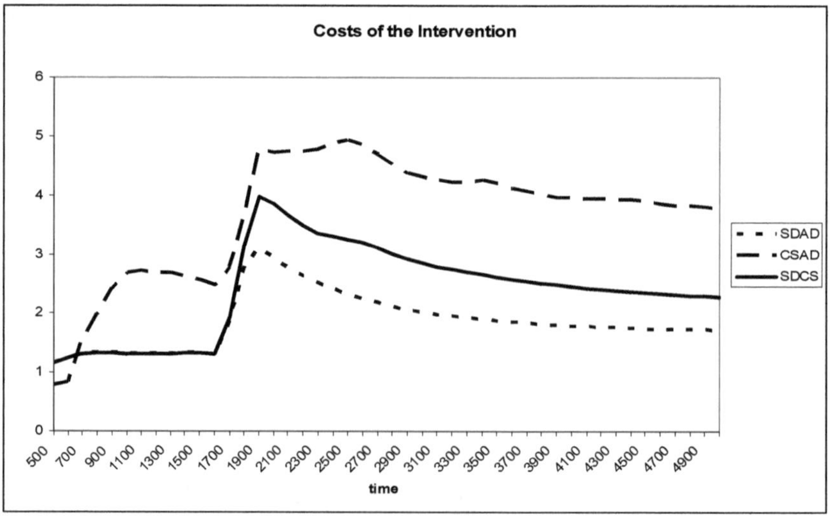

Fig. 2.31 Additional costs cc_t^{SDAD}, cc_t^{CSAD} and cc_t^{SDCS}

References

Ausiello G, Feuerstein E, Leonardi S, Stougie L, Talamo M (2001) Algorithms for the On-Line Travelling Salesman. Algorithmica 29: 560-581

Bertsimas DJ, van Ryzin G (1989) The Dynamic Traveling Repairman Problem. MIT Sloan School Working Paper No. 3036-89-MS

Bianchi L, Birattari M, Chiarandini M, Manfrin M, Mastrolilli M, Paquete L, Rossi-Doria O, Schiavinotto T (2005) Hybrid Metaheuristics for the Vehicle Routing Problem with Stochastic Demands. Technical Report No. IDSIA-0605

Bramel J, Simchi-Levi D (1997) The Logic of Logistics. Springer, New York

Brotcorne L, Laporte G, Semet F (2003) Ambulance location and relocation models. European Journal of Operational Research 147: 451-463

Chen Z-L, Xu H (2006) Dynamic Column Generation for Dynamic Vehicle Routing with Time Windows. Transportation Science 40(1): 74-88

Clifton C, Iyer A, Uzsoy R (2002) A Prototype Integrated Transaction Data Analysis and Visualization Environment for the Transportation, Distribution and Logistics Sector. Purdue University, Indiana, USA

Crainic T, Laporte G (1997) Planning models for freight transportation. European Journal of Operational Research 97: 409-438

Crainic T, Laporte G (1998) Fleet Management and Logistics. Kluwer

Fiat A, Woeginger G J (1998) Online Algorithms – The State of the Art. Springer, Berlin Heidelberg

Fleischmann B, Gnutzmann S, Sandvoß E (2004) Dynamic Vehicle Routing Based on Online Traffic Information. Transportation Science 38(4): 420-433

Gayialis S, Tatsiopoulos I (2004) Design of an IT-driven decision support system for vehicle routing and scheduling. European Journal of Operational Research 152: 382-298

Gendreau M, Guertin F, Potvin J-Y, Taillard E (1999) Parallel tabu search for real-time vehicle routing and dispatching. Transportation Science 33 (4): 381-390

Gendreau M; Potvin J-Y (1998) Dynamic Vehicle Routing and Dispatching. In: Crainic T G; Laporte G.: Fleet Management and Logistics: 115-126

Ghiani G, Guerriero F, Laporte G, Musmanno R (2003) Real-time vehicle routing: Solution concepts, algorithms and parallel computing strategies. European Journal of Operational Research 151(1): 1-11

Gutenschwager K, Niklaus C, Voß S (2004) Dispatching of an Electric Monorail System: Applying Metaheuristics to an Online Pickup and Delivery Problem. Transportation Science 38(4): 434-446

Hiller B, Krumke S O, Rambau J (2005) Reoptimisation Gaps versus Model Erros in Online-Dispatching of Service Units for ADAC. Submitted for publication in Discrete Applied Mathematics

Holzer M (2003) Hierarchical Speed-up Techniques for Shortest-Path Algorithms. Diploma Thesis, University of Konstanz, Germany

Ibaraki T, Nonobe K, Yagiura M (2005) Metaheuristics: Progress as Real Problem Solvers. Springer, Heidelberg

Irani, S, Lu, X, Regan, A (2004) On-line Algorithms for the dynamic traveling repairman problem. Journal of Scheduling 7: 243-258

Jaillet P (1988) A Priori Solution of a Traveling Salesman Problem in Which a Random Subset of the Customers Are Visited, Operations Research, 36(6): 929-936

Jaillet P, Odoni A R (1988) The Probabilistic Vehicle Routing Problem. In: Golden B L, Assad A A (eds.) Vehicle Routing: Methods and Studies, North-Holland: 293-318

Jensen M T (2001) Robust and Flexible Scheduling with Evolutionary Computation. PhD Dissertation, University of Aarhus, Denmark

Kopfer H, Schönberger J (2006) Adaptive Optimierung: Selbststeuernde Anpassung einer operativen Transportkostenoptimierung an strategische Qualitätsziele. In: Jacquemin M, Pipernik R, Sucky E (eds.) Quantitative Methoden der Logistik und des Supply Chain Management, Verlag Dr. Kovac: 321-337

Krumke, S (2001) Online-Optimisation – Competitive Analysis and Beyond. Habilitation Thesis, Technical University Berlin, Germany.

Lackner, A (2004) Dynamische Tourenplanung mit ausgewählten Metaheuristiken. Cuvilier Verlag Göttingen.

Makowski, M. (1994) Design and Implementation of Model-based Decision Support Systems. Working Paper WP-94-86, IIASA

Michalewicz Z, Fogel D B (2004) How to Solve It: Modern Heuristics. 2nd Edition. Springer, Heidelberg

Mitrović-Minić S, Krishnamurti R, Laporte G (2004) Double-horizon based heuristic for the dynamic pickup and delivery problem with time windows. Transportation Research B 38: 635-655

Pankratz G (2002) Speditionalle Transportdisposition. DUV

Psaraftis H (1988) Dynamic vehicle routing problems. In: Golden B L, Assad A A (eds) Vehicle Routing: Methods and Studies, North-Holland: 223-248

Psaraftis H (1995) Dynamic vehicle routing: status and prospects. Annals of Operations Research 61: 143-164

Sandvoss E (2004) Dynamische Tourenplanung auf Basis von Online-Verkehrs-Informationen. ProBusiness

Savelsbergh M, Sol M (1998) DRIVE: Dynamic Routing of Independent Vehicles. Operations Research 46:474-490

Schönberger J, Kopfer H (2007) On Decision Model Adaptation in Online Optimisation of a Transport System. Accepted for publication in the proceedings of the conference Supply Networks and Logistics Management – Decision Support, Information Systems and OR Tools

Schönberger J (2005) Operational Freight Carrier Planning. Springer

Séguin R, Potvin J-Y, Gendreau M, Crainic T G, Marcotte P (1997) Real-time decision problems: an operational research perspective. Journal of the Operational Research Society 48(2):162-174

Slater A (2002) Specification for a dynamic vehicle routing and scheduling system. International Journal of Transport Management 1: 29-40

Solomon M (1987) The vehicle routing and scheduling problems with time window constraints. Operations Research 35 (2): 254-265

Williams H P (1999) Model Building in Mathematical Programming. 4th Edition. Wiley, Chichester

3 Autonomous Control Methods for the Managment, Information and Communication Layer

3.1 Approaches to Methods of Autonomous Cooperation and Control for the Management-, Information- and Communication-Layer of Logistics

Michael Hülsmann[1], Katja Windt[2]

[1] Management of Sustainable System Development, Institute for Strategic Competence-Management, Faculty of Business Studies and Economics, University of Bremen, Germany

[2] Department of Planning and Control of Production Systems, BIBA, University of Bremen, Germany

Recent changes like short product life cycles, mass customization as well as a decreasing number of lots with a simultaneously rising number of product variants and higher product complexity have led to an increase of complexity of logistics in transportation systems as well as in production systems (Hülsmann / Scholz-Reiter / Freitag / Wycisk / De Beer 2006). Therefore, all participants in supply networks and their processes have to develop new planning and control methods for their logistics in order to cope with these requirements (Scholz-Reiter / Windt / Freitag 2004). To achieve an ability to adapt to these new challenges, the approach of decentralized planning and control by intelligent logistic objects in autonomously controlled production systems is discussed in different disciplines that deal with logistics (e.g Hülsmann / Grapp 2005; Windt 2006). In general, it is postulated that autonomous cooperation and control is one possible approach to cope with rising dynamics and complexity and to increase the level of emergence and robustness in supply chains and networks (Hülsmann / Windt / Wycisk / Philipp / Grapp / Böse 2006).

However, for the implementation of the principles of autonomous cooperation and control into the organisation domain of supply chains and supply networks, approaches will be needed that allow to transfer the general idea of autonomy in decision making into functional rules of cooperation and control for the different layers of logistics, i.e. the management-, the information-, and the communication layer (Scholz-Reiter / Windt / Freitag 2004). That addresses the need for appropriate methods that allow to reflect the major constitutive characteristics of autonomous cooperation and control – which are e.g. autonomy, heterarchical structures, interaction and interrelations, non-determinism (Hülsmann / Wycisk 2006) – in the modelling, measurement, and management of logistic processes. Therefore, this chapter would like to develop and describe approaches on how existing methods of logistics – like for example the layout of communication networks, the strategizing of logistic service providers, the management of distributed knowledge – can be adopted to the paradigm of autonomy. Additionally, new methods, based solely on the concept of autonomous cooperation and control, should be introduced – such as adaptive business processes within the production logistics domain.

The authors of **"Self-Organization in Management Science"** – Michael Hülsmann, Jörn Grapp, Ying Li, Christine Wycisk – aim to develop a general understanding of self-organization in management science so as to contribute to the establishment of a framework for studying self-organization. The concept of self-organization has its roots in various natural sciences and has the potential for enrichment of management theories by giving new interpretations to key aspects of conventional management approaches. However, in management science, research on self-organization is dispersed, with different angles of observation and a variety of terms used as synonymies. Consequently, this lack of an overarching framework for studying self-organization may impede the recognition and application of this concept in management science. Therefore, this chapter introduces selected concepts using self-organization from management science, compares the characteristics of self-organization implied by these concepts according to selected criteria (organizational structure, behaviour and abilities), and forms a general framework for studying self-organization in management science.

In the chapter **"Autonomous Cooperation – A way to vitalize organizations?"** Michael Hülsmann and Christine Wycisk deal with the question as to how far autonomous cooperation can provide a tool to unlock organizations. A locked organization is the organization trapped in a dilemma between need of a large amount of information, and limited capacity for obtaining and analyzing information, the consequence of which are fewer alternatives for organizational development. Such a dilemma results from the complexity and dynamics in the environment. As being locked implies the risk of collapse of organizations, ways to unlock organizations need to be studied. This chapter proposes autonomous cooperation as an alternative way and analyzes its contribution for the flexibilisation of organizations. The theory applied is competence-based view.

The chapter **"Self-organization Concepts for the Information- and Communication Layer of Autonomous Logistics Processes"** by Markus Becker, Andreas Timm-Giel, Carmelita Görg describes the application of self-organization approaches in the information and communication layer of autonomous logistic processes: the self-organized selection of communication networks, gateway discovery and ad hoc routing. The purpose of applying self-organization approaches is to improve communications between logistic objects. Among others, concepts of Autonomic Communication, Autonomic Computing and Self-Star are most relevant for logistics processes. In this chapter, self-organized selection of communication networks and services regarding communication selection and time are presented. Next, different methods and implementations of service discovery are introduced, followed by description of principles of ad hoc routing and different scenarios.

Hagen Langer, Jan D. Gehrke, Otthein Herzog introduce with **"Distributed Knowledge Management in Dynamic Environment"** an approach to the agent-based modeling of logistic processes which makes use of an explicit knowledge management system and hence enables agents to fulfill complex logistic tasks in dynamic environments. This task is driven by the realization that conventional optimization models neglect the important role of knowledge and communication in real-world process and the dynamics of parameter values. This chapter introduces agents as basic components as a framework for modeling, discusses agent-based approaches to logistics, and depicts distributed knowledge management for multi-agent systems (agent roles, decision parameters, and an interaction protocol).

The chapter **"Proactive Knowledge-based Risk Management"** – contributed by Martin Lorenz, Boris Bemeleit, Otthein Herzog and Jens Schumacher – presents the description of new possibilities in reducing damage, lateness and other aberrations to given goals for autonomous logistic objects through the usage of a suitable risk management concept. The increased complexity of logistic systems is followed by a more complicated planning and control of logistic systems and of the related processes in combination with an increased sensitivity of the total system to disturbances and malfunctions. This fragility calls for a risk management system to ensure successful realization of autonomous logistic objects. Therefore, this chapter gives an overview of different levels of risk and risk management for planning and controlling the logistic processes by agent based autonomous objects. Besides, the basic risk management concept and technical realization of a local risk management system is introduced and discussed regarding the requirements for agent based logistic objects.

The chapter **“Autonomy in Software Systems”**, written by Ingo J. Timm, Peter Knirsch, Hans-Jörg Kreowski, and Andreas Timm-Giel, presents one of the main characteristics of autonomous cooperation and control – i.e. autonomy – as a core property of innovative software systems, like agents and autonomous units. Therefore, the ideas of agency in software systems are sketched. That is the basis for analysing how communities of autonomous units deal with autonomous decision makers in comparison with multi-agent-systems. The relationship between autonomous units as a graph-transformation-based approach to handling autonomous decisions in a rule-based formal framework and agents as a widely used logical structure in artificial intelligence is discussed in this article regarding environmental states, transformation steps, perception, and decision-making.

Bernd Scholz-Reiter, Jan Kolditz, and Torsten Hildebrandt present in their article **“Specifying Adaptive Business Processes within the Production Logistics Domain — A new Modelling Concept and its Challenges”** the idea of autonomous logistic processes and focus on a concept for modelling such processes. Today, enterprises are exposed to an increasingly dynamic environment. Last but not least, increasing competition caused by globalisation more and more necessitates gaining competitive advantages by improved process control, within and beyond the borders of production companies. One possibility to cope with increasing dynamics is the autonomous control of logistic processes. This chapter gives a short overview of the concept of autonomous logistic processes, presents the overall system development cycle and discusses process modelling under the paradigm of autonomy.

References

Hülsmann M, Grapp J (2005) Autonomous Cooperation in International-Supply-Networks – The Need for a Shift from Centralized Planning to Decentralized Decision Making in Logistic Processes. In: Pawar KS et al (eds) Proceedings of the 10th International Symposium on Logistics (10th ISL). Loughborough, United Kingdom, pp 243-249

Hülsmann M, Scholz-Reiter B, Freitag M, Wycisk C, De Beer C (2006) Autonomous Cooperation as a Method to cope with Complexity and Dynamics? – A Simulation based Analyses and Measurement Concept Approach. In: Proceedings of the International Conference on Complex Systems (ICCS 2006). Boston, MA, USA. Web-publication. http://necsi.org/events/iccs6/viewpaper.php?d=339. 8 pages

Hülsmann M, Windt K, Wycisk C, Philipp T, Grapp J, Böse F (2006) Identification, Evaluation and Measuring of Autonomous Cooperation in Supply Networks and other Logistic Systems. In: Baltaciglu T (ed) Proceedings of the 4th International Logistics and Supply Chain Congress. Izmir, Turkey, pp 216-225

Hülsmann M, Wycisk C, (2006) Selbstorganisation als Ansatz zur Flexibilisierung der Kompetenzstrukturen. In: Burmann C.; Freiling J.; Hülsmann M. (Hrsg) Neue Perspektiven des Strategischen Kompetenz-Managements. Deutscher Universitätsverlag, Wiesbaden, pp 323-350

Scholz-Reiter B, Windt K, Freitag M (2004) Autonomous logistic processes: New demands and first approaches. In Monostori L (ed) Proceedings of the 37th CIRP International Seminar on Manufacturing Systems, Budapest, pp 357-362

Windt K (2006) Selbststeuerung intelligenter Objekte in der Logistik. In: Hütt M., Vec M., Freund A. (Hrsg) Selbstorganisation: Ein Denksystem für Natur und Gesellschaft. Böhlau Verlag, Köln Weimar Wien, pp 271-314

3.2 Self-Organization in Management Science

Michael Hülsmann, Jörn Grapp, Ying Li, Christine Wycisk

Management of Sustainable System Development, Institute for Strategic Competence-Management, Faculty of Business Studies and Economics, University of Bremen, Germany

3.2.1 Introduction

Today's real-time economy is characterized by three phenomena: hyper-linking, hyper-competition and hyper-turbulence (Tapscott 1999; Siegele 2002). As a result, management, which is responsible for designing social systems (Remer 2003), is confronted with high complexity and dynamics. However, conventional management seems not to be capable enough to cope with highly complex and dynamic situations (Hülsmann and Berry 2004) due to limited ability of human beings to obtain and analyze information (Simon 1957). The concept of self-organization might contribute to social systems' competence and thus to managing complexity and dynamics (Hülsmann and Wycisk 2005). The concept of self-organization has its roots in various natural sciences and has been studied by quite a few natural scientists (Foerster 1960, Prigogine 1971, Haken 1983, Maturana and Varela 1982). The original idea that self-organization could enable spontaneous formation of order (Prigogine and Glansdorff 1971; Maturana and Varela 1987) inspires interests of researchers from management science. Probst claims that the idea of self-organization enriches management theories by giving new interpretations to key aspects of the conventional management approaches such as planning, organizing and motivating (Probst 1984). However, in management science, research on self-organization is dispersed, with different angles of observation and a variety of terms used as synonymies. Consequently, this lack of an overarching framework for studying self-organization may impede the recognition and application of this concept in management science.

The primary aim of this paper is therefore to develop a general understanding of self-organization in management science so as to contribute to the establishment of a framework for studying self-organization. To fulfill this aim, this paper tries to answer the following questions: 1) how do different researchers understand self-organization in management science and what are the major aspects of self-organization in their view? 2) What are the commonness and differences between these concepts? To answer these questions, at first selected concepts using self-organization from management science will be introduced. Next, characteristics of self-organization implied by these concepts will be compared according to selected criteria in order to form a general framework to study self-organization in management science. Finally, future research needs will be proposed.

3.2.2 Selected concepts using self-organization in management science

In this section, selected concepts using self-organization from management science will be presented. However, this list of concepts relevant to self-organization is not exhaustive, that is, there are more concepts brought forward by other researchers besides those introduced in the following (e.g. Knyphausen-Aufseß 1993; Kieser 1994; Ulrich 1984; Dachler 1984). Factors taken into consideration during the selection process are primarily systematization, explicitness and citation frequency. Besides, among similar approaches (e.g. concept of evolutionary management raised by Knyphausen-Aufseß 1993, Kieser 1994, Kirsch 1992 and Malik 2000 respectively) those which might have a more comprehensive understanding are chosen (Malik 2000; Kirsch 1992).

Order as the result of human action (F. A. von Hayek)

Von Hayek (1899-1992), economist and Nobel Price winner, works with core problems in social theories and social policies. He is especially interested in topics of how structures of human society develop and how a variety of humans together build a society. His main statement is that social systems do not result from consciously steered actions, but come into being spontaneously (von Hayek 1994). In order to explain this phenomenon, he draws analogies between phenomena in social fields (e.g. development of a relationship net in social systems) and those in fields of natural science such as physics and biology (e.g. natural evolution processes) (von Hayek 1981). Göbel sees his work as the original business concept of self-organization in economics (1998).

Von Hayek attributes the formation of a society's ordered structure to a self-organizing process and calls such a structure self-organized order. (von Hayek 1994), which he characterizes as polycentric and spontaneous (von Hayek 1969). As an example, he frequently uses the image of the "invisible hand" on economic markets raised by Adam Smith (1723-1790). In this understanding, the process bringing a balance between demand and supply is not consciously controlled by any entity. Such self-organizing processes are based on the evolvement of a relationship net (Caldwell 2003). According to von Hayek, relationship net is a constitutive characteristic of human society. It is shaped by mutual adjustment of actions between humans. With the establishment of relationships, humans might anticipate their fellows' possible behavior, which will be considered when deciding their own actions. During the development process of the relationship net, some relationships will be sustained and become stable while some others are up to individual choices and unstable. At the same time, new relations will be generated and existing ones will adapt to changing situations. Thereby, the interpersonal relationships and expectations of each other's behavior lead to an ordered structure, which unifies a variety of humans into a society (von Hayek 1994). It has to be stated that characteristic of self-organization in the formation of social structures is reflected in the absence of conscious human design. The prerequisite for such a self-organizing creation of ordered structure in a system is the elements' adherence to abstract rules, which are embedded in generally accepted norms, cultural aspects, traditions and customs (von Hayek 1980). Individuals are not necessarily aware of these general rules, as education and influence of society can implicitly shape individuals' rationality of behavior without his or her consciousness (e.g. the behavior rule of respecting others' properties, which is gained through education). Only if all individuals adhere to the same rules, they can anticipate other system members' behavior and adapt themselves accordingly in order to attain their goals (Caldwell 2003). As each system element reacts in its individual environment according to generally accepted rules, a social order comes into being. In the above mentioned example of economic markets, all market participants follow the same rationality: to produce and distribute enough goods with a price capable of gaining profit (von Hayek 1984).

In contrast to the order created by a self-organizing process, there is also order resulting from conscious planning and building (e.g. in organizations), which is called by Hayek as taxonomic order (von Hayek 1969). Similarly to order out of self-organization, such order is also a consequence of behavior rationality and rules. However, these rules (e.g. organizational rules) are established with awareness. Due to the limited ability of human beings in recognizing and analyzing problems (bounded rationality) (Simon 1957), deliberately designed structures are of a simple nature, which means that they could hardly reach states as complex as those found in self-organizing structures. Von Hayek points out that the knowledge about general principles of self-organization could help to generate complex order by creating accordant conditions (von Hayek 1994).

Self-organization in social systems (N. Luhmann)

Luhmann (1927-1998), jurist and sociologist, is one of the founders of system theory. He does interdisciplinary research in the fields of economics, jurisprudence, theology, history, literature and communication science. His aim is to apply the conceptual instrument of social system theories to describe all objects in the field of sociology (Kneer and Nassehi 1993). In 1984, he published his major work "Social Systems", in which such a conceptual instrument is described (Luhman 1984). His social system theory has become one of the most famous theory models in the German-speaking area applied to sociology as well as psychology, management theory and literature theory. Luhmann sees a paradigm shift in the research results of Maturana and Varela and tries to transfer their approach of autopoiesis to social systems (i.e. principles of self-organization). In his work, he regards social systems as autopoietic with the characteristics of emergence and structural coupling (Brans and Rossbach 1997).

According to Luhmann, autopoiesis in social systems means that social systems are closed operating entities, which sustain and regenerate themselves through recursive production of communications (Luhmann 1984). He interprets communications as smallest elements in social systems, which are unable to be divided. Every communication produces another succeeding communication, which is explained by Luhmann as a chain effect. After a person X hears or reads what another one has said or written, his or her words might further be heard or read by a third person. As this process keeps going, new communications are produced one after another (Luhmann 1990). Consequently, social systems keep reproducing themselves, which reflects self-organizing processes. However, Habermas criticizes that Luhmann portrays social systems as consisting of only communications (Luhmann 1990) without taking into consideration humans (Christodoulidis 1991) involved in social interaction. The consequence might be weak transferability to real life, as social interaction might disrupt cultural reproduction (Habermas 1987).

Emergence refers to the generation of a new order level, which cannot be explained by the material and energy foundations (Luhmann 1984). In a psychic system, though the generation of thoughts depends on activities in the brain for necessary material and energy supply, this process is going on without the influence of the brain. The reason is that thoughts to be produced cannot be inferred by observing the activities of the brain. At the same time, from certain thoughts the processes in the brain cannot be inferred, either. Therefore, the psychic system is an emergent order level for the brain. Similarly, communications in social systems cannot be inferred from organic, neural and psychic processes. Consequently, communications build a new order level over other systems (Luhmann 1985), which in this case describes a major principle of self-organization.

Two structural coupled systems constitute environments for each other but are closed operating systems, like the psychic system and the brain described above (Luhmann 1984). Though psychic systems have the possibilities to disturb, inspire or irritate communications (e.g. a person is happy so that he wants to tell others his story), it is impossible to conclude from a communication how the involved psychic systems think. For example, even though one party of the communication is confident that he clearly knows what his partner thinks, his thinking occurs only in his own psychic system and this is not a process of communication (Luhmann 1985). In this sense, psychic systems and social systems operate independently while having certain influence on each other (Kneer and Naasehi 1993). As a consequence, psychic and social systems are structurally coupled (Luhmann 1984), which describes another major principle of self-organization.

Though being autopoietic like the living system studied by Maturana and Varela, psychic and social systems differ from other systems in that they exist for some "meaning" (Luhmann 1984). The meaning they pursue is decided by both reality and possibilities on hand. When the risk of instability has to be faced, the possibilities for systems' further development are considered under constrains of reality. In other words, psychic and social systems are constantly choosing between possibilities to update their actual status. In psychic systems, every thought is accompanied by certain intentions and could lead to further possible intentions, which update the original thought (e.g. specify a decision on increased information over time). This selection process between possibilities in thinking is also applied to communication processes, which contain intentions and could be connected by a number of possible communications (i.e. process of self-organization). In this sense, a meaning always points to another meaning through a selection process of thoughts and communications, as meanings are embedded in thoughts and communications (Kneer and Nassehi 1993).

Evolutionary management (F. Malik)

Malik (1944-) from St. Gallen belongs to the evolutionary management school, who is dedicated to the development of his own concept of evolutionary management. The central theme of his approach to evolutionary management is the configurability and tractability of complex and dynamic systems. He bases his theory mainly on works of Beer (Beer 1972), Drucker (1974) and von Hayek (1984). Especially cybernetics (Wiener 1948; Aschby 1974) and general system theory (Bertalanffy 1969) lay foundation for the development of his thoughts.

The evolutionary management school considers complexity and dynamics as causes for uncertainty of system behavior and thus recognizes the limits of organizational planning and controlling (Malik 2000). Malik points out the complexity in social systems which means social systems could have a number of possible states due to numerous interactions between system elements (Malik 1984, 1993). Similar to cybernetics (Ashby 1974), the evolutionary management school focuses on the central assumption that only complexity can absorb complexity (Malik 1984). In his approach, Malik tries to identify the general principles of applying complexity as well as its opportunities, limitations and consequences for management practice (Malik 2000).

The evolutionary approach claims that complete control of company systems is impossible due to a high level of complexity, which means unpredictability for the system development (Malik and Probst 1981). This recognition is reflected in systems' objectives from the perspective of evolutionary management. Unlike classic approaches arguing that profit maximization is the systems' objective, the evolutionary approach regards viability as systems' objectives (Malik 2000). A certain degree of control could only be achieved by influencing general structures and rules. Detailed rules are abandoned, because the conception and implementation of these rules are not realistic for limited human knowledge in face of high complexity in firms. Instead, Malik recommends abstract rules for guiding complex systems towards the desired direction (Malik 1993, 2000). This exhibits a self-organizing order-building process (Kieser 1994).

As the development of social systems is driven by decisions and actions of problem solving processes, Malik analyzes different problem solving processes (analytic constructivist approach vs. evolutionary cybernetic approach) combined with distinctive order building processes raised by Hayek (taxonomic order vs. self-organized order, see Section 2.1) (von Hayek 1969, Malik 2000).

The analytic constructivist approach tries to build order through detailed formulation of processes. An optimal solution would be chosen through rational evaluation of alternatives and implemented in organizational practice. This problem solving process in this sense could be considered as planned and conscious. It can often be found in tightly hierarchical organizations for ensuring organizational functions. Therefore, the rationality of the constructivist approach is to design a taxonomic order in advance and control its further development. However, this approach does not work for self-organized order. The reason is that the establishment of such order needs flexibility and adaptability of system elements, but the constructivist problem solving process might hinder the self-organization tendency by imposing pre-defined solutions (Malik 2000).

In contrast, evolutionary cybernetic approach claims that order building depends on system structures, certain general behavior rules as well as interaction patterns of elements (Malik and Probst 1987). Malik claims that the evolutionary problem-solving process is a "blind" variation and selection process (Malik 2000). "Blind" refers to the fact that "right" strategies for solving problems could only be obtained through trial and error processes when an organization faces complex situations (Malik 1984). Variation means the generation of specific actions, which are based on some basic behavior patterns but are adapted to specific environmental conditions. Selection refers to the retention of effective behavior alternatives after a number of trials. However, an evolutionary problem-solving process does not mean leaving freedom of decision and action totally to employees, because they have to behave under general objectives and rules given by management (Malik 1984). Combined with taxonomic order building, this approach introduces ideas like job-enrichment and job-enlargement as well as cooperative leadership style (Malik 2000). However, Malik points out that this approach could not be fully realized in a taxonomic order form. As the taxonomic order form is oriented to planning and optimization, it tends to offer few possibilities for an organization's development. The combination of this approach with self-organized order could be an important component of today's evolutionary theory in both biological and social development, because this combination might work out a variety of alternatives and thus make a system adaptable (Malik 2000).

Concept of the progressive organization (W. Kirsch)

Kirsch (1937-) is professor of management science at Ludwig Maximilians University of Munich, Germany. He works in the field of leadership and management. He focuses on the limits of managing complex dynamic systems like firms. In 1992 he published his major work "communicative action, autopoiesis and rationality", which contains his concept of the progressive organization (Kirsch 1992).

Kirsch's approach conforms to the understanding of the evolutionary management school and claims that firms are evolving systems, which are capable of adapting themselves to the changing environment by changing their own structures and processes (Kirsch 1992). In his work, Kirsch brings forward the hypothesis that organizations change with low predictability over time. Due to complexity and dynamics, organizations' objectives could hardly be achieved by management's deliberate design (Kieser 1994).

Though his approach has many similarities with that of Malik, there are two major differences in this conception. On the one hand, while Malik attributes complexity of firms to a variety of unknown data and events (Malik 1984), Kirsch credits complexity to collision of different people within different contexts, needs and goals (Kirsch 1992). Therefore, an important component of Kirsch's research conforms to Habermas' theory of communicative action, which studies the communication between system members (Harbermas 1981). On the other hand, while in his research Malik sees the firms' goal in survival on the market, the major concern of Kirsch's theory is to create a goal based on consensus (Kirsch 1992). In order to meet different goals, needs and motives of firm members, decisions made by individual members should benefit the whole organization, because progress of the whole organization is the prerequisite for individual development (e.g. all employees could get satisfactory compensation or training opportunities only if the organization is operating smoothly and efficiently). As the theory of autopoiesis mainly deals with system development, it conforms to Kirsch's core idea of progressive organization (Kirsch 1997). Therefore, Kirsch studies what knowledge of using autopoiesis to deal with complexity could be transferred to social systems and to what extent external forces could be relied on to attain the systems' goal (Ringlstetter and Aschenbach 2003).

With the concept of the "progressive organization", Kirsch emphasizes that in their evolutionary process organizations could develop some capabilities which enable organizational development and problem solving with a certain degree self-organization (Kirsch 1992). The extent of self-organizing ability depends on three system capabilities (Kirsch 1997). One capability is the action capability, meaning that an organization has enough resources for further organizational development and necessary changes due to certain impetuses. This capability helps to retain an organization's identity, as it ensures that the system could respond appropriately to perceived problems. The second capability is the learning capability, which means that an organization is able to master and apply knowledge. Organizational learning builds a common knowledge base, which puts together individual employees' knowledge. However, an important premise for organizational learning is that all employees should have access to the knowledge base and have the opportunity to make use of it. Besides, the learning capability implies that an organization can filter irrelevant and redundant information (e.g. by distributed decision-making so as to reduce information overload for management). Besides absorbing new knowledge, the learning capability also means that an organization can learn from its own behavior (e.g. the failure of formal rules leads to management's decision on giving employees more power for decision-making). Therefore, this capability is self-referential. The third capability is responsiveness capability, which means that an organization is sensible to the needs and interests of its stakeholders (Kirsch 1992). A responsive organization always undertakes actions which address the needs of relevant parties. However, appropriate responses are preconditioned by the organization's ability to recognize such needs. Therefore, the organization should be sensitive to individual contexts and life styles of the concerned parties, which articulate their needs. According to Kirsch, if the above three capabilities are well developed, an organization could reach a high development level. However, he points out that complete self-organization has to be seen only as an ideal model (Ringlstetter and Aschenbach 2003).

Kirsch describes a self-organizing process like this: when a system member perceives a problem, he or she can establish his or her own hypotheses about who else is involved in this problem and who can contribute to solving the problem. Then this member sets up contact with other concerned members, who again produce hypotheses regarding concerned parties and interact with them. Kieser calls this process "self-organizing snowball process" because through this process a chain of members are connected without the influence of external forces. However, this process could only be realized when the framework for action given by external forces (e.g. management) allows members to make independent decisions (i.e. regarding concerned parties) (Kieser 1994). In this way, system elements could have a wide scope for independent decision-making, which might lead to a self-organized problem solving without the external intervention resulted from hierarchy. However, to which extent problems can be solved by system elements in a self-organization process depends on the level of action, learning and responsiveness capability. If such a self-organization process fails to generate consensus, one of the members could take the role of leadership and impose formal rules, which would bring self-organization to an end (Kirsch 1992).

Order building processes in social systems from an integrated view (G. Probst)

Probst (1950-), professor of organization and management science at the University of Genf, Switzerland, understands self-organization as the consequence of interaction and exchange processes of organization members (Probst 1992b). He claims that the order pattern does not solely result from actions of managers, organizers and planners, but is constructed and developed by all organization members in self-organizing processes (Probst 1987). Therefore, the result of deliberate management design can not be predicted and may deviate from the original goal (Kieser 1994).

Probst regards self-organization as the prerequisite for survival of systems (Probst 1992b). He points out that a social system has a relationship of mutual exchange with its environment (e.g. a system gets resources from the environment and offers its output to the environment). When the environment changes, the system also has to change in order to retain its identity, e.g. through absorbing new technology to meet higher requirements of consumers so as to stay on the market (Probst 1992b). However, a social system is difficult to plan and control due to both external complexity (e.g. new technologies) and internal complexity (e.g. variety in attitudes towards introduction of a new technology). Due to management's inability of planning in such situations, self-organization is assumed to endow social systems with the ability to appropriately respond to changes (Probst 1987).

Probst identifies several characteristics of self-organization, namely self-reference, complexity, redundancy and autonomy. Self-reference builds a system's border and differentiates the system from its environment. It means a system makes decisions and implements actions based on its current state (e.g. to produce more due to low inventory). Due to its function of offering information for decision-making, this self-reference is the starting point for system behavior, for taking measures against disturbance and for realizing internal synergy (Probst 1992a). In this way, a social system develops its own logic and thus gains its identity.

Complexity is reflected in the fact that a self-organizing system comes into being through a high density of interactions between a variety of elements. Therefore, system order could take a variety of forms, which depends on the system history and its elements (Probst 1992a). Social systems have objectives, which might be different. Therefore, individuals and departments inside systems have to cooperate to realize common objectives. However, cooperative relationships keep changing (e.g. the cooperating relationships are only temporary inside a project within a company). Therefore, system structures change constantly and swing between order and disorder. In self-organizing processes, the whole system and the elements have to be oriented towards finding new equilibria to retain the system's identity (Probst 1987).

Self-organizing systems are redundant, because their structure and behavior are not designed by a single designer but developed by all system elements (Probst 1987). In this view, the systems' functions instead of system elements are redundant. This redundancy in functions results from a heterarchical structure, where a number of people could have the same capabilities. Therefore, it is possible that some organization members can fulfill several roles and functions. Redundancy ensures the normal organizational operation even when systems are exposed to disturbance. Therefore, the development of a firm might be based on a design of heterarchical structure, where all system members are empowered to manage the firm (Probst 1992b).

Autonomy means that elements, relationships and interactions within a system are independent of external forces. Though a system has a loose relationship with the environment for more options in the future (e.g. to absorb talents from the environment to develop new products), it can establish its own goals as well as means to attain the established goal (Probst 1992a). An example in a firm can be that each department only follows the guiding principle based on the goal of the firm and can decide its own objective and actions.

Though self-organization has potential to be applied to organizations and facilitates the organizational development, Probst stresses that self-organization has to be separately studied in specific contexts (e.g. different industries in which firms are situated, different sizes of firms) (Probst 1987). Thus, the optimal degree of every characteristic should be studied (e.g. how much autonomy should be given to each department in a firm). Moreover, an instrument measuring single characteristics of self-organization under the consideration of cost-benefit relationship is still lacking.

Autogenous and autonomous self-organization (E. Göbel)

Göbel (1956-) finished "Theorie und Gestaltung der Selbstorganisation" in 1997 (Göbel 1998). She claims that the effect of a selected structure on organizational performance can only be evaluated by observing the interaction of organization members within the formal organizational structure. According to Göbel, the structure is the result of external organization and self-organization. External organization refers to goal-oriented structure design while self-organization is based on individual as well as systemic behavior (Göbel 1993). Göbel's goal is to assess limitations of deliberate structure design and develop suggestions for implementing self-organization (Göbel 1998).

Self-organization could be understood as the removal of bureaucracy and formality as well as the reduction of hierarchy and specialization. At the same time, a new structure should be established in the form of teams and processes (Göbel 1998). The formation of such a structure means more self-decision power for organization members, which is given by the management. Göbel stresses that management should set itself as an example and be the motivator in the learning process while being the initiator of self-organization processes (Göbel 1993). The application of concrete management concepts like divided management (Mintzberg 1990), rotating management (Peters 1993) and collective management (Heintel and Krainz 1990) however should take the specific context into consideration (Göbel 1998).

Göbel sees self-organization as a phenomenon which manifests itself in different aspects: micro-organization (autonomous complementary organization), informal organization (autonomous alternative organization), interpretation of organizational reality (autogenous alternative organization) and momentum of systems (autogenous complementary organization) (Göbel 1998). Self-organization as micro-organization means that system elements can use options given by an external organizer like management to build their internal structures. But management has limited influence on this kind of order building, which depends on some factors hardly visible to management such as personal capabilities and habits of organizational members (Göbel 1993). Self-organization as informal organization refers to the situation that formal and informal rules exist in parallel. As a result, there are both formal and informal communications. However, whether such self-organization contributes to organizational performance remains unknown. For example, informal communication could be regarded as positive for performance, as it might speed up information flow (e.g. directly between employees instead of through a complex hierarchy). However, due to the uncontrollability, informal communications could negatively impact organizational performance by deviating it from the desired state (e.g. departments work together to hide problems). Self-organization as interpretation of organizational reality exhibits an individual psychological perspective, meaning that organization members can construe reality by themselves. In this context, members evaluate and process reality by using their own experience and approaches. Consequently, it is possible that similar processes and structures could lead to very different organizations. The risk lies in the incongruity in perception, which results in conflicts for organizations (Göbel 1993). Self-organization as momentum of systems stresses that a system should be regarded as a whole. Individual elements' behavior which is totally independent from the management is regarded as harmful for the development of the whole system, as elements might misuse full autonomy and pursue their own benefits in conflict with the system's goal. Therefore, management predefines a number of actions and system elements have the freedom to choose and combine these actions.

In general, self-organization is assumed to have positive influence on organizational efficiency, as it might help to fulfill the requirements of environment concerning time (e.g. timely response by fast information flow through direct communication in heterarchical structures) and resources (e.g. employees' creativity resulting from more autonomy) (Göbel 1993; Staehle 1991). Besides, employees' satisfaction and motivation might be enhanced through gaining more power for decision-making (Göbel 1997; Laux and Liermann 1993; Ulrich 1991). However, possible negative effects of self-organization on organizational efficiency could be seen in potential conflicts (e.g. due to different perception of autonomous elements) (Göbel 1993; Rosenstiel 1985) and excessive demands and overload for employees (Göbel 1993; Jung 1985). Other problems could be resistance by rooted routines and habits, management's unwillingness to give up power (Göbel 1993) as well as organization members' opportunism for self-interest. As a consequence, a combination of external organization and self-organization might be required (Göbel 1998).

Self-organization as evolutionary process (A. Remer)

Remer (1944-) is professor of management and organization science at the University of Bayreuth, Germany. His conception of organization has human beings as its focus. He claims that the personnel in an organization have double functions: as system members they design an organization's structure; as system participants, they interact with each other to play the defined roles and realize the desired structure which they have designed as system members (Remer 1985). This process is covered by the concepts of self-organization or organizational self-structuring.

One possibility to enhance self-organized structure building could be more frequent interaction with the concerned social systems (Remer 1994). In this context, management's role is not restricted to the realization of ideas and goals. It is regarded rather as intermediate between environmental conditions and employees in a constant process of adaptation (e.g. improve products to fit consumers' needs) and selection process (e.g. absorb necessary technology to improve products) (Weick 1985).

Remer understands self-organization as an evolutionary and learning process (Remer 1994), in which a system acquires its structure through its capabilities of "structural learning" (Pautzke 1989). The existing knowledge of a social system is regarded as "genes" or "comps" (competences) (Segler 1985). The system's ability to survive depends on processes of "self-observation" and "self-selection", where genes could adapt themselves to the environment (Remer 1994). The prerequisite for structural learning consists of feedback on actions (Argyris and Schön 1978) and variation (Hedberg 1981). Feedback on actions means that new actions should be based on existing problems (e.g. a firm's decision on updating technology, because they have recognized that the existing technology cannot fulfill consumers' needs). Variation means that the variety of comps should be facilitated, because variety contributes to evolutionary success of a system by giving more possibilities for the system's development (Remer 1987). Variation could be achieved by taking into consideration ideas of all system members and participants as well as other institutions (e.g. formal rules). A means to realize variation is decentralization of organization processes, which enhances the capacity of a whole system in problem solving (e.g. overload of information for management could be replaced by an appropriate amount of information for a number of elements) as well as diversity of perspectives. Remer calls the process of achieving structural learning as "organizational reflexivity" (Remer 1997), which refers to Luhmann's concept of "reflexive mechanisms" (Luhmann 1973).

According to Remer, the progress in thinking of organizational problems by including the concept of self-organization could be considered as a shift in perspectives (i.e. from a mechanic perspective to a biological perspective) (Remer 1994). A biological perspective emphasizes the generation and evolution of organizations with the recognition of their dynamic nature, which the rule rather than the exception in real life. Nevertheless,

Remer also points out that the introduction of self-organization means the loss of opportunities to design an organization according to certain cause-effect patterns (Remer 1997). The reason is that the endowment of organization members with a certain degree of autonomy means that management might not predict the behavior of individual members and the aggregate effect of their behavior (behavior of the whole organization) (Remer 1987).

3.2.3 Major characteristics of self-organization in management science

Criteria for comparison

In order to develop a general understanding of self-organization in management science, a comparison of the concepts introduced in Section 2 shall be carried out according to the following criteria: "organizational structure" "organizational behavior" and "organizational abilities". Characteristics of self-organization classified under "organizational structure" depict the context of self-organization, that is, organizations themselves. Characteristics under "organizational behavior" indicate how an organization develops. Characteristics under "organizational abilities" represent what an organization is capable to do.

There are two reasons for choosing such criteria. One reason is concerning system analysis. In the comparison, a system-oriented view of organizations will be adopted, which sees organizations as systems adaptive to changing environment (Hicks and Gullett 1975). It might contribute to the generalization of research results (Ulrich 1984) while enabling an interdisciplinary observation and analysis of concepts (Remer 1982) like self-organization. The other reason is concerning system design. The above criteria stress different dimensions of system design. They are relevant for studying management problems, because management is seen as dealing with the design of organizations as social systems (Remer 2000).

Results of comparison

Organizational structure

Complexity is a common characteristic of organizations discussed in the concepts from Section 2. From a system-oriented perspective, complexity is based on the number and variety of elements, the number and variety of connections between elements (Patzak 1982) as well as aggregated characteristics of the system (Dörner 2001). However, there are also differences in emphasis among the concepts discussed in the last section, despite their common recognition of complexity as a characteristic of self-organization. For example, Malik regards complexity as a number of possible states due to numerous interactions between organization participants (Malik 1984) while Kirsch stresses that complexity results from collision of various needs and goals of participants (Kirsch 1992).

Dynamics is another common characteristic. According to Hill et al, dynamics refers to changing of a system's state over time (Hill et al. 1994). Dynamics manifests itself in various forms among business approaches. For example, in von Hayek's approach dynamics is the evolvement of the relation net (von Hayek 1980). In Probst's approach dynamics means organizations' swinging between order and disorder, which is the result of ever changing cooperative relationships between participants (Probst 1992a).

System openness is also common to the organizations with which the above business concepts are dealing. Openness means that a system and its environment interact with each other and mutually adapt to each other. Therefore, failure to adapt will endanger a system's survival. Among the business concepts presented above, Probst explicitly points out that an organization should change according to the environment's requirements so as to retain its identity. Due to the bounded rationality (Simon 1957), self-organization ensures that an organization can timely and appropriately respond to changes (Probst 1992b). Göbel also claims that the requirements of the environment underline the importance of self-organization for the purpose of efficiency in terms of time and resources (Göbel 1993). Luhmann recognizes the interdependence between social systems and their environment. However, he lays more emphasis on the aspect of system closure, as he sees the environment mainly as the source of material and energy supply without substantially influencing the system's operations (Luhmann 1984).

Organizational behavior

Concerning the organizational behavior, non-determinism is a common characteristic. Non-linearity in this context means that effect is disproportional to cause (Sterman 2001), which refers to the behavior of the system can not be causally predetermined and thus is not predictable (Haken 1983). Among the presented business concepts, this characteristic is embodied in a number of alternatives an organization has during its process of development. For example, Luhmann claims that social systems are constantly choosing between possibilities to update their actual status (Luhmann 1984). Probst points out that a social system can have a variety of forms depending upon its past and the interaction of participants (Probst 1992a).

Autonomy is another characteristic which can be observed in the presented business concepts. It refers to the freedom of rendering decisions by individual organizational units (Probst 1987). Examples could be variation from basic behavior patterns in specific contexts (Malik 2000), organization participants' own determination of concerned parties for solving a certain problem (Kirsch 1992), participants' options to build internal structure given by an external unit (Göbel 1993) and participants' following only general ideas and goals set by the management (Remer 1994).

Self-reference is also frequently talked of by the business concepts discussed above. Probst gives self-reference as an organization's decision making and action implementation based on its current state (Probst 1992a). Some other concepts use either different terms or elaborate the meaning of self-reference indirectly. For example, Luhmann uses the term self-reflexivity to describe social systems' analysis of themselves and optimization of their own actions based on this analysis (Luhmann 1984). Remer mentions self-reference in his understanding of self-organization as a learning process, stating that feedback on actions should be the reference for future behavior (Remer 1992).

Organization abilities

Emergence is one of the organizational abilities within a self-organizing organization, which is identified in the above business concepts. Emergence means the generation of new qualitative characteristics of a system resulting from synergy effects of interacting elements (Haken 1993). Luhmann explicitly deals with emergence, stating that a psychic system is an emergent order for the brain, whose function is merely supply of materials and energy. In contrast, other concepts implicitly address this characteristic (Luhmann 1984). For example, Göbel underlines that the order of an organization should be evaluated by observing the interaction of organizational members instead of focusing on single members (Göbel 1993). Besides, von Hayek uses the example of "the invisible hand" originally studied by Adam Smith to illustrate emergence, that is, the market order comes into being as a result of interaction of market participants who follow the same rationality (von Hayek 1984).

Dynamic equilibrium is another organizational ability found in those business concepts. It means that an organization can swing between different stable states instead of sticking to a single one (Carver and Scheier 2002). For example, Probst points out that a self-organizing organization keeps looking for new equilibria for retaining its identity and developing itself (Probst 1992a). Other researchers like Malik (2002) and Kirsch (1992) implies this idea by stressing the evolution of an organization by adapting to the changing environment for the purpose of viability. With the changing process, the organization keeps moving to a new equilibrium by changing its structure and behavior.

Another identified organizational ability, which is common to those business concepts, is self-control. This means that an organization can steer itself towards its objective with no or little external influence. The above discussed business concepts address this ability rather by explanation. For example, Kirsch sees this ability as dependent upon three capabilities, namely action capability, learning capability and responsiveness capability (Kirsch 1992). These capabilities ensure that an organization can pertain to its objective while responding appropriately to the environment. Remer points out personnel inside an organization design the organizational structure while interacting with each other to realize the structure. During this process, the steering towards the established goals is realized through the interaction of organizational participants rather than through any external forces (e.g. a central planning unit) (Remer 1985).

3.2.4 Conclusions

To establish an overarching framework for studying self-organization, this paper has compared different approaches using self-organization and deduced common characteristics classified into three groups, namely:

- Organizational structure: complexity, dynamics and system openness;
- Organizational behavior: non-determinism, self-reference and autonomy;
- Organizational abilities: emergence, dynamics equilibrium and self-control

As such a framework offers a unified terminology, it may enable clear description instead of a mess of terms; as such a framework combines different dimensions of system design, it may simplify analysis by focusing on every single dimension at each time. As a result, these categorized characteristics might allow an easier comparison and integration of different perspectives from existing literature on the one hand and might be used as directions for further research on self-organization on the other hand. However, one remark should be made that these nine characteristics are only superordinate terms. This means that different concepts might have slight differences in understanding a certain characteristic, which is explained by the examples shown above.

Self-organization might contribute to strategic competence management, because it could simultaneously increase flexibility and stability in complex and dynamic environments (Hülsmann and Wycisk 2005). Thus, it might be worthwhile to apply self-organization in practice by the management, which is responsible for the conscious and goal-oriented structuring of purposive social systems (Remer 2003). A framework with categorized characteristics might be helpful for management to implement self-organization along different dimensions (i.e. structure, behavior and abilities) with key aspects.

Finally, there are some requirements for further research on self-organization. In this paper, a relatively small number of concepts are selected, where authors with different focus are dealing with self-organization in management science. However, due to the specific context of their research, only a limited number of aspects of self-organization are studied in respective works. Therefore, a task of future research could lie in the absorption and evaluation of more concepts. Besides, during the process of aggregating characteristics of self-organization in this paper, there is potential risk of information loss. As a result, another task might be the generation of a more detailed categorization, as this prevents the sacrificing of seemingly unimportant information which might be proved significant in practice. Finally, a third requirement on further research could be empirical studies in organizations. The verification of the existence of self-organization as well as its effect on order building might be enhanced by some real-life observation and measurement (e.g. through interviews with organizational members).

References

Argyris C, Schön D (1978) Organizational Learning: A Theory of Action Perspective. Klett-Cotta, Stuttgart

Ashby WR (1974) Einführung in die Kybernetik. Suhrkamp, Frankfurt am Main

Beer S (1972) Brain of the firm – the managerial cybernetics of organization. Lane, London

Bertalanffy v L (1969) General system theory: foundations, development, applications. Braziller, New York

Brans M, Rossback S (1997) The autopoiesis of administrative systems: Niklas Luhmann on public administration and public policy. Public Administration 75(3): 417-439

Caldwell B (2003) Hayek's Challenge. University of Chicago Press, Chicago

Carver CS, Scheier MF (2002) Control Processes and self-organization as complementary Principles underlying behavior. Personality and Social Psychology Review 6: 304-315

Chmielewicz K (1994) Forschungskonzeptionen der Wirtschaftswissenschaft. Schäffer-Poeschel, Stuttgart

Christodoulidis EA (1991) A case for reflexive politics: challenging Luhmann's account of the political system. Economy and Society 20(4): 380-401

Dachler P (1984) Some explanatory boundaries of organismic analogies for the understanding of social systems. In: Ulrich H, Probst G (eds) self-organization and management of social systems. Springer, Berlin Heidelberg New York Tokyo, pp 132-147

Dörner D (2001) Die Logik des Misslingens, 14th edn. Rowohlt, Hamburg

Drucker P (1974) Management-tasks, responsibilities, practices. Heinemann, London

Forster v H (1960) On Self-Organizing Systems and their Environment. In: Yovits MC, Cameron S (eds) Self-Organizing Systems. Pergamon, London, pp 31-50

Glaserfeld v E (1981) An epistemology for cognitive systems. In: Roth G, Schwegler H (eds) Self-organizing Systems. Campus Verlag, New York, pp 121-131

Grapp J, Wycisk C, Dursun M (2005) Systematischer Ansatz zur Identifikation von transdisziplinären Diffusionsprozessen-Entwicklung einer Literaturdatenbank. In: Hülsmann M (ed): Forschungsbeiträge zum Strategischen Management 4. University of Bremen

Göbel E (1993) Selbstorganisation – Ende oder Grundlage rationaler Organisationsgestaltung. Zeitschrift Führung und Organisation 62: 389-393

Göbel E (1998) Theorie und Gestaltung der Selbstorganisation. Duncker and Humblot, Berlin

Habermas J (1981) Theorie kommunikativen Handelns. Suhrkamp, Frankfurt am Main

Habermas J (1987) The theory of communicative action-systems and life-world: a critique of functionalist reason. Beacon Press, Bosten

Haken H (1983) Erfolgsgeheimnisse der Natur, Synergetik: Die Lehre vom Zusammenwirken. Dt. Verl.-Anst., Stuttgart

Haken H (1993) Eine Zauberformel für das Management? In: Rehm W (ed) Synergetik: Selbstorganisation als Erfolgsrezept für Unternehmen; ein Symposium der IBM. Expert, Stuttgart, pp 15-43

Hayek v FA (1969) Freiburger Studien. Mohr, Tübingen

Hayek v FA (1981) Recht, Gesetzgebung und Freiheit, vol 1: Regeln und Ordnung. Verl. Moderne Industrie, München

Hayek v FA (1981) Recht, Gesetzgebung und Freiheit, vol 2: Die Illusion der sozialen Gerechtigkeit. Verl. Moderne Industrie, München

Hayek v FA (1994) Freiburger Studien. Mohr, Tübingen

Hedberg B (1981) How Organizations learn and unlearn. In: Nystrom PC, Starbuck WH (eds) Handbook of Organizational Design. Oxford University Press, New York

Heintel P, Krainz EE (1990) Projektmanagement: eine Antwort auf die Hierarchiekrise? Gabler, Wiesbaden

Hicks HG, Gullett CR (1975) Organizations: theory and behavior. McGraw-Hill, New York

Hill W, Fehlbaum R, Ulrich P (1994) Organisationslehre. Haupt, Bern Stuttgart

Hülsmann M (2004) Das Dilemma mit dem Dilemma-Management. In: Hülsmann M, Müller-Christ G (ed) Modernisierung des Managements. Gabler, Wiesbaden

Hülsmann M (2005) Contributions of the concept of self-organization for a strategic competence-management. The 7th International Conference on Competence-based management

Hülsmann M, Berry, A (2004) Strategic management dilemma – its necessity in a world of diversity and change. The SAM/IFSAM 7th World Congress

Jung RH (1985) Mikroorganisation: eine Untersuchung der Selbstorganisationsleistung in betrieblichen Führungssegmenten. Haupt, Bern Stuttgart

Kieser A (1994) Fremdorganisation, Selbstorganisation und evolutionäres Management. Zeitschrift für betriebswirtschaftliche Forschung (ZfbF) 46(3): 199-228

Kirsch W (1992) Kommunikatives Handeln, Autopoiese, Rationalität, Sondierung zu einer evolutionären Führungslehre. Herrsching, München

Klimecki RG (1995) Self-organization as a New Paradigm in Management Science? Management Forschung und Praxis, Diskussionsbeitrag 10: 1-29

Kneer G, Nassehi A (1993) Niklas Luhmanns Theorie sozialer Systeme – Eine Einführung. Fink, München

Knyphausen D zu (1991) Selbstorganisation und Führung – Systemtheoretische Beiträge zu einer evolutionären Führungskonzeption. Die Unternehmung 45(1): 47-63

Laux H, Liermann F (1993) Grundlagen der Organisation, Die Steuerung von Entscheidungen als Grundproblem der Betriebswirtschaftslehre. Springer, Berlin Heidelberg New York Tokyo

Luhmann N (1973) Zweckbegriff und Systemrationalität. Über die Funktion von Zwecken in sozialen Systemen. Suhrkamp, Frankfurt am Main

Luhmann N (1984) Soziale Systeme: Grundriß einer allgemeinen Theorie. Suhrkamp, Frankfurt am Main

Luhmann N (1985) Die Autopoiese des Bewusstseins. Soziale Welt 36: 402-446

Luhmann N (1990) Die Wissenschaft der Gesellschaft. Suhrkamp, Frankfurt am Main

Malik F (1984) Strategie des Managements komplexer Systeme: ein Beitrag zur Management-Kybernetik evolutionärer Systeme, 2nd edn. Haupt, Bern

Malik F (1993) Systemisches Management, Evolution, Selbstorganisation: Grundprobleme, Funktionsmechanismen und Lösungsansätze für komplexe Systeme. Haupt, Bern

Malik F (2000) Strategie des Managements komplexer System: ein Beitrag zur Management-Kybernetik evolutionärer Systeme, 6th edn. Haupt, Bern

Malik F, Probst G (1981) Evolutionäres Management. Die Unternehmung 35: 303-316

Maturana HR, Varela F (1982) Autopoietische Systeme. In: Maturana HR (ed) Erkennen: Die Organisation und Verkörperung von Wirklichkeit. Vieweg, Braunschweig, pp 180-235

Maturana HR, Varela F (1987) Der Baum der Erkenntnis: Die biologischen Wurzeln des menschlichen Erkennens. Scherz, Bern

Mintzberg H (1990) The manager's job: Folklore and Fact. Havard Business Review 68(2): 163-177

Patzak G (1982) Systemtechnik. Springer, Berlin

Pautzke G (1989) Die Evolution der organisatorischen Wissensbasis: Bausteine zu einer Theorie des organisatorischen Lernens. Herrsching, München

Peters T (1993) Jenseits der Hierarchien, Liberation Management. Econ, Düsseldorf

Prigogine I, Glansdorf P (1971) Thermodynamic Theory of Structure, Stability and Fluctuations. Wiley, London New York Sydney Toronto

Probst G (1987) Selbst-Organisation: Ordnungsprozesse in sozialen Systemen aus ganzheitlicher Sicht. Parey, Berlin

Probst G (1992a) Selbstorganisation. In: Frese E (ed) Handwörterbuch der Organisation. Schäffer-Poeschel, Stuttgart, pp 2255-2269

Probst G (1992b) Organisation: Strukturen, Lenkungsinstrumente und Entwicklungsperspektiven. Moderne Industrie, Landsberg Lech

Probst G, Siegwart H (1985) Integriertes Management – Bausteine des systemorientierten Managements. Haupt, Bern Stuttgart

Remer A (1982) Instrumente und instrumentelles Dilemma der Verwaltungsführung. In: Remer A (ed) Verwaltungsführung: Beiträge zu Organisation, Kooperationsstil und Personalarbeit in der öffentlichen Verwaltung. De Gruyter, Berlin New York

Remer A (1985) Vom Produktionsfaktor zum Unternehmensmitglied. Grundlagen einer situations- und entwicklungsbewußten Personallehre. In: Bühler W, Hofmann M, Malinsky AH, Reber G, Pernsteiner AW (eds) Die ganzheitlich verstehende Betrachtung der sozialen Leistungsordnung. Springer, Wien New York

Remer A (1987) Führung als Managementinstrument. In: Kieser A, Reber G, Wunderer R (eds) Handwörterbuch der Führung. Schäffer-Pöschel, Stuttgart

Remer A (1994) Organisationslehre, 2nd edn. REA-Verl, Bayreuth

Remer A (1997) Organisationslehre, 4th edn. REA-Verl, Bayreuth

Remer A (2002) Managementsystem: System und Konzept. REA-Verl, Bayreuth

Ringlstetter MJ, Aschenbach M (2003) Perspektiven der strategischen Unternehmensführung: Theorien, Konzepte, Anwendungen. Gabler, Wiesbaden

Rosenstiel v L, Molt W, Rüttinger B (1995) Organisationspsychologie. Kohlhammer, Stuttgart Berlin Köln

Segler T (1985) Die Evolution von Organisationen. Lang, Frankfurt am Main Bern New York

Simon H (1957) Models of man. Wiley and Sons, New York

Staehle WH (1991) Redundanz, Slack und lose Kopplung in Organisationen: Eine Verschwendung von Ressourcen? In: Staehle WH, Sydow J (eds) Managementforschung 1. de Gruyter, Berlin New York, pp 313-345

Sterman (2001) System dynamics modelling: tools for learning in a complex world. Irwin/McGraw-Hill, Boston

Ulrich H (1984) Management – A Misunderstood societal function. In: Ulrich H, Probst G (eds) self-organization and management of social systems. Springer, Berlin Heidelberg New York Tokyo, pp 80-93

Ulrich E (1991) Arbeitspsychologie. Schäffer-Poeschel, Zürich

Weick KE (1985) Der Prozess des Organisierens. Suhrkamp, Frankfurt am Main
Wiener N (1948) Cybernetics or control and communication in the animal and the machine. MIT Press, Cambridge MA

3.3 Autonomous Cooperation – A Way to Vitalize Organizations?[5]

Michael Hülsmann, Christine Wycisk

Management of Sustainable System Development, Institute for Strategic Competence-Management, Faculty of Business Studies and Economics, University of Bremen, Germany

3.3.1 Complexity and dynamics of social systems – the problem of unlocking

In the age of information technology the rising amount and the permanent alteration of information will cause a rise of complexity and dynamics (Hülsmann and Berry 2004). The fast development and spreading of the internet and new communication services are well known examples of these technological changes, which imply new possibilities of interaction for organizations and customers (Pflüger 2002).

In terms of the complexity of a system, not the quantity of elements is decisive but the existence of multiple interrelations between the elements of the system as well as between the system and its environment (Dörner 2001; Malik 2000). According to Dörner (2001), a complex system can be understood as „the existence of many interdependent characteristics in a section of reality [...]“. When this definition is transferred to an example in the field of information technology, the amount of available information based on the innovations in those technologies represent the rising amount of elements in this section of reality.

The term dynamics describes the accelerated variation of the system`s status over time. Here, the internet can be quoted as a technological example: dynamics mean the permanent alteration of available information on the internet. In this case, the elements (pieces of information) themselves

[5] A former version of this paper was presented at the 22nd EGOS Colloquium “Unlocking Organizations“ in 2005.

change and thus the relations between them and other systems (e.g. companies) alter.

This development in turn leads to a higher complexity of the firm's environment. As a result, firms have to cope with this complex information to maintain their capacity of reacting to timely to changing demands. In order to handle complexity and dynamics, there is a need for a flexible adaptation of the system, which is realized through processes belonging to system theory: system openings and system closures.

Processes of system openings (Luhmann 1973) enable the system to communicate with the environment through mutual inter-relations. Thereby the system it sustains the existential exchange process of resources (Staehle 1999; Böse and Schiepek 1989). During these system openings, the system absorbs a part of the environmental complexity (e.g. information) to incorporate necessary resources. In order to avoid the risk of an information overload, system openings have to go along with system closures. This means that the system does not absorb the entire complexity of the environment but only the portion that, in terms of the ability of solving specific problems, corresponds to the system's identity (Luhmann 1994) and ability to handle it. System closure therefore ensures that the system does not absorb more information than needed or than manageable by the system's capacity.

The challenging task of the management, keeping the best possible balance between those system processes, implies a dilemmatic decision-making situation. Since the degree of necessary information to solve specific problems rises along with the increased complexity and dynamics of the environment, the decision maker has to absorb more complexity (information) through system openings, while still possessing the same ability of handling this piece of information. At the same time, the management faces the difficult selection of information in terms of quality and quantity and has to take into account the dynamics of information and the risk of an information overload caused by system closure (e.g. Hülsmann 2005; Gebert and Boerner 1995; Gharajedaghi 1982).

A possible outcome of this dilemmatic situation is a limited ability of decision-making (Hülsmann and Berry 2004). In this state of being caught in its own complexity the organization is called a locked organization. The environmental complexity outgrows the organization's capability of handling it and the immanent lack of information of a decision called the problem of bounded rationality (Simon 1972: a manager cannot have the complete information about his problem of decision) renders the situation suboptimal.

Since the system will then be unable to continue its exchange of vital resources with the environment, the event of locking will have negative effects on the continuity of the organization. The latter will lose its flexibility and will not be able to respond to the requested resources of the environment in time, quality, quantity, or place (e.g. products of the company which are needed by the environment but cannot be provided). In the worst case, a locked system may result in the risk of a collapse of the organization. The notion of a "locked organization" describes a dysfunctional and suboptimal situation with a limited choice of possible decisions (Schreyögg, Sydow and Koch 2003). The adjective "dysfunctional" in this context describes the limited ability of a rational decision-making. The immanent lack of information for a rational decision (the problem of bounded rationality (Simon, 1972: a manager cannot have the complete information about his problem of decision) is connotated with the adjective "suboptimal". This leads to the question of how durable flexibility can be generated and integrated in the organizational structure. For the research of the generation of flexibility the concept of autonomous cooperation is of interest whereas the idea of competence-management may offer a tool to integrate flexibility into the organizational structure.

The aim of this paper is to analyze to which extent autonomous cooperation can provide a tool to unlock organizations. For this purpose, the approach of the competence-based perspective is used to apply the concept of autonomous cooperation to business science and to identify its contributions to a flexibilization of the organization.

In the following, the concept of autonomous cooperation will be analyzed from a competence-based perspective. Section 2 describes autonomous cooperation in its history of development (2.1), its core statements (2.2) and its understanding in business science (2.3) to establish common background knowledge as well as an analytical basis. Section 3 analyzes the role of flexibility from a competence-based perspective to point out its relevancy in this context. For this purpose, the approach of the competence-based perspective is presented in a short introduction of its main statements and the role of flexibility from a competence-based perspective is analyzed. In section 4 the attributes of the concept of autonomous cooperation are combined with their contributions to a flexibilization of the company structure to discuss possible effects of autonomous cooperation on flexibility. A conclusion of the results of the paper can be found in section 5.

3.3.2 The concept of autonomous cooperation

Origins of autonomous cooperation

The concept of autonomous cooperation belongs to the field of complexity science. It deals with the problem of complex and dynamic systems in natural science and analyzes how these systems generate system adaptiveness, robustness, and emergent order. The basic idea derives from the science of self-organization, whose intention is to study, explain, and identify general principles on how complex systems autonomously create ordered structures. This concept was originated in the 70s by separate scientists of different disciplines, e.g. von Foerster (1960) (cybernetics), Prigogine and Glansdorff (1971) (chemistry), Haken (1973) (physics), Maturana and Varela (1980) (biology). After recognizing a common background of the notions complexity and order at the end of the 70s, a basis for a comprehensive interdisciplinary theory was established. Until now this young science is still at a stage of forming and developing. Initial results of different approaches of self-organization have already diffused into other fields of science. The approach of autopoiesis of Maturana and Varela (1980), for instance, appears in different scientific fields, such as sociology with reference to Luhmann's systems theory (Luhmann 1994), as well as in psychology in the area of family therapy (e.g. Hoffmann 1984).

Classification of autonomous cooperation

Before the main statements of the concept of autonomous cooperation are presented, a short classification of the concept and a distinction from similar terms will follow. A clearly defined usage of the notions 'self-management', 'self-organization' and 'autonomous cooperation' has not been established yet. The specifications of the terms could be categorized in the following way.

Fig. 3.1 Classification of the terms 'self-management', 'self-organization' and 'autonomous cooperation'

The term self-management comprises the most widespread concept of the mentioned terms. It describes the ability of a system to organize itself autonomously. This means the system determines its own objectives, autonomously chooses its strategies and organizational structure and also raises the necessary resources itself (Manz and Sims 1980). Therefore, a self-managed system is able to design and to vary its own management system. Self-organization as a part of management describes the way of autonomously creating an emergent order. It focuses on the autonomous formation of structures and processes (Bea and Göbel 1999; Probst 1987). Finally, the term autonomous cooperation as the narrowest perspective of the mentioned terms describes processes of decentralized decision-making in heterarchical structures. It presumes interacting elements in non-deterministic systems which possess the capability and possibility to render decisions independently (Hülsmann and Windt 2005).

Main statements of autonomous cooperation

Autonomous cooperation aims at achieving an increased robustness and a positive emergence of the total system resulting from distributed and flexible coping with dynamics and complexity (Hülsmann and Windt 2005). As self-organization and autonomous cooperation have the same scientific roots, they share the main attributes such as autonomy, interaction, emergences, and non-determinism (Von Foerster 1960; Prigogine and Glansdorff 1971; Haken 1973; Maturana and Varela 1980). Among other attributes the named ones were chosen for an analysis in the following as they feature the characteristic of reflecting the process of self-organization.

Autonomy

A system or an individual is autonomous if its decisions, relations, and interactions are not dependent on external instances and therefore are operationally closed (Probst 1987). However, a complete independence of other systems cannot be assumed (Varela 1979; Malik 2000), as each system only represents a part of a wide-ranging total system which it is to a certain extent dependent on and influenced by. Therefore, we have to speak of a relative autonomy of the individual or the system in relation to certain criteria (Varela 1979; Probst 1987). In the organization these criteria are defined by the given scope of action and decision-making of the autonomous subject. For this reason autonomy manifests itself in the organization as a result of the processes of decentralization and delegation. (Kappler 1992).

Interaction and emergences

The core statement of the concept of self-organization is that open, dynamic and complex systems (natural or social systems) develop a self-organized order within a system (von Foerster 1960; Prigogine and Glansdorff 1971; Maturana and Varela 1980), which is the result of various interactions of the individual system elements (Haken 1987). From this process of interaction new qualitative characteristics of a system arise, namely emergences (Haken 1993). These emergences are not related to individual system components but result from the synergistic effects of the interacting elements. It is not yet clarified how these synergetic effects arise from the interacting elements and how they may be analyzed and explained. According to Haken (1987), the system reaches a new increased level of quality through the emergences as they enable the system to better cope with environmental demands.

Non-determinism

Another feature that can be found in all self-organizing systems is non-determinism. In autonomous, cooperating systems general rules of decision-making are predetermined (Hülsmann and Windt 2005) and the desired final state of the system may be predicted but not the way of how to achieve this. Since the system elements are able to autonomously take decisions, the system behaviour is casually not predetermined and thus not predictable (Haken 1983; Prigogine 1996).

3.3.3 Flexibility out of a competence-based-view

From a system theoretical point of view, flexibility can be seen as a driver for unlocking organizations. Flexibility describes the ability of a system to open its boundaries for required resources (e.g. information) and thereupon to change the system structures according to the demands of its relevant environment if needed. Through processes of system openings the border to the system's environment becomes increasingly indistinct. Therefore, it is all the more important to compensate the degree of flexibility through processes of stabilization (system closure) to maintain the system's identity in the permanent processes of adaptation. Consequently, organizational flexibility is needed to cope with internal and external dynamics and complexity and to avoid the risk of locked organizations.

According to the strategic management, achieving sustainable competitive advantage should be the aim of an organization. The literature of the strategic management argues that there are two essential sources of competitive advantages – one from the market position (market-based view) and one from competencies (competence-based view).

The concept of the competence-based view started with articles and books by Prahalad and Hamel beginning in the late 1980s (Hamel and Prahalad 1989; Prahalad and Hamel 1990; Hamel and Heene 1994; Sanchez et al. 1996). The main statement of the theory of the competence-based view is that companies focus on their competencies to achieve competitive advantages. According to Sanchez et al. (Sanchez and Heene 1996; Sanchez 2004) competences can be understood as „[...] the ability to sustain the coordinated deployment of assets in ways that help a firm achieve its goals." In the theory of the competence-based view a firm is seen as a learning organization that builds and deploys assets, capabilities, and skills to achieve strategic goals (Hamel and Heene 1994).

Flexibility plays an important role in the competence-based management. Representing particular forms of activeness and processes within the organization, competence-building and competence-leveraging go hand in hand with a certain degree of alteration and consequently require organizational flexibility. In strategic management literature, for instance, the work of Sanchez covered the topic of flexibility, which underlines its importance. Sanchez (2004) defined five "modes" of competences, each of which stands for a different kind of flexibility that all respond to changing environmental conditions.

On the one hand, organizations have to develop flexibility to ensure their survival in the long run by adapting to changing environmental demands. But on the other hand, a basic flexibility should be present within the organization's predisposition to enable a continuous competence building and leveraging. Consequently, a dualistic role of organizational flexibility can be identified (Hülsmann and Wycisk 2005), which leads to two basic challenges for management: the basic requirement of flexibility has to be assured while flexibility and stability also have to be balanced (Hülsmann and Wycisk 2005).

3.3.4 The contribution of autonomous cooperation to a flexibilization of social systems from a competence-based perspective

Autonomous decision-making as a tool to cope with complexity

In the context of business science autonomy is characterized by processes of delegation and decentralization (Kappler 1992), which can be understood as the degree of autonomous decision-making among the organization's employees. Therefore, those processes will be analyzed in their effects on flexibility and stability as well as in relation to qualitative, quantitative, temporal and spatial aspects from a competence-based perspective.

Delegation empowers the elements (members) or sub-units of the system to freely develop various patterns of competences and to make autonomous decisions, which are spatially closer to the operational level of work (Mullins 2005). Thus, the system can partially react towards changing environmental demands while the rest of the organizational structure remains unaffected. Moreover, there is a link between the spatial closeness of decision-making and the temporal effect of flexibility in autonomous, cooperating organizations. Ways of decision-making become shorter and easier as information on the level of the sub-units flow faster so that the total system's ability of problem solving quantitatively as well as qualitatively increases.

Through processes of decentralization, the entire complexity of an organization (consisting of the system's as well as the environment's complexity) can be distributed among its diverse sub-units and elements so that a reduction of the quantitative level of complexity can be achieved. These processes may be coupled with an increase of system flexibility. Instead of controlling and focusing on all of the required competences of each individual element and its system interrelations, the organization now merely has to consider the sub-units in its processes of planning, designing, and developing competences.

However, processes of delegation and decentralization always imply the risks of intransparency and moral hazard as well as autogenous self-organization (Göbel 1998) and intergrouping conflicts (Staehle 1999), which the management needs to consider. Furthermore, it has not yet been ascertained which degree of empowerment proves to be effective and provides the most valuable contribution to a flexibilization.

Interaction as a tool to obtain redundancy and emergences

The interaction processes of autonomous, cooperating systems involve the effect of redundancy. According to the concept of autonomous cooperation, each element or subsystem of the complete system is equipped with the same assets and abilities by nature as shown for example by the individual light waves of Haken's laser light (1983) or the atoms of the dissipative structures of Prigogine (1996). Applied to social systems, it could be assumed that with a high degree of interaction and exchanged information the elements learn about each other's capabilities and know-how through organizational structures, such as job rotation or job enlargement (e.g. Schreyögg 1998; Mullins 2005). With a high degree of autonomous cooperation, each member could undertake every function of the system. This redundancy, which could be understood as a competence of the system itself, feeds the system with flexibility because its employees are able to react flexible wherever needed and even if some members turn out. However, a disadvantage of redundancy could be a lack of expertise within the system. Due to the learning of different functions, the knowledge of the employees is mainly characterized by diversity, which may cause higher costs in case expertise is needed.

Resulting from the interaction of the various system elements, the effect of emergences represents new qualities of the system. From a competence-based perspective, the latter would be defined as a competence arrangement that is characterized by an improved ability to cope with complexity and dynamics and therewith by a better fit of system structure and environmental demands. Through interaction of the system elements, for instance, a bundling of company-specific resources as core competences could evolve (Hamel 1994), which sustain competitive advantages.

Non-determinism as a tool to promote creativity

Based on the ability of autonomous decision-making, the members of an organization initially do not act in a predetermined way. As a result, a wide range of alternatives of action for the system elements is preserved, which assumably involves an increased flexibility of action and thus reaction to sudden environmental demands. By authorizing the system elements to use innovative strategies of problem-solving their creativity will be stimulated so that eventually more effective ways of organizational acting will be generated. This evolutionary process provides a basis for retention (Wolf 2003), which in this case stands for the firm maintenance and stabilization of profitable competences within the system. In Addition, the creativity will amount to context-conditional changes in the competence structure, which from an evolution-theoretic perspective would be conceptualized under the term of variation (Macharzina 2003). The formation of variation patterns bears the opportunity of selection (Wolf 2003), i.e. the opportunity of sorting out ineffectual action alternatives.

However, the organization's way of acting is not completely indetermined. One reason for this is the openness of social systems meaning that they are in a permanent process of exchange (e.g. of information and material) with their surroundings, which goes along with a permanent affection by environmental influences. Another reason can be found in the system's history. According to the theory of path dependencies, a grown system is always predetermined by its formerly made decisions. Thus, an unlimited amount of acting alternatives cannot exist (Schreyögg et al. 2003).

3.3.5 Conclusions

In the previous specifications we described the situation of a locked organization as a suboptimal situation with a limited choice of possible decisions (Schreyögg et al. 2003), meaning that the organization is caught in its own complexity and thus not longer able to make rational decisions. Organizational flexibility was identified as a means to unlock this dilemmatic situation of decision-making (Hülsmann 2005). To obtain organizational flexibility – which may be understood as a competence itself or as a basic requirement of the whole company structure (Hülsmann and Wycisk 2005) – the concept of autonomous cooperation was analyzed to determine the extent of its contribution to a flexibilization of the company from a competence-based perspective. In doing so, several links and starting points for a flexibilization through autonomous cooperation were found.

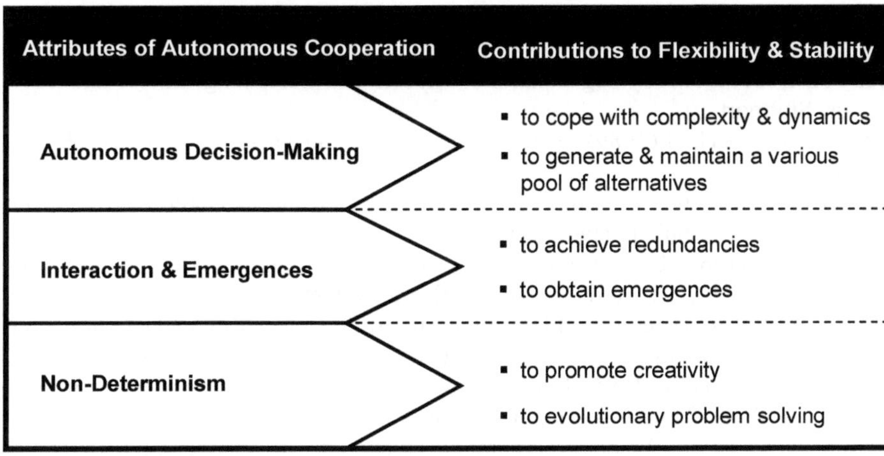

Fig. 3.2 Contributions of the concept of autonomous cooperation to generate organizational flexibility and stability

Since the previous discussion gives a rough insight into possible contributions of the concept of autonomous cooperation to flexibilize a company's structure and processes, further research tasks arise out of a scientific and pragmatic perspective.

From the scientific perspective, the achieved results of this discussion could be regarded as assumptions about the correlation of autonomous cooperation and flexibility within organizations. Unless those assumptions become a status as established statements or even a part of a theory, they need to be examined more detailed in their logical explanatory power. Further the logical statements should be verified in an empirical way, to raise their factual validity (Raffée 1995).

From the pragmatic perspective the concept of autonomous cooperation needs to be more examined regarding its manageability. For a targeted appliance of autonomous cooperation, its measurement, control and steering abilities are necessary. The process of measuring presumes visibility as well as predetermined goals of achievement. One task will therefore be to detail the concept of autonomous cooperation in its constitutive attributes to gain higher visibility. Another research requirement will be to generate a measuring system which is able to quantify the level of autonomous cooperation in a system and to evaluate these results in comparison to the desired achievements. These questions are part of the work of the CRC 637 "Autonomous Cooperating Logistic Processes: A Paradigm Shift and its Limitations".

References

Bea FX, Göbel E (1999) Organisation – Theorie und Gestaltung. Lucius and Lucius, Stuttgart

Böse R, Schiepeck G (1989) Systemische Theorie und Therapie. Asanger, Heidelberg

Dörner D (2001) Die Logik des Misslingens: Strategisches Denken in komplexen Situationen. Reinbek, Hamburg

Foerster v H (1960) On Self-Organizing Systems and their Environment. In: Yovits MC, Cameron S (eds) Self-Organizing Systems. London

Freiling J (2002) Strategische Flexibilität und Strategiewechsel als Determinanten des Unternehmenswertes. Deutscher Universitätsverlag, Wiesbaden

Gebert D, Boerner S (1995) Manager im Dilemma – Abschied von der offenen Gesellschaft. Campus-Verlag, Frankfurt am Main et al

Gharajedaghi J (1982) Social Dynamics (Dychotomy or Dialectic). In: General Systems 27: 251–268

Göbel E (1998) Theorie und Gestaltung der Selbstorganisation. Duncker and Humblot, Berlin

Haken H (1973) Synergetics: cooperative phenomena in multi-component systems (Proceedings of the Symposium on Synergetics from April 30 to May 6, 1972, Schloß Elmau Stuttgart)

Haken H (1983) Erfolgsgeheimnisse der Natur: Synergetik, die Lehre vom Zusammenwirken. Deutsche Verlags-Anstalt, Stuttgart

Haken H (1987) Die Selbstorganisation der Information in biologischen Systemen aus Sicht der Synergetik. In: Küppers BO (ed) Ordnung aus dem Chaos. Piper, München, pp 35–60

Haken H (1993) Synergetik: Eine Zauberformel für das Management ?. In: Rehm W (eds) Synergetik: Selbstorganisation als Erfolgsrezept für Unternehmen. Ein Symposium der IBM, Stuttgart, pp 15–43

Hamel G (1994) The concept of core-competences. In: Hamel G, Heene A (eds) Competence-based competition. Wiley, Chichester, pp 11–33

Hamel G, Heene A (1994) Competence-based competition. Wiley, Chichester

Hamel G, Prahalad CK (1989) Strategic intent. Harvard Business Review

Hoffman L (1984) Grundlagen der Familientherapie. ISKO Press, Hamburg

Hülsmann M (2005) Dilemmata im Krisenmanagement. In: Burmann C, Freiling J, Hülsmann M (eds) Management von Ad-hoc-Krisen – Grundlagen - Strategien – Erfolgsfaktoren. Gabler, Wiesbaden, pp 401–422

Hülsmann M, Berry A (2004) Strategic Management Dilemma: It's necessity in a World of Diversity and Change. In: Wolff R et al (eds) Conference Proceedings of SAM and IFSAM VII World Congress: Management in a World of Diversity and Change, Göteborg

Hülsmann M, Windt K (2005) Selbststeuerung – Entwicklung eines terminologischen Systems. Universität Bremen FB Wirtschaftswissenschaften, Bremen (forthcoming)

Hülsmann M, Wycisk C (2005) Contributions of the concept of self-organization for a strategic competence-management. The Seventh International Conference on Competence-Based Management: Value Creation through Competence-Building and Leveraging, 2.4. Juni 2005. Antwerpen, Belgien. (forthcoming)

Kappler E (1992) Autonomie. In: Frese E (ed) Handwörterbuch der Organisation. Schäffer-Poeschel, Stuttgart, pp 272–280

Luhmann N (1973) Zweckbegriff und Systemrationalität. In: Reihe Suhrkamp-Taschenbuch Wissenschaft 12, Frankfurt am Main

Luhmann N (1994) Soziale Systeme: Grundriss einer allgemeinen Theorie. In: Reihe Suhrkamp-Taschenbuch Wissenschaft 666, 5th edn. Frankfurt am Main

Macharzina K (2003) Unternehmensführung – Das internationale Managementwissen. Gabler, Wiesbaden

Malik F (2000) Strategie des Managements komplexer Systeme: Ein Beitrag zur Management-Kybernetik evolutionärer Systeme, 6th edn. Haupt, Bern.

Manz C, Sims HP (1980) Self-management as a substitute for leadership: A social learning theory perspective. In: AMR 5and1980, pp 361–367

Maturana HR, Varala F (1980) Autopoiesis and cognition: the realization of living. Reidel

Mullins LJ (2005) Management and organizational behaviour, 7th edn. Harlow et al

Pflüger, M (2002) Konfliktfeld Globalisierung : Verteilungs- und Umweltprobleme der weltwirtschaftlichen Integration. Heidelberg

Prahalad CK, Hamel G (1990) The Core Competencies of the Corporation. Harvard Business Review 68 (3): 79–91

Prigogine I, Glansdorff P (1971) Thermodynamic Theory of Structure, Stability and Fluctuation. Wiley, London et al

Prigogine I (1996) The End of Certainty: Time, Chaos, and the New Laws of Nature. The Free Press, New York

Probst GJB (1987) Selbstorganisation, Ordnungsprozesse in sozialen Systemen aus ganzheitlicher Sicht. Parey, Berlin et al

Raffée, H (1995) Grundprobleme der Betriebswirtschaft in Reihe Betriebswirtschaftslehre im Grundstudium der Wirtschaftswissenschaft, Bd 1, Göttingen

Sanchez R (2004) Understanding competence-based management: Identifying and managing five modes of competence. Journal of Business Research 57: 518–532

Sanchez R, Heene A (1996) A systems view of the firm in competence-based competition. In: Sanchez R, Heene A, Thomas H (eds) Dynamics of Competence-Based Competition. Pergamon, Oxford, pp 39–42

Sanchez R, Heene A, Thomas H (1996) Dynamics of competence-based competition: Theory and practice in the new strategic management. Pergamon, Oxford

Schreyögg G (1998) Organisation – Grundlagen moderner Organisationsgestaltung, 2nd edn. Gabler, Wiesbaden

Schreyögg G, Sydow J, Koch J (2003) "Organisatorische Pfade – Von der Pfadabhängigkeit zur Pfadkreation?. In: Schreyögg G, Sydow J (eds) Strategische Prozesse und Pfade. Managementforschung 13, Wiesbaden

Simon HA (1972) Theories of Bounded Rationality. In: McGuire CB, Radner R (eds) Decision and Organization. North-Holland Publications, Amsterdam, pp 161–172

Staehle WH (1999) Management - eine verhaltenswissenschaftliche Perspektive, 8th edn. Vahlen, München

Varela FJ (1979) Principles of biological autonomy. North Holland, New York

Wolf J (2003) Organisation, Management, Unternehmensführung – Theorien und Kritik. Gabler, Wiesbaden

3.4 Self-Organization Concepts for the Information- and Communication Layer of Autonomous Logistic Processes

Markus Becker, Andreas Timm-Giel, Carmelita Görg

Communication Networks, Otto-Hahn-Allee - NW1, University Bremen, 28359 Bremen, Germany

As described in Chapter 2.2 on Historical Autonomy in ICT, self-organisation is a well-known concept in information- and communication technology. In this chapter it is shown how self-organisation can be employed in the Information- and Communication Layer of Autonomous Logistic Processes.

3.4.1 Autonomic communication, autonomic computing and self-star

Especially the concepts Autonomic Communication, Autonomic Computing and Self-Star, already mentioned in Chapter 2 are of interest to be applied in Autonomous Cooperating Logistic Processes.

The components involved in the processes, in particular the vehicles, need very reliable communication systems due to their mobile nature. The communication equipment fixed to the vehicle has to be highly reliable to avoid frequent reset/repair times. The time the transport vehicle spends in the garage for resetting, reconfiguring or repairing the communication system is necessarily downtime and hence extremely costly. Resistance of the shipping companies against unreliable IT systems has become obvious with the introduction of the TollCollect system (Bundesverband Güterkraftverkehr Logistik und Entsorgung e.V. 2005).

The next paragraphs introduces example applications of Autonomic Computing and Self-Star such as Self-organized Selection of Communication Networks, Service Discovery, Gateway Discovery and Ad hoc Routing that are needed for always best connected communication services of sensor and mobile nodes.

Self-organized selection of communication networks and services

To show the application of the concepts, a self-organized selection of communication networks has been implemented, called Autonomous Communication Gateway. The autonomous communication gateway is documented in (Becker et al. 2006a) and has been demonstrated as part of an autonomous logistic support system (Becker et al. 2006b; Morales et al. 2006). The device has the ability to communicate using three different communication networks, namely Wireless Local Area Network (WLAN), Universal Mobile Telecommunication System (UMTS) and General Packet Radio Service (GPRS).

The independence of the different systems can be established by means of MobileIP and a Virtual Private Network (VPN).

For the autonomous logistic processes case the permanent addressability is established by the use of a Virtual Private Network (VPN), as shown in figure 3.3. The VPN provides static Internet Protocol addresses for use in the applications, although the IP addresses change, when changing the communication network. This approach includes encryption of the transmitted data, which is crucial to business data. The mechanism of VPN can be used in this case, because only connectivity to one node is needed. In the general case when more than one node needs to be connected, a MobileIP solution should be preferred.

The standardized MobileIP as described in (Montenegro et al. 2003) provides Internet Protocol (IP) layer mobility. The components enabling the mobility are the Home Agents (HA) and Foreign Agents (FA). The mobile node needs to register with one of the agents. When in a foreign network, it registers with a foreign agent, which will inform the home agent of the current Care of Address (CoA). Several implementations of MobileIP exist and are tested in more general scenarios. Furthermore, Mobile IP combined with mobile ad hoc networks can be used for gateway discovery, as described in a later section.

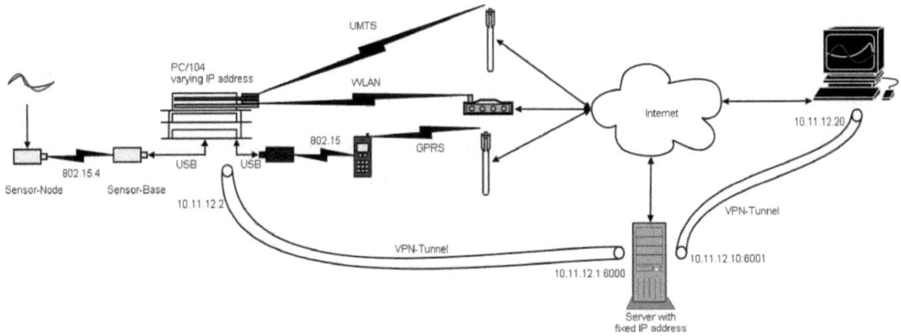

Fig. 3.3 Autonomic Network Selection as Support for Autonomic Logistic Proc esses

Communication Selection

The communication selection process includes the service availability and additional selection criteria, e.g. cost and required data rate of the application. Currently the application that is in use is the surveillance of sensor data using the system as described in Chapter 4.6

Communication timing

A self-organizing instance needs to evaluate the advantage of transmitting the data against the costs of the transmission. This instance might decide to postpone the transmission of the data until a cheaper system is available or the importance of the data has increased. Additionally, in future it is planned to give feedback to the autonomous logistics application on the available networks, their bandwidth, the costs, changes in availability and security. Based on this information the application can then decide to delay its transmission, aggregate its data or to signal the gateway to use a different system.

This approach yields an economic usage of the scarce and expensive wireless bandwidth.

3.4.2 Service discovery and gateway discovery

Another area in which self-organization plays a major role is Service Discovery. Service discovery is the process of detecting which services are available and where they are available.

There are many different methods and implementations of Service Discovery. Standardized Protocols such as ZeroConf (Cheshire 2006), SLP (Gutmann et al. 1999), SDP (Handley et al. 2006), Sun Microsystem's Jini (Jini 2006) or Microsoft's Universal Plug and Play (UPnP Forum 2006) as well as research protocols such as GSD (Chakraborty et al. 2002) and Konark (Helal et al. 2003) are available, to name only a few.

For autonomous logistic processes there are at least two service discovery types: Sensor Discovery and Gateway Discovery.

Sensor discovery is the discovery of sensor nodes with specific sensors as discussed in 4.6. The mentioned protocols cannot be used in Wireless Sensor Networks because of their dependence on IP networks, instead specialized WSN solutions are needed that take into account the low memory space, low data rates, limited energy sources, for example (Raluca Marin-Perianu et al. 2006).

Gateway discovery in Internet Protocol based wireless networks such as the one depicted in Fig. 2 means discovering the node that provides a connection to the infrastructure network, e.g. the Internet. Usually only one node provides this service to all the other nodes. There are several solutions to autonomic gateway discovery. In this project two approaches are being considered. The first approach is called Proxy Gateway. The second approach is an integration of the already mentioned Mobile Internet Protocol (MobileIP) and Ad-hoc Networks (MANETs).

The gateway discovery in the Proxy Gateway approach uses a modified gateway (in the vehicle) that replies to route requests to an unknown network node with its own address and forwards the data. This approach provides connectivity of the logistic ad-hoc network (e.g. in the vehicle) to the Internet. Addressability of the logistic ad-hoc network from the Internet is not included.

The second approach to gateway discovery integrates MobileIP (Montenegro et al. 2003) and MANETs (Macker and Chakeres 2006). The Foreign Agent of MobileIP, which handles the registration of Mobile Nodes, advertises its gateway functionality into the MANET. The nodes receiving the advertisement can use the address specified in the advertisement for identifying the gateway to the Internet. Implementations of MobileIP and AODV for embedded devices such as Access Points have been developed by the authors (Becker et al. 2006c).

3.4.3 Ad hoc routing

The working principles of ad hoc routing have already been discussed in Section 2.2 on Historical Autonomy in ICT. Here the usage of ad hoc routing in logistical environments is presented.

Fig. 3.4 depicts the usage of ad hoc networks in logistics. The nodes and packages have communication units attached to them. Those communication units set up a mobile ad hoc network. The networks might be divided into sub-networks depending on the spatial distance between them. When for example a packet is loaded into a vehicle, existing radio links will be discontinued and new links will be established.

A commonly investigated usage scenario of ad hoc networks in the Collaborative Research Centre (CRC) is the intelligent transport good, which decides itself with which transport vehicle to start negotiations for transport services. Factors considered are for example: expected time of arrival at the destination, risk not to arrive in time, suitability of transport vehicle for the transport, costs, and sensors available in the transport vehicle to monitor the transport conditions.

When the packet is loaded onto the transport vehicle the agent representing the transport good can be copied from the RFID tag into the board computer of the transport vehicle. The agents of more intelligent goods can communicate themselves with the vehicle's agent (see figure 3.4). It is also possible that the transport good is just identified by the transport vehicle, which then downloads the transport requirements from the Intra- or Internet. Among other things the transport good is advising the vehicle's agent on how to configure the sensors to monitor the right transport conditions.

Other scenarios foresee that a sufficient number of sensors are in the transport vehicle and also on the transport goods themselves. Each transport vehicle configures a sensor network out of the available sensors suitable for monitoring the required transport conditions.

Fig. 3.4 Ad hoc routing in logistic applications

3.4.4 Conclusions

This chapter has shown examples of the application of self-organisation approaches in the information and communication layer of autonomous logistic processes: the self-organized selection of communication networks, gateway discovery and ad hoc routing. The self-organisation methods applied here aim at improving the communications of logistic objects.

References

Becker M, Wenning BL, Timm-Giel A, Görg C (2006a) Usage of Mobile Radio Services in Future Logistic Applications. In: Proceedings of Mobilfunktagung 2006, pp. 137-141. Osnabrück, Germany

Becker M, Jedermann R, Behrens C, Sklorz A, Westphal D, Timm-Giel A, Görg C, Lang W, Laur R (2006b) Sensoren und RFIDs für selbststeuernde logistische Objekte und deren mobile Vernetzung. Bremen, Germany. Berichtskolloquium des SFB 637

Becker M et al. (2006c) Port of MobileIP (uob-nomad) and AODV (aodv-uu) to an embedded platform. https://dev.openwrt.org/browser/packages/net

Bundesverband Güterkraftverkehr Logistik und Entsorgung e.V. (2005) LKW-Maut in Deutschland - Wahrung von Schadenersatzansprüchen im Zusammenhang mit dem Aufspielen der neuen OBU2-Software. http://www.bgl-ev.de/web/service/nutzungsausfall-tollcollect.htm

Chakraborty D, Joshi A, Yesha Y, Finin T (2002) GSD: a novel group-based service discovery protocol for MANETS, Proc. MWCN 2002

Cheshire St (2006) DNS Service Discovery (DNS-SD). http://www.dns-sd.org/
Guttman E, Perkins C, Veizades J, Day M (1999) IETF RFC 2608: Service Location Protocol, Version 2
Handley M, Jacobson V, Perkins C (2006) IETF RFC 4566: SDP: Session Description Protocol
Helal S, Desai N, Verma V (2003) Konark - a service discovery and delivery protocol for ad-hoc networks, Choonhwa Lee, Proc. IEEE WCNC
Jini (2006) http://www.jini.org/wiki/Main_Page
Macker J, Chakeres I (2006) http://www.ietf.org/html.charters/manet-charter.html.
Marin-Perianu RS, Scholten J, Havinga PJM, Hartel PH (2006) Energy-Efficient Cluster-Based Service Discovery in Wireless Sensor Networks. Technical Report TR-CTIT-06-43 Centre for Telematics and Information Technology, University of Twente, Enschede. ISSN 1381-3625
Montenegro G, Roberts P, Patil B (2003) http://www.ietf.org/html.charters/OLD/mobileip-charter.html
Morales E, Ganji F, Jedermann R, Gehrke JD, Lorenz M, Wenning BL, Be cker M, Behrens C, Congil J, Ober-Blöbaum C (2006) SFB-Demonstrator "Intelligenter Container". Bremen, Germany. Newsletter des SFB 637 1/2006
UPnP™ Forum (2006) http://www.upnp.org/

3.5 Distributed Knowledge Management in Dynamic Environments

Hagen Langer, Jan D. Gehrke, Otthein Herzog

TZI - Center for Computing Technologies, University of Bremen, Germany

3.5.1 Introduction

Logistic processes are inherently dynamic and hence require the ability to plan and re-plan in complex situations, under rigid time constraints, and in light of uncertain, incomplete, and false information. Standard scenarios of logistic processes typically have been modeled on the basis of static graph-theoretic representations. The well-known traveling salesman problem (TSP), the vehicle routing problem (VRP), or the pickup and delivery problem (PDP) reduce the complex task of transportation to a route optimization problem. They neglect both the important role of knowledge and communication in real-world logistic processes (cf. (Hult et al. 2003)) and the fact that relevant parameters, e.g., traffic flow, incoming orders, etc. change over time.

In this paper we will describe an approach to the agent-based modeling of logistic processes which makes use of an explicit knowledge management system and hence enables agents to fulfill complex logistic tasks in dynamic environments.

This paper is organized as follows. We introduce agents as basic components of our framework in Section 3.5.2 and discuss agent-based approaches to logistics (3.5.3). Section 3.5.4 presents our approach to distributed knowledge management for multiagent systems. We discuss agent roles, decision parameters, and an interaction protocol for the two most important knowledge management roles of our framework. In Section 3.5.5 we summarize the main conclusions of this work.

3.5.2 Intelligent agents

Agents are currently one of the most prominent paradigms for creating autonomous software systems. A broad variety of agent architectures have been proposed in the past. One extreme in the spectrum of agent architectures are reactive agents which are not necessarily much more autonomous in their decision-making than standard software components, but share other important properties with prototypical agents (e.g., the existence of sensors and actuators). The other extreme is established by cognitive agents, which mimic our assumptions on human cognitive processes as close as possible. Cognitive agents are often implemented as BDI agents (belief, desire, and intention), cf. (Rao and Georgeff 1991). BDI agents possess an autonomous knowledge base (the beliefs) and this knowledge base is modified whenever the agent interacts with its environment or when the agent updates its knowledge base by inferring new knowledge from its existing background knowledge. In the simplest case, this interaction means that the agent receives percepts via its sensors. The behavior of a BDI agent is also determined by its desires and its intentions. The desires are long-term goals which, together with the beliefs, determine the agent's intended actions. The beliefs and actions of a BDI agent depend not only on the agent's environment but also on the agent's existing knowledge base and its desires.

3.5.3 Agent-based logistics

Previous research on applying multiagent systems in the logistics domain has put a strong emphasis on price negotiations and auctions. In these approaches the inter-agent communication is often reduced to bidding (cf., e.g., (Zhengping et al. 2001)), or the internal structure of an agent is defined by a set of equations (e.g., (Bos et al. 1999)). Scholz et al. (2004) apply MAS to shop floor logistics in a dynamic production scenario. It aims at a flexible and optimal scheduling of production plans in a heterogeneous shop floor environment. Hofmann et al. (1999) aim at replacing conventional tracking and tracing in the logistics domain based on sending (i.e., pushing) EDIFACT messages by an agent-based pull mechanism. Smirnov et al. (2003) present a prototype of a multi-agent community implementation and a constraint-based protocol designed for the agents' negotiation in a collaborative environment. Most of the previous approaches to multi-agent-system-based logistics, however, employ simplified models of logistic processes which do not involve any explicit knowledge management.

Our approach is based on a system of autonomous agents which represent logistic entities. Besides its primary logistic functionality, each agent can adopt a role as part of a distributed knowledge management system.

Our framework makes the following assumptions about real-world logistic scenarios

- Real-world logistic scenarios are never static, but highly dynamic.
- Agents involved in logistic processes have to plan and act on the basis of incomplete, uncertain, and rapidly changing knowledge.
- Optimal decision making under the circumstances sketched above presupposes an appropriate knowledge management framework.
- Knowledge is a valuable resource and can also be a tradable good.

We can envision a scenario in which agents are used to represent real-world entities such as trucks and containers, abstract objects such as weather or traffic services, or even human decision makers, such as a ramp agent at a loading dock. We believe this kind of autonomous, decentralized decision-making can help make the operational processes more efficient, cost-effective, and allow the participating enterprise to stay competitive. It is also a major improvement over traditional centralized approaches in which individual agents are ill-equipped to deal quickly with sudden events since control usually resides with the entities that are removed from the scene of the event and thus have only delayed access to the relevant information. In addition, agents must be able to negotiate, form coalitions, and thrive in the presence of competition, for example, for customers (orders) or resources, and are also subject to unpredictable changes in their environment.

In contrast to standard approaches to the computational modeling of transportation processes[6], we do not presuppose that there is a central omniscient unit which plans, coordinates, and controls the activities of logistic entities (e.g., vehicles, depots). We, on the contrary, assume that these logistic entities are autonomous and control themselves. This setting requires that there is a robust and flexible knowledge management system which is able to provide the necessary knowledge for each agent.

[6] By standard approaches we mean settings such as the well-known traveling salesman problem (TSP), the vehicle routing problem (VRP), or the pickup delivery problem (PDP).

3.5.4 Knowledge management based on roles and parameters

Agent-based knowledge management has been studied under different assumptions, but the main focus of previous research has been on single agents, as opposed to multiagent systems (MAS), which we employ, and on knowledge management by agents for human users, as opposed to our approach which is not only by agents but moreover for agents. Another important difference between our framework and previous agent-based knowledge management systems is that we do not presuppose a one-to-one mapping between agents, on the one hand, and knowledge management functions, on the other.

Three main components are ingredients of our framework: agents, knowledge, and roles. Agents represent process owners (e.g., decision makers) or real-world entities in the logistics domain (e.g., cargo transport centers, vehicles, transport containers, or even single packages).

In addition, an agent has specific properties (e.g., speed, weight, enterprise affiliation), capabilities (e.g., transportation or storage capabilities, or sensors for measuring humidity), desires (e.g., minimizing delay of a shipment or maximizing the utilization ratio), and intentions (i.e., tactical plans). The set of beliefs forms an agent's knowledge base and is associated with specific inferential capabilities.

We envision that these agents, which must act in a rational fashion, can be implemented as goal-oriented agents following the BDI (belief, desire, intention) approach as discussed above. The BDI approach is well suited for this purpose since it provides the appropriate concepts and structures for representing our agents. For example, the strategic layer of agents may be modeled within the desires, operational aspects within beliefs, and tactical features within intentions or plans. Furthermore, the BDI approach attempts to closely mimic human decision-making (Bratman 1987) and represents one of the dominant approaches for modeling intelligent behavior within the agent research community (d'Inverno et al. 2004). For a comprehensive discussion of the applicability of BDI to represent rational, autonomous agents see also (Timm 2004).

The second component of our framework provides knowledge management functionalities including knowledge representation, storage, and manipulation. In our framework, the terminological domain knowledge is organized in associated ontologies for transportation and production logistics which include, e.g., a representation of the transportation network as an annotated graph, together with a two-dimensional map-like representation (similar to geographic information systems) enabling spatial reasoning (e.g., inferring properties of proper sub-regions using a part-of relation), the basic types of agent and their properties (e.g., for a vehicle, its average and maximum speed, the types of routes in the network it can use, its load capacity, and its corporate affiliation), and the properties of `inactive' objects, such as highways, traffic hubs, depots, etc.

The visibility of the ontology is determined by an agent's predefined tasks and capabilities. For example, in contrast to a shipment agent, an agent representing a navigation system must have complete access to all relevant details of the transportation network part of an ontology.

Knowledge management enables agents to request new or missing knowledge, or update existing knowledge. Intuitively, our approach is similar to peer-to-peer knowledge management. Agents have the ability to form dynamic knowledge networks and to share knowledge. Hereby knowledge management becomes a secondary task orthogonal to the primary logistic tasks.

The third component of our framework integrates the multiagent approach with knowledge management functionalities using roles. Examples of these roles are knowledge acquisition, brokerage, and processing. Depending on their capabilities and tasks in the logistics domain, agents may assume any one of these roles, which may change over time. For example, an agent representing a ship may assume the role of a knowledge provider reporting weather information to other ships. At a different point in time, the same agent may also assume the role of a knowledge consumer requesting information about its cargo and destination from a dock agent after loading is complete. Communication among agents is implemented by the already existing agent communication infrastructure.

In contrast to conventional knowledge management systems, our approach is inherently distributed. In particular, it focuses on knowledge management performed by agents and for agents as decision makers in logistic processes. Nevertheless, humans remain an important factor because they need the capability to monitor the logistic processes and the agents therein.

As a prerequisite to apply our framework we are tacitly assuming the existence of standard information technologies to provide the proper support such as networking, document storage, retrieval, metadata annotation, etc. Despite potentially existing connections by corporate affiliation, we do not presuppose initial structures in the knowledge management network. In contrary, as argued above, we emphasize the necessity of dynamic situation- and location-dependent interactions. In a sense, the structure of the knowledge management system emerges from the interaction of agents by virtue of implementing specific roles autonomously and in dynamic change.

Knowledge management as it is proposed in this framework is one key enabling factor to the envisioned autonomy of logistic processes. Autonomous entities need to make decisions based on a technically implemented decision theoretic process. In order to achieve this they not only need knowledge about their environment, but also have to assess possible future states of this environment and judge alternative options. In (Lorenz et al. 2005) we propose a mechanism for assessing the risk associated with an option based on knowledge the agent has about its current environment. This risk management is very closely related to knowledge management (cf. (Bemeleit et al. 2006) in this volume). On the one hand it can trigger the acquisition of additional knowledge. On the other hand it may be necessary to evaluate the risk linked with a KM decision, e.g., giving away certain information or asking an expensive but reliable source instead of a free but inaccurate one.

It is important to note that distributed knowledge management is restricted by various sociological and technological boundaries. For example, on a sociological level, agents may represent competing enterprises, which may lead to inconsistent or even incompatible desires. In addition, there is the important issue of trust. Low trust levels could prevent agents to assume certain roles (e.g., that of a knowledge broker or provider). High trust levels strengthen the connections between certain agents, causing an increase in traffic over time. As far as technological boundaries are concerned, the presence of embedded computational entities, which are partially moving in the physical world, leads to hard restrictions on network availability and computational power.

Agent roles

The agent-oriented approach, which advocates decomposing problems in terms of autonomous agents that can engage in flexible, high-level interactions (Jennings 2000), employs a multitude of agents to solve the knowledge management problem. In our approach to distributed knowledge management the agents have a special primary task, e.g., self-organization of a logistic entity. Managing and sharing knowledge becomes an optional secondary capability orthogonal to their primary logistic task. Thus, in contrast to previous approaches to agent-based knowledge management (van Elst et al. 2004a), there is no one-to-one correspondence between agents and knowledge management functions, such as providing knowledge or brokering knowledge.

In order to cope with this system characteristic we map knowledge management functions onto agent roles. Herrmann et al. (2004) report on a number of case studies which show that in sociologically inspired systems (in that case a collaborative learning environment) users "attempted to take different roles and tried to change their roles dynamically in being able to structure their communication." They give an overview on the application of sociological role concepts in computer supported collaboration and state a need for role development in computer-supported knowledge management. In a sociologically inspired computer system, e.g., a MAS, it seems therefore straightforward to apply the role metaphor from computer-supported KM for humans to KM for agents. This is especially true as human agents are explicitly included in the overall concept.

Within our framework a knowledge management role includes certain reasoning capabilities, a visibility function on an agent's beliefs, a deliberation pattern (i.e., a plan how to accomplish the KM task), and a communication behavior with interacting roles. The aim of KM roles is to provide a formal description of knowledge management tasks that eases the development of agents and reduces the computational complexity by means of a minimum set of processed knowledge and applied reasoning capabilities. One agent can assume different roles and may switch them over time. The minimum role model includes the roles of a provider offering information and a consumer being in need of information. The next extension would be a broker mediating between the two (van Elst et al. 2004b). Taking the agent-based approach, our claim to fully automate knowledge management raises new reasoning demands especially on the brokering and maintenance of knowledge, which have not been addressed so far. For example, in classical KM approaches, knowledge brokering and maintenance are performed by human actors (cf. (Maurer 2003)). We propose an extended role model that incorporates all knowledge management functions we identified as needed for autonomous logistic processes.

We distinguish internal and external roles. The latter ones are interactive and presuppose at least two involved agents, the former ones do not require inter-agent communication, but refer to intra-agent processes. Both types of roles are independent from the primary logistic task of an agent and define a complex behavior which results in a modification of an agent's knowledge base.

Figure 3.5 depicts a conceptual overview of the most important external roles in our framework together with the corresponding communication acts. Figure 3.6 shows the internal roles' operations with respect to an agent's knowledge base. We briefly describe the resulting role set and the respective tasks in our proposed framework:

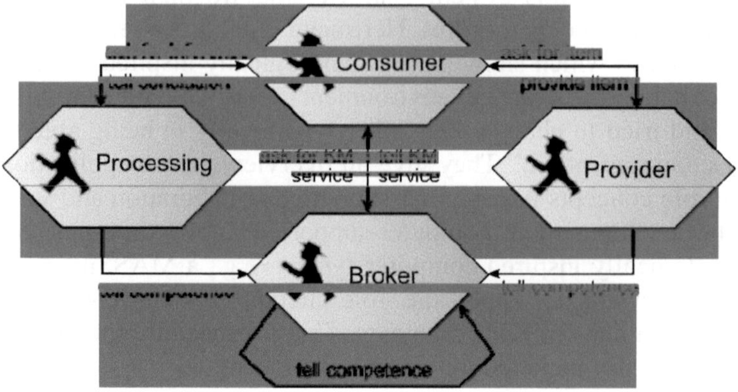

Fig. 3.5 External roles

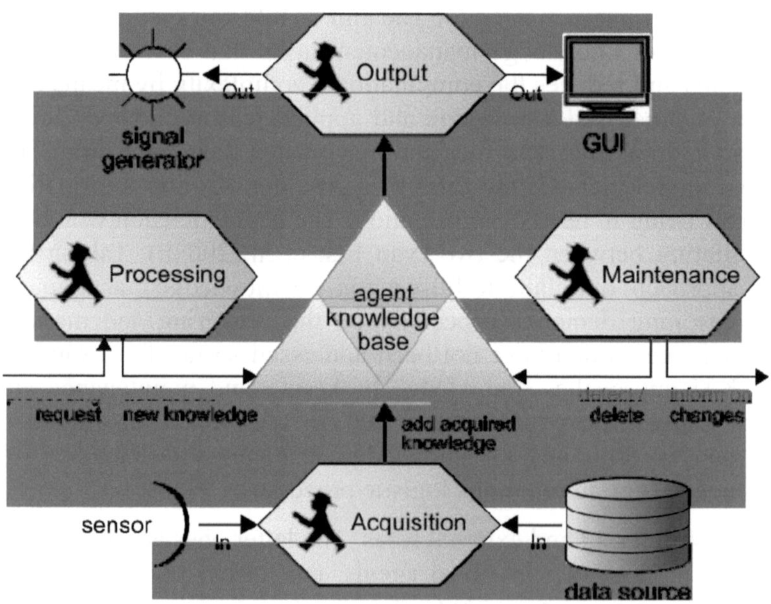

Fig. 3.6. Internal roles

Knowledge consumer: An agent acts as a knowledge consumer the moment it discovers a lack of its own local knowledge. Which knowledge it considers to be most important depends both on the agent's current knowledge base and its environment, and will be explained in more detail in Sect. 0. In order to determine the most appropriate knowledge source the agent uses its meta-knowledge on its inferential abilities, own sensors, available data sources, and provider agents. If the agent decides to ask another agent and its meta-knowledge on adequate services is considered insufficient the agent may consult a broker (see below) as for who would be able to provide the needed service. Successful direct transfers with providers will strengthen the relationship to them and decrease the necessity to use brokers.

Knowledge provider: An agent assuming the knowledge provider role in a knowledge transfer process provides parts of its internal knowledge repository either on demand or as part of some pro-active behavior. To be able to provide knowledge pro-actively this role has to implement a publisher/subscriber mechanism. An agent that aims at providing knowledge (including trade with costs) will tell other agents, particularly esteemed brokers, about the kind of knowledge it is willing to offer. When asked for knowledge the provider weighs up whether or not to consent to the transfer depending on the importance and potential confidentiality of the requested knowledge and the social (e.g., organizational) relationship to the asking knowledge consumer.

Knowledge broker: The knowledge broker acts as a yellow pages service within the system. It collects meta-knowledge on KM services (e.g., providers, other brokers, and processing services) and points a knowledge-seeking consumer to the right service. The broker also maintains a reputation list. Therefore it can rule out answers from unreliable partners upon request (Quality of Service enforcement). A broker may also act as a coordinator for adequate knowledge distribution within a legal organization or any other group of cooperating agents.

Knowledge processing: This role provides services that generate or reveal new knowledge based on knowledge already available. This comprises semantic mediation and integration, learning, and inference which may be regarded as sub-roles, respectively. Inference is the KM function that reveals knowledge as a conclusion by logical deduction. Learning analyzes the knowledge base for generalization rules that may, e.g., allow deletion of inferable knowledge or prediction of recurring situations. The mediation function translates and possibly integrates knowledge from different sources and ontologies. In general, this service will be used within an agent by request of the consumer role or other sub-systems. In some cases it may also be offered as a service to other agents. Knowledge processing is the most complex role and demands for sophisticated reasoning and learning capabilities. Thus, only some agents will implement this role entirely.

Knowledge acquisition: This internal role is intended to provide an interface to external data sources including sensors. Therefore it needs the capability to query a specific source and build up an internal representation of it. Changes in the source might trigger the generation of new knowledge items.

Knowledge maintenance: This internal role incorporates tasks needed to keep the knowledge base manageable and to monitor changes. If required the role informs an agent's sub-systems of relevant changes which may trigger an update of situation assessment and planning.

Knowledge output: An internal role providing an interface to the external environment through signal generators and user interfaces. The communication for this role is unidirectional toward the external interface. Possible responses from the environment are handled by knowledge acquisition which can of course be implemented within one agent.

It is important to reiterate that one instantiated agent can incorporate more than one role (e.g., an agent representing a truck can first act as a knowledge provider and later as a knowledge consumer). Hence, the incorporation of roles is a decomposition of the KM problem, which is in essence orthogonal to the mapping of organizational entities to agents. Furthermore, since different roles of agents need different reasoning capabilities, the encapsulation of roles can reduce the complexity of tasks which have to be performed by an agent at any given time.

Consumer provider interaction

A minimal role interaction model requires one agent A in the knowledge consumer role and another agent *B* in the knowledge provider role. In this section, we will discuss in more detail under which circumstances an agent assumes the role of a knowledge consumer and provider, respectively. We will describe how different parameter settings determine the decisions and actions associated with these roles.

The parameters are involved in the knowledge transfer process between consumer and provider. Adequate providers are identified by the consumer agent itself or by asking broker agents. Agent interactions in the transfer process may be modeled as an (iterated) contract net protocol with the consumer as initiator and at least one participating provider.

Consumer parameters In general, the role of a knowledge consumer presupposes a situation which meets the conditions listed below:

- The agent A intends to obtain a knowledge item k specified by a knowledge item description d. A knowledge item can be thought of as the truth value of a statement or the value of a variable. Its description d can, in principle, be provided by a query, possibly in combination with additional constraints, e.g., the definition of a minimum precision for the intended response and a response deadline. These additional constraints make a knowledge item description different from a query or a sentence form which subsumes the intended knowledge item.
- The knowledge item described by d is not already part of A's knowledge base.
- The knowledge item described by d cannot be inferred from A's knowledge base, given A's inferential capabilities.
- A believes that the knowledge item described by d is, in principle, available now or later.

These prerequisites closely resemble the Gricean maxims on rational cooperative discourse (Grice 1975) in many respects. They have to be combined with additional criteria, e.g., the (estimated) cost of obtaining k, in order to cover situations where knowledge is a tradable good and the knowledge consumer has a limited budget for acquiring knowledge from other agents. In its interaction with other agents an agent in the knowledge consumer role has to make many decisions, including the following:

- It has to assign a rank or weight to k in comparison to other knowledge items. This weight depends on how important k is for achieving its current goals, and if there are alternative knowledge items which might serve the same or very similar purposes.
- The knowledge consumer has to choose among different knowledge sources, e.g., its sensors and knowledge provided by other agents. This decision can make use of the agent's own experiences from earlier knowledge transfers, or it can be made solely on the basis of a general trust/reputation mechanism.
- The agent has to decide upon the maximum acceptable price being paid for k and the required response time.
- After each finished knowledge transfer, A has to assess its quality, e.g., if the actually delivered knowledge item deviated significantly from what the agent expected in advance.
- Finally, the agent has to decide upon the next steps, e.g., if k implies that other knowledge transfers are necessary.

These decisions of an agent assuming the knowledge consumer role are governed by the parameters importance, confidence, cost, availability, compliance, and value which are now discussed in more detail.

The importance parameter gives the (subjective) importance of a knowledge item k. The range of this parameter is between 0 (i.e., completely irrelevant) and 1 (maximum importance). As most other parameters, too, importance depends on the agent in question and time. Hence, we write Imp(A,k,t) for the importance an agent A assigns to a knowledge item k at time t. The importance parameter thus reflects an agent's point of view at a particular time which may differ significantly from the 'true' importance. The process of determining the importance of a knowledge item can be based on the agent's planning or risk management component (cf. (Lorenz e al. 2005) for details).

Conf(A,B,d,t) describes the confidence of the knowledge-consuming agent A at time t that knowledge-providing agent B will answer the knowledge request d correctly. The parameter value ranges from -1 to 1. -1 means A feels certain that B is lying or just has incorrect beliefs, whereas a confidence of 1 corresponds to absolute confidence in B's answer. 0 stands for neutral confidence, i.e., agent A has no clue whether B's answer will be rather right or wrong. The parameter determines provider selection.

The cost parameter determines the maximum cost an agent is able and willing to accept for obtaining a knowledge item. Since cost, again, depends on an agent and time, we write Cost(A,k,t). This parameter includes costs arising in the communication process and possible costs to obtain k as payment to the knowledge provider or knowledge brokers. The maximum accepted costs are closely related to k's importance and the agent's budget. In general, the accepted costs do not correspond to the price actually communicated to the provider.

A successful knowledge transfer presupposes that the knowledge item intended by the knowledge consumer agent is available, in principle. Hence, it is required that an agent assumes that there is a non-zero probability to obtain the intended item. This probability is given by the availability parameter. A zero availability, Avail(A,k,t)=0, means that the agent does not believe that there is any chance to obtain k at time t and, hence, will not make any further attempts into that direction.

Availability of knowledge is based on prior experiences and background knowledge. It is used, for example, in deciding which knowledge items should be acquired (in case there are choices).

Compl(k',d) denotes the degree of compliance of the obtained knowledge item k' with respect to the intended knowledge specified by d. The value ranges from 0 to 1. If the value is 0 the whole knowledge transfer has to be reiterated with another provider. The value 1 means that k' perfectly matches d. Item k' may differ in terms of spatial and temporal validity or precision of measurement. The consumer needs to evaluate kind and scale of a potential deviation in order to plan and execute appropriate actions to finally get the knowledge needed.

After a completed knowledge transfer, an agent determines how successful the transfer has been, i.e., its (net) value. This parameter depends on the initial importance of (the higher the initial importance, the higher its value), the compliance of the actually obtained knowledge item with the intended one (divergence decreases the value), and cost (the higher the cost the lower the value). The value of a knowledge transfer will affect the future behavior of an agent. Successful knowledge transactions with a particular provider agent, for example, will strengthen the connection between the involved agents and increase the likelihood of future transactions between them.

Provider parameters Similar to the knowledge consumer role, the role of a knowledge provider incorporates multiple decisions during the transfer process which are characterized by a set of decision parameters to achieve a rational behavior. These parameters may be engaged in decisions that determine whether the provider is willing to transfer knowledge at all, or which transfer conditions the provider will propose and accept.

The proposed distributed knowledge management framework also considers agents being in competition or just belonging to different organizations. Thus, a provider agent needs to deliberate whether requested information may be propagated to some other agent or not. This confidentiality parameter is either defined by an explicit, predefined classification of the requested type or item of knowledge or may be determined by an intelligent estimation of the possible impact this knowledge may have on the providing agent and its organization if once revealed. Confidentiality is always specified with respect to an (agent) group of interest.

If the requested knowledge item is classified the provider agent will refuse to perform the answer action, except the asking agent has a sufficient security clearance which is organization-dependent. If the requested knowledge is considered basically confidential to some minor extent the provider's willingness to answer a given query strongly depends on the asking agent. This decision is influenced by the trust parameter. This parameter ranges from -1 to 1 and describes whether the consumer agent is supposed to comply with a non-disclosure agreement for the requested knowledge item. A trust value of 0 is neutral, i.e., the provider does not know anything about the consumer's trustworthiness.

Irrespective of confidentiality, trust, and transfer cost, an agent may have the disposition to agree to a knowledge transfer due to a social relation to the consumer agent. In this case, a transfer is motivated by a common organizational background, a current or past cooperation, or the aim to initiate a new (long-term) cooperation and to increase mutual trust (cf. (Alam et al. 2005)). The affinity parameter describes this disposition w.r.t. a specific agent and time. The parameter ranges from 0 (minimum affinity) to 1 (maximum). A high affinity decreases the minimum accepted price. If affinity is 1 the price is 0, i.e., the knowledge is provided as a gift.

In order to meet the consumer agent's requirements on precision and certainty the provider has to compute Comp(k',d). Due to different knowledge and/or inferential capabilities the value computed by the provider is not necessarily identical to the compliance value computed by the consumer agent after the completed knowledge transfer.

Whether the provider agrees to a knowledge transfer and under which conditions (minimum accepted price, response time) also depends on the expected expenses arising due to the transfer. This is represented by the provider's cost parameter. It may include a temporal and financial dimension consisting of, e.g., communication costs and reasoning costs.

If the provider is in general willing to answer an agent's query it determines the minimum accepted price for this service. The price is determined by the common value (if any), the provider's private value, the expected transfer cost, and the affinity to the consumer agent. The minimum accepted price, of course, may and in general will differ from the communicated price.

Interaction Protocol The knowledge transfer process between knowledge consumer and provider incorporates an informative act. Unfortunately, the FIPA Query Interaction Protocol is not sufficient to model all aspects of the depicted transfer process. As described above, the consumer request is more than just a query but a more complex description of the needed knowledge with constraints on that knowledge (e.g., precision) and the corresponding transfer process (e.g., response deadline and payment). Furthermore, the process may include a negotiation. Thus, we think of the actual knowledge transfer as a special action. In a consumer-initiated knowledge transfer process, the action with the preceding consumer's call for proposals, the provider's proposals, and possible negotiations are part of an (iterated) contract net protocol. The corresponding FIPA Interaction Protocol was adapted or interpreted for this special purpose (see figure 3.7). A provider-initiated process would follow the FIPA Propose Interaction Protocol.

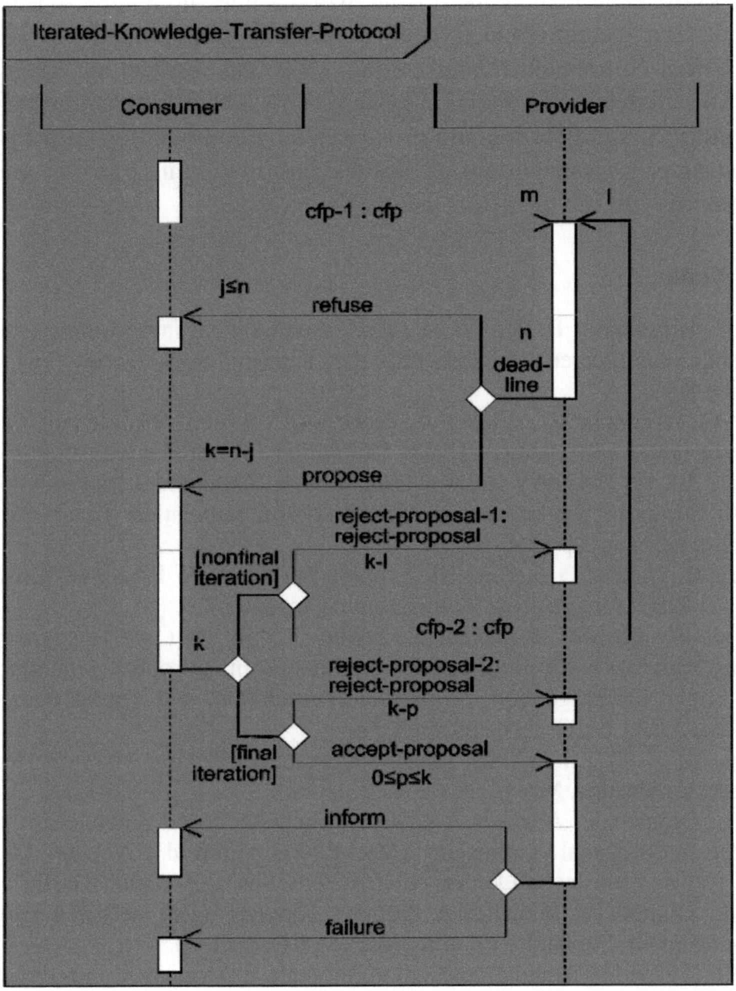

Fig. 3.7 Knowledge Transfer Interaction Protocol

3.5.5 Conclusions

We presented a framework for distributed knowledge management for modeling complex and dynamic scenarios from the logistics domain on the basis of multiagent systems. We introduced a set of knowledge management roles, decision parameters for them, and an interaction protocol for the two most important roles of our framework, the knowledge consumer and the knowledge provider.

This knowledge management framework is part of an agent-based logistic simulation system (Becker et al. 2006) and forms, together with a risk management component (Lorenz et al. 2005; Bemeleit et al. 2006), a dynamic knowledge-based decision system. This KM infrastructure offers the opportunity to simulate logistic processes as a combination of primary logistic tasks (e.g., transportation), knowledge processing/transfer, and decision-making.

References

Alam SJ, Hillebrandt F, Schillo M (2005) Sociological implications of gift exchange in multiagent systems. Journal of Artificial Societies and Social Simulation 8(3)

Becker M, Wenning BL, Goerg C, Gehrke JD, Lorenz M, Herzog O (2006) Agent-based and Discrete Event Simulation of Autonomous Logistic Processes. In: W. Borutzky and A. Orsoni and R. Zobel (eds) Proceedings of the 20th European Conference on Modelling and Simulation, Bonn, Sankt Augustin, Germany, pp 566-571

Bemeleit B, Herzog O, Lorenz M, Schumacher J (2006) Proactive Knowledge-Based Risk Management. In: this volume, Chapter 3.6

Bos A, de Weerdt MM, Witteveen C, Tonino J, Valk JM (1999) A dynamic systems framework for multi-agent experiments. In: European Summer School on Logic, Language, and Information. Foundations and Applications of Collective Agent Based Systems workshop

Bratman ME (1987) Intentions, Plans, and Practical Reason. Harvard University Press, Cambridge, MA

Elst v L, Dignum V, Abecker A (2004a) Agent-Mediated Knowledge Management: International Symposium AMKM 2003, Stanford, CA, USA, March 24-26, Revised and Invited Papers, LNCS 2926. Springer-Verlag, Berlin

Elst v L, Dignum V, Abecker A (2004b) Towards agent-mediated knowledge management. In: van Elst L et al. (eds) (2004), pp 1-30

Grice HP (1975) Logic and conversation. In: Cole P, Morgan JL (eds) Syntax and Semantics, vol 3, Speech Acts. Academic Press, New York, pp 41-58

Herrmann T, Jahnke I, Loser KU (2004) The role concept as a basis for designing community systems. In: Darses F, Dieng R, Simone C, Zacklad M (eds) COOP. IOS Press, pp 163-178

Hofmann O, Deschner D, Reinheimer S, Bodendorf F (1999) Agent-supported information retrieval in the logistics chain. In: HICSS '99: Proceedings of the Thirty-second Annual Hawaii International Conference on System Sciences, vol 8, IEEE Computer Society, p 8028

Hult GTM, Cavusgiland ST, Calantone RJ (2003) Knowledge as a strategic resource in logistics and purchasing. Tech. Rep. Working Paper Series, No. 03-001, Marketing Science Institute (MSI), Cambridge, Massachusetts

d'Inverno M, Luck M, Georgeff MP, Kinny D, Wooldridge M (2004) The dMARS architecture: A specification of the distributed multi-agent reasoning system. Autonomous Agents and Multi-Agent Systems 9(1-2):5-53

Jennings NR (2000) On agent-based software engineering. Artificial Intelligence 177(2):277-296

Lorenz M, Gehrke JD, Hammer J, Langer H, Timm IJ (2005) Knowledge management to support situation-aware risk management in autonomous, self-managing agents. In: Czap H, Unland R, Branki C, Tianfield H (eds) Self-Organization and Autonomic Informatics (I), Frontiers in Artificial Intelligence and Applications, vol 135. IOS Press, pp 114-128

Maurer M (2003) Important aspects of knowledge management. In: Klein R, Six HW, Wegner L (eds) Computer Science in Perspective: Essays Dedicated to Thomas Ottmann, LNCS 2598. Springer-Verlag, Berlin, pp 245-254

Rao AS, Georgeff MP (1991) Modeling rational agents within a BDI-architecture. Tech. Rep. 14, Australian AI Institute, Carlton, Australia

Scholz T, Timm IJ, Woelk PO (2004) Emerging capabilities in intelligent agents for flexible production control. In: Katalinic B (ed) Proceedings of the International Workshop on Emergent Synthesis (IWES 2004)

Smirnov A, Pashkin M, Chilov N, Levashova T (2003) Multi-agent knowledge logistics system "KSNet": Implementation and case study for coalition operations. In: Mařík V et al. (eds) CEEMAS 2003, LNAI 2691, pp 292-302

Timm IJ (2004) Dynamisches Konfliktmanagement als Verhaltenssteuerung Intelligenter Agenten. Dissertationen in Künstlicher Intelligenz, vol 283, Akademische Verlagsgesellschaft, Berlin

Wooldridge M, Lomuscio A (2001) A computationally grounded logic of visibility, perception and knowledge. Logic Journal of the IGPL 9(2):257-272

Zice S, Zhengping L, Runtao Q, Mansoor S (2001) Agent-based logistics coordination and collaboration. Tech. Rep. SIMTech (AT/01/011/LCI), Singapore Institute of Manufacturing Technology

3.6 Proactive Knowledge-Based Risk Management

Martin Lorenz[2], Boris Bemeleit[1], Otthein Herzog[2], and Jens Schumacher[1]

[1] Bremer Institut für Betriebstechnik und angewandte Arbeitswissenschaft(BIBA), Bremen, Germany

[2] TZI - Center for Computing Technologies, University of Bremen, Germany

3.6.1 Introduction

Globally distributed production networks accompanied by the reduction of the vertical range of manufacturing, customer-driven markets, decreasing product life-cycle times and increasing information flows alter the requirements for the management of logistic systems and processes. The reduction of the size of goods that have to be transported and as a consequence thereof an increasing amount of transports are main reasons for a relative shortage of logistic infrastructure and lead also to rising utilization of existing logistic processes and to more complex logistic systems. These developments for example are caused through the evolution of virtual organizations and the increasing maturity of new information and communication technologies (ICT) technologies like RFID and ubiquitous computing.

To coordinate all these processes, an increasing demand of required information for just in time deliverables is needed. These requirements exceed the abilities of existing standard logistic processes. Dynamic development of modern ICT (e.g. telematics, mobile data transfer, and transponder technology) open new possibilities for the development and emergence of intelligent logistic systems which can fulfill the requirements of rising utilization and relative shortened logistic infrastructure. An approach to face the challenges on existing and upcoming problems in logistics is the concept of autonomous logistic processes represented by autonomous logistic objects.

The autonomous control of logistic processes can be realized through decentralized control systems, which select alternatives autonomously or logic based semi-autonomously and decide within a given framework of goals. Coming along with the autonomy of the logistic objects is a shift from the responsibility for the realization of the decisions from a central deciding system to the single logistic object. This has to be regarded by developing a concept for the management of autonomous logistic objects and the complexity of the total system which is an after-effect of the high number of logistic objects which are needed in such a system.

The complexity of logistic systems depends on the amount of the embedded logistic objects. The amount and the character of the relations within logistic systems affect also the complexity of the logistic system. The third factor, which is an important influencing factor for logistic systems, is the dynamic of the system. This dynamic is displayed by the number of system states and changes in the amount of system elements. However, the complexity of a logistic system allows still no conclusions regarding the sensitivity of the system in relation to the malfunction of individual objects or relations between them. The integration of strategic planning may enable the system to compensate a temporary or unlimited mal-function of an object or a system relevant relation between two or more objects. The increased use of modern ICT doesn't necessarily assure the constant availability and high quality of data and information to plan and control the logistic processes. A malfunction or a loss of information- and communication systems can lead to substantial negative consequences.

Risk in autonomous logistic processes

The increased complexity of logistic systems is followed by a more complicated planning and control of logistic systems and of the related processes in combination with an increased sensitivity of the total system to disturbances and malfunctions. The hazard of delayed delivery in transportation, latency in manufacturing and reduced adherence to delivery dates are results of complex system structures and increased customer requirements. All these numerated disturbances and changed conditions clarify that logistic systems and the related logistic processes are very fragile and the contained hazards and chances have to be managed to ensure the success of the logistic processes. These circumstances show that the development of a management system for risks is essential for a successful realization of autonomous logistic objects. Direct disturbances of the processes caused by risks which exist impartial from the logistic objects and risks which result from the interaction of the logistic processes. Traditional literature on risk management (RM) knows six strategies to handle risk: (1) acceptance, (2) avoidance, (3) reduction, (4) transfer, (5) compensation, and (6) diffusion (e.g. Finke 2005). Not all of them are applicable for an autonomous system. The possibilities of avoiding, reducing and partly compensating risks by a proactive risk management system are to identify and analyze risk which could be dangerous for the fulfillment of goals given to the autonomous logistic object in advance. Such a risk management system will be developed in the sub project "Risk Management" of the Collaborative Research Centre 637.

The consequence from the shift of responsibility from a central instance to an autonomous logistic object is a different situation of risk which could endanger the success by reaching of the goals of the logistic process. In classic logistic systems a malfunction of the centralized, deciding instance is a danger for the success of all logistic processes. Other problems are, that central systems are suitable to only a limited extent in reacting on changing local conditions and that a local lack of information affects the total sys-tem. By contrast to a central deciding instance there are other risks to be considered in logistic system which is based on autonomous logistic objects. For an autonomous logistic object it has to be kept in mind that there are additional risks which result from the required communication between the involved objects and that the interaction between them which leads to non calculable states on local and global level. It is also important to consider that contradictory information generated from different objects is another source of risk for the logistic processes in relation to their specific goals and that an optimization object level can compromise the goal of the total system. These flexible characteristics of disturbances can be categorized in 3 types of risk:

- External risk, which is caused by an event, that exists independently from the autonomous processes and may affect them.
- Internal risk, which is a result of the interaction between autonomous processes, the reasoning within an autonomous process and
- Information risk, which is related to the information which are available but may be inconsistent, contradictory fuzzy, incomplete or unreliable.

An overview about the different characteristics of risk which could influence the logistic objects is given in the figure 3.8:

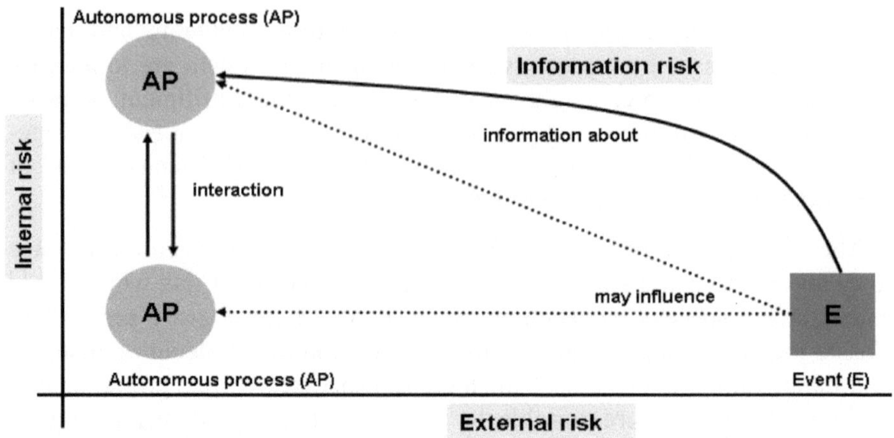

Fig. 3.8 Risks induced by events.

Managing the types of risk mentioned and shown above is essential to understand the meaning of risk for autonomous logistic objects and their environment. To handle existing and new risks for autonomous processes and autonomous objects a proactive risk management has to be established as a part of the whole system, because it helps to develop logistic processes which are robust and insusceptible to existing and occurring risks: A risk management system supports the autonomous objects in decision making and realizing these decisions considering the risk which is related to the whole logistic processes. For this reason the development of a proactive risk management system can be considered as a relevant success factor for autonomous logistic processes.

An additional advantage caused by the use of a proactive risk management in comparison to a traditional reacting risk management system is the gain of auxiliary scopes. Avoiding needless hazards and getting the chance of using these scopes can be made available by:

- Evaluation of all available information;
- Examination of new Information regarding the validity and relevance for the processes;
- Interpretation of new Information in relation to given risk factors
- Analysis of risk factors;
- Evaluation of the overall risk for the whole process.

As shown above the existence of a proactive risk management system leads to more opportunities and more calculable risk for the autonomous processes. The question about the gained opportunities and the more calculable risk to answer is: How to manage the risk? A declaration of defining an effective risk management was made by Kenney:

"The principal element involved in managing risks can be boiled down to a single sentence: Good process risk management results in perfect containment and safe handling of the hazard." (Kenney 1993)

This single sentence has to be enhanced for autonomous cooperating logistic processes: ...in perception of existing and future options for the autonomous objects. Kenney exemplifies the fundamentals and principles for a functioning risk management system in three predications (Kenney 1993):

- The hazards of a process must be capable of being defined at any time.
- The risks resulting from these hazards must be controllable by equipment, by procedures, or by some combination thereof.
- Management must uncompromisingly maintain control over the equipment and procedures that are identified to control the risks. (Kenney 1993).

A possible approach about how these tasks can be realized in a risk management system is to develop a suitable comprehension of risk for autonomous logistic processes and objects, examine existing risk concepts and determine the requirements for their realization.

Definition of risk for autonomous logistic processes

To develop a suitable comprehension of risk, existing definitions of risk and approaches of risk management have been analyzed. As part of this analysis many definitions of risk, hazard and uncertainty have been examined. The first step was the differentiation of the terms risk, chance, uncertainty and hazard because in some cases risk and uncertainty are used in the same context and the terms chance and hazard are not well differentiated.

In established literature nescience of the future and future developments are called uncertainty in wide sense. If an impartial or pseudo-impartial (subjective) occurrence probability can be allocated to a future event or development of events it is called risk. If it cannot be allocated to a future incidence, it is called uncertainty. This differentiation was developed by Frank Knight and is deemed to be the economical standard approach (Schwarz 1996). This differentiation is also used by Motsch. Motsch describes that decisions fraught with risk exist if the deciding instance has clear knowledge about the occurrence probability. If this instance knows only the amount of possible and relevant future conditions but can not give full particulars about the occurrence probability the decisions are made under uncertainty (Motsch 1995).

The possibility to assess risk during the planning phase and the accomplishment of logistic processes is a necessary feature for a proactive risk management which shall be able to modify the original plan which was developed after a process oriented risk assessment if necessary. All of these differences between the diverse risk definitions and concepts of risk lead to the next part of this chapter which contains the requirements for a risk term to be developed and the development of the risk term itself.

For the development of an adequate definition of risk in autonomous logistic processes additional requirements have to be considered. The requirements for a suitable risk term for an application in a pro active risk management system are:

- The total risk and the individual risks are connected to the system "autonomous logistic processes". This is important, because the risk assessment will be done by an autonomous object, which is part of this system and not an external element which does not influence the system by its decisions.
- The risk term includes upside risk and downside risk. The consideration of both characteristics of risk is necessary because there is also the possibility to outperform the given goals of a specific process.
- Risk is connected to the goals and /or aims of the system (and the process as part of the system). This fact is important for an automated evaluation and assessment of risk accomplished by single agent.
- Risk has to be regarded in connection with endogenous and exogenous influences or malfunctions. To consider internal and external developments is important because the system "autonomous logistic objects" is not a closed or self-contained.

From these requirements and the examined definitions and approaches of risk the following definition for the CRC 637 was developed:

"Risk is the contingency that the result does not correspond to the goals of the system due to differences."

This definition includes uncertainty about the future and future events by using the terms risk and contingency. Upside and downside risk represented by chance and hazard are contained by using the term "differences" which allows a positive or negative deviation in relation to given goals and does also apply on internal and external risk. This deviation can be of technical and economic origin. It is applicable for the use in autonomous logistic objects because of its simplicity and reduction on terms which may be used in a dynamic system on their own.

The definition of risk is the basis for the development of a pro-active risk management system for robust logistic processes. To develop this risk management system the research on methodical concepts on risk analysis in the context of autonomous objects in a complex and dynamic system is the next step in realization.

To realize the development of a proactive risk management system it is also essential to implement a suitable mechanism of risk identification and risk analysis into the logistic objects. For this reason existing methods of risk management have to be evaluated considering some requirements which are essential for the implementation into an autonomous logistic object. The next step in developing a pro active Risk Management is to choose a risk concept which contains a methodological approach which can be integrated into an autonomous logistic object.

3.6.2 Risk management for autonomous decision-making

The main difference between engineering oriented and other approaches is the declaration of the meaning of the term risk and the understanding in relation to the possible impact(s). Most engineers consider risk as a negative term, where only a possibility of loss or a negative development is included. Two examples for engineering oriented approaches are:

- Risk is the hazard of the negative deviation between plan and reality (Hess and Werk 1995).
- Operational Risk is the risk of loss resulting from inadequate or failed internal processes, people and systems or from external events (Nash 2003).

These are the so called asymmetric approaches of risk because the appearance of risk is only expected in a way with consequences which characteristics show only in one direction (positive or negative development). Most of these approaches are used in different forms of safety analysis like FMEA (Failure Method Effect Analysis, developed in the 1960's) or for example HAZOP (HAZard and OPerability Studies, developed in the early 1970's and extended to software development in the 1990's). These kinds of risk oriented safety analysis were originally developed to reduce only the error probability in engineering or chemical research and development. An exception in relation to the other approaches which are mostly focused on engineering tasks is the approach of Haindl. Haindl exemplifies that risk (especially delivery risk) is the hazard of loss caused by external disturbances within the field of the supplier as well as in communication between supplier and customer (Haindl 1995). If a definition of risk comprises additional positive possibilities it can be allocated to the symmetric approaches of risk. The differentiation between symmetric and asymmetric can be found in (Pfohl 2002).

Financial and entrepreneurial approaches as well as approaches on project management used in the majority of cases are symmetric approaches and differentiate between downside risk (negative development) and upside risk (positive development). Downside risk is also called hazard while upside risk is referred to as chance. An overview on these differentiations can also be looked up in (Pfohl 2002). The mathematical approach on risk (Risk = probability * impact) can also be treated as a symmetric approach because the impact can be positive or negative.

Another differentiation of risk concepts in relation to the definition of risk is the differentiation between action risk, which may result from a wrong decision and precondition risk which results from changing conditions of the relevant environment. A determination of these two risk differentiations was made by Haller and can be found in Mikus "Risikomanagement" (Mikus 2001). The insufficiency and the problems by using action risk oriented concepts or definitions will be discussed in the paragraph "Risk as a possibility of a wrong decision" (Härterich 1987).

To integrate a suitable risk term for autonomous logistic processes it is also important to analyze existing concepts of risk and risk management for that the interdependencies between definition of risk and a risk concept can be considered by developing a CRC specific risk term. This consideration is necessary because both are bearing columns of a proactive risk management system and affect each other.

Haerterich (Härterich 1987) divides risk in three main areas:

- Risk as goal deviation;
- Risk as a possibility of a wrong decision;
- Risk as a deficit of information;
- Risk as a combination of deficit of information and goal deviation.

These concepts have a different orientation and understanding of risk and risk management. They will be shortly introduced and analyzed on their advantages and disadvantages. The first approach is "Risk as goal deviation". Risk comprises the possibility and not the realized goal deviation. This concept has a high fit with respect to complex system structures with different impacts and probabilities. The goal deviation is a neutral factor which contains hazard and chance. Part of the goal deviation approach is "Risk as a possibility of a wrong decision". This concept also includes a correlation to given goals, because a decision can not be assessed as wrong without goal analysis. It is difficult to measure decision oriented risk, because the risk assessment can be conducted after analyzing what really happened and how other decisions would have influenced the result under the existing conditions. This relation between deciding in a situation fraught with risk and examining this decision afterwards is also a problem by using the action risk oriented concept following Haller. The next approach is the "Risk as a deficit of information" concept. Risk is here characterized as a lack of information in situations where a decision has to be made. The disadvantage of this concept is the limitation to situations where decisions have to be made. Risk always exists and it is not limited to selected situations. The last concept is "Risk as a combination of a deficit of information and goal deviation". This concept follows from the combination of the goal deviation approach and the information deficit approach. The risk is divided into two components:

- Description through objective and subjective probability distribution;
- A goal deviation for symmetric or asymmetric risk.

The approach of a risk concept in a logistic environment has to fulfill several requirements. The first requirement is the measurability of the risk and the contained risk factors. In the approach that considers risk as a goal deviation this problem can be solved by splitting the total risk. Chosen examples for risk are:

- Time risk (early, in time or delayed delivery or production);
- Cost risk (within monetary restrictions, overpriced);
- Quality risk (quality related to the input data and related to the object quality; this can also be enhanced by regarding sustainability of the accomplished process steps).

It is possible to measure the relevant risk factor for a sufficient risk assessment with this idea. The "Risk as a deficit of information" approach is not able to fulfill the requirement of measuring risk adequately, because risk is reduced to a probability distribution but the flexible characteristics (additional cost, delay in delivery, damaged object) remain unconsidered.

After consideration of these facts we have the highest fit for autonomous logistic objects by usage of the goal oriented approach or the approach where risk is defined as a combination of a deficit of information and goal deviation. Regarding these facts concerning risk concepts supports the definition of risk developed for the CRC 637 because it fulfils the requirements shown above and fits into the risk concepts chosen above. The subset "risk as a possibility of a wrong decision" of the goal oriented approach is not sufficient for a risk management approach which fulfils the requirements for future oriented logistics; because in this approach risk is limited to the decision points and can not occur during the realization of a decision. Another reason which constricts this concept for an application in a logistic environment is the fact that the real risk can only be assessed after a logistic process has finished and all states and decision that lead to an optimal result are known. Yet, another reason for the refusal of the subset "risk as a possibility of a wrong decision" is the difficulty in allocating unexpected events and certain decisions.

There are different possibilities to assess the risk in complex logistic systems and for autonomous logistic objects. One possibility is to analyze potential nonconformities and malfunctions in relation to their cause and the other possibility is to examine process relevant events in relation to their impact on the logistic system or on the logistic objects. This leads to a classification of methods into forward oriented methods which evaluate occurring events and backward oriented methods which analyze the causes for malfunctions. Another important element for developing a risk management system is the ability to manage nearly all risk afflicted situations without external help. This can only be realized if the method(s) used for the risk management do not need abilities which are used by human (supported) instances like associativity, because autonomous objects do not posses such abilities but shall be able to asses the risk in the logistic processes. For this reason it is obvious that the method of risk management integrated in a in a complex and dynamic logistic system has to consider the potential fuzziness of the information which are essential for the decisions. This can only be realized if the chosen approach of risk management is able to examine the consistency of the information and act in case of need without them if they are not fully available by using a methodological approach which also uses components of plausibility and decomposes complex problems into parts to assess the risk.

To fulfill the requirements for the development of a proactive risk management system in complex logistic systems or for autonomous logistic objects it is required that the method is forward oriented and can be well integrated into an ICT supported system architecture because the application of such environment has many advantages compared to a central, human controlled system and may be realized based on requirements shown above. It is also important that the method which will be used is able to assess risk as a permanent factor during the whole process and has the ability to regard:

- Uncertainty;
- Upside und downside risk;
- Internal and external risk.

How these abilities of the risk management can be realized and which requirements have to be regarded concerning the realization in an autonomous logistic object will be presented in the next part of this chapter.

3.6.3 Requirements for risk management for autonomous systems

As shown above goal fulfillment is the defining characteristics of a risk concept for autonomous logistic entities. In the logistics domain this goal might be to reach a given destination in shortest possible time or with lowest possible fuel consumption. But primary goal fulfillment is only one aspect of risk management within an autonomous system. The autonomous entities aim to maximize its local utility will usually subsume primary goal fulfillment but aspects like system continuance or contribution to a global utility of the enterprise the entity belongs to induce different risks.

Collectives of autonomous systems in the way they are modeled in our work (i.e. all logistic entities in a transportation network are regarded as a collective—in itself subdivided into enterprises, trucks, loads, etc.) have a close relationship to social systems. The autonomous entities are self contained and follow an individually rational goal. In the basic assumption they are individually rational decision-makers in the sense of game theory, each aiming at maximizing their individual utility function (Tumer 2004). Following (Weigand and Dignum 2004) intelligent entities in a collective must above all be seen as autonomous in that they can't be directly manipulated neither by a "governing authority" nor by other members of the collective. This autonomy of individual agents implies that the collectives' performance highly depends on the individual "willingness" of its members to contribute to the global goal.

In case of a pure technical system one could argue that it is the designer's responsibility to ensure the "willingness" of the autonomous entities. This can be achieved as long as we deal with closed systems. In open systems the benevolence of an entity cannot be a priori assumed. Therefore it is crucial for an autonomous entity in an open system to assume the autonomy and hence the possibility of malevolence of its counterpart be it artificial or human. The autonomous system therefore needs to acquire and maintain an internal model of its environment and the processes therein. Using a "foretelling" mechanism can than enable the assessment of situations that will be occurring. Such a mechanism has of course to be of technical nature and thus needs to calculate future states of the world based on probabilities. Most classical methods of risk management employ brainstorming and experts assumptions to assess the possibilities of events that can have an influence on a process (Seidel 2005) prior or during a structured process. In a technical autonomous system one can either employ these methods in advance (the "design time") or find a computer implementable method to assess risks.

The former is simply a matter of completeness of the design process. The disadvantage of design-time assessment is obviously that new situations in which risks occur cannot be handled by the autonomous system. In conventional control tasks a human operator will be responsible and able to intervene. In the autonomous decision-making case this task is delegated to the system itself. Therefore enabling autonomous risk assessment is the only remaining alternative.

Engineering risk aware autonomous processes

Engineering autonomous processes in logistics includes three perspectives: material, information, and management. The challenge for the implementation of autonomous decision behavior is to enable distributed systems, where the different levels gain the ability to interact autonomously and flexibly. For the design and implementation of autonomous entities as autonomous decision-makers this challenge includes high-level decision-behavior which may not be realized by simple reactive architectures. Therefore, we assume that intelligent entities with deliberative decision behavior and explicit knowledge representation and reasoning capabilities are needed to meet these requirements.

We believe this kind of autonomous, decentralized decision-making can help make the operational processes more efficient, cost-effective, and allow the participating enterprise to stay competitive. It is also a major improvement over traditional centralized approaches in which individual entities are ill-equipped to deal quickly with sudden events since control usually resides with the expeditor who is removed from the scene of the sudden event and thus has only delayed access to the relevant information. A decision within a computer implemented autonomous entity always is a decision among previously known alternatives. So the decision process will have to calculate and assign some kind of value to all known and accessible alternatives in order to choose for exactly one.

Enabling this type of autonomous decision-making is challenging given the potentially large number of entities that could be involved as well as the dynamic and sometimes even competitive environment in which the entities operate. In principle, enabling a technical system, to make decisions that are designed to impact real-world entities delegates the assessment of consequences of the decisions to the agent. Economical management interests therefore require the technical system to be dependable in terms of awareness of hazards, competitor malevolence, malfunctions, etc.

The special challenge in logistics arises from the different interests within the system. On the interaction level, entities should maximize their utility. Each entity is a representative of an enterprise and, therefore, its local decision behavior should improve the performance of the corresponding enterprise. However, on a global level, we hope to achieve a better performance of the overall logistics resp. the optimization of the global system. For practical applications, it still has to be proven, that optimization is realized at least on the enterprise level, as the enterprises have to invest into this innovative technology and transfer competence on the entities level. So dependability of the technical system is of utmost significance to the principal and implies that it behaves as ordered. Thus, the conclusions of straightforward emergence of macroscopic optimality from microscopic autonomy has to be questioned especially in this domain.

The engineering task therefore involves the provision of mechanisms for local autonomous decision capabilities as well as for dependability from a (human) principal's prospect. Regarding decision-making based on local knowledge as the core ability for an autonomous entity we have to focus on how it can be enabled to identify, assess and regard risks in its decision process.

Knowledge and uncertainty

To the same extent as the future is perceived as decision-dependent, any decision to be made by the technical system must be regarded as risky (Luhmann 2003). The goal of risk management (RM) is to attempt to optimize the entities decisions in the presence of incomplete, imprecise, or debatable information by reducing the uncertainty about future events.

Thus, context-based, situation-aware, and local decision-making, which in turn supports autonomous, self-managing behavior of logistic entities, calls for the integration of knowledge management functions with the entities planning and situation assessment.

Knowledge is and evolves locally in different entities and organizations. Only the ability to represent, organize and communicate knowledge enables the deliberative decision making of an autonomous entity as well as its collaboration with others and thus the emergence of distributed problem solving. It is obvious that knowledge is a core element of an approach to autonomous logistics, as it is constitutive for sophisticated decision-making within an autonomous entity (Langer et al. 2006).

As uncertainty is the major source of risk in decision-making the autonomous entity will need a mechanism to evaluate the knowledge it has regarding the expected state of the environment that might influence the current goal. To achieve this is a challenging task for a technical system. It involves not only to have knowledge but also to generate hypotheses about future states of the environment and to evaluate the amount of knowledge it has regarding this hypothesis.

A logistic entity's environment is inherently unpredictable. While the degree of uncertainty of well structured environments such as container stowage is relatively low, others especially open world logistics involving multi modal routing and road traveling are highly dynamic and in many ways unpredictable. In this many issues that arise in autonomous robotics are also applicable to autonomous logistics (e.g. Thrun et al. 2005).

Internal models of the environment are abstractions of the real world. As such they only partially model the underlying physical processes of the logistic entity and its environment. Furthermore the capability of acting of a logistic entity is limited depending on its kind. On the one hand a self-steering trolley on a shop floor has all actuators it needs to fulfill its task of getting its payload from one place to another. Uncertainty arises only from control noise or mechanical failure. A single parcel on the other hand has no physical actuators at all and will therefore be inherently unsure weather its intended action is going to be carried out.

What Thrun et al. state for robotics is also very true for autonomous logistics: "Managing uncertainty is possibly the most important step towards robust real-world robot systems." (Thrun et al. 2005)

Planning and predicting

Decisions are subject to changing conditions. The dynamics of the environment requires a number of short- and mid-term goal oriented decisions to be taken during every process. In order to fulfill a given goal an autonomous entity will have to use its knowledge of its environment to formulate a plan. Thus planning is a crucial capability for autonomous systems.

The complexity of a planning task increases with the amount of uncertainty in the environment. In a simple and static world the autonomous entity can formulate a complete model and thus calculate definite plans. With increasing complexity the model on which a plan can be based must be more abstract thereby introducing a source of risk namely incomplete knowledge.

Furthermore the dynamics of the environment interferes with the attempt to execute a plan. Thus the autonomous entity will have to possess the capability to observe processes occurring in the environment and extrapolate them into the future.

The planning capability therefore depends on the accurateness of the model not only of the world and its entities but also of the processes the entity can trigger, observe or endure.

Components of autonomous risk management

Thus for a proactive risk management within an autonomous logistic entity we need 5 technically implementable components. (1) An internal local model of the environment, which will contain static elements that are common to all entities and inherently subjective parts originating from local perception and communication with other entities. To fulfill a given goal it will (2) need to make plans using the knowledge it has and (3) generate hypotheses about future states of the environment. The subjective part of the knowledge needs (4) a mechanism to assign a certainty value to each item and evaluate its contribution to hypotheses, triggering the acquisition of additional information as necessary. Finally it will need to (5) evaluate plans it made and predicted states of the environment for their potential of risk.

3.6.4 Implementation of proactive risk management for autonomous logistic entities

The goal oriented risk concept chosen above is destined to enable a risk management strategy for autonomous entities such that they achieve robust behavior supporting a global goal. We employ the agent metaphor to model autonomous logistic entities and to support autonomous decision-making. Agents seem to be adequate due to their inherent autonomy and flexible interaction which enables them to interact dynamically in open systems.

Software systems implementing autonomous logistic processes (i.e. agents) need to share information on a continuous basis, for example, product specifications, manufacturing capabilities, delivery schedules, etc., and are required to make decisions which are consistent with the policies and overall economical situation of the enterprise they represent. In this context (Langer et al. 2006) introduce autonomous knowledge management (KM) to support the agent in improving its decisions in the presence of incomplete, imprecise, or debatable information as well as the inherent uncertainty that results from the dynamic of the domain.

In conventional research on multiagent systems, it is claimed, that the local interaction of autonomous systems (microscopic behavior) should lead to an optimized behavior on the global level (macroscopic behavior) (Langer et al. 2005). However current agent architectures are not designed to model this complex decision-making process which requires agents to process knowledge about internal structures and organizations, show awareness of other agents and communicate or even cooperate with them, and perceive changes in their environment. In the BDI (belief, desire, intention) approach as introduced by Rao (Rao and Georgeff 1998), the strategic layer of agents may be modeled within desires, operational aspects within beliefs, and tactical features within intentions or plans. The BDI approach also attempts to closely mimic human decision-making1 and is the currently most widely used approach for modeling intelligent behavior within the agent research community (Inverno et al. 2004).

The major shortcoming of current agent deliberation cycles is the relatively simple discovery and evaluation of alternatives. The standard approach to creating consistent subsets (goals) for action selection is not sufficient for dynamic environments, as the agent must often conduct multi-criteria optimization, which may also be based on competing goals. (Timm 2004) introduces a dynamic conflict resolution scheme for an agents options which in turn are derived from its goals.

An important challenge for this project is to augment the agent's deliberation cycle with the ability to identify and assess the underlying risks that are associated with the options that determine the next course of action. If necessary, the agent must be able to augment its knowledge base with missing or updated knowledge, for example, from other agents, to be able to properly assess and evaluate the feasible options. In an abstract sense this could mean to equip the agent with meta knowledge and meta reasoning capabilities, which is considered impossible for an artificial system as it would mean to engineer consciousness—a claim that AI has finally identified as unrealizable. For our approach we don't aim at a universal meta reasoning ability but add one meta layer to an agents reasoning capabilities, which can be realized by modal logics (cf. (Fagin and Halpern 2003)).

In (Langer et al. 2005) we proposed a framework for an enhanced agent deliberation process. This framework is being developed as a common basis for risk- and knowledge-management in agent decision-making (Langer et al. 2006). It includes explicit risk and knowledge management, termed decision-support in the figure, which may work in an inter-leaved fashion to augment the deliberation cycle of the agent. Generally speaking, we use risk management to identify and assess the risks associated with one or more options, and knowledge management to acquire missing knowledge, for example, to improve risk assessment or to generate additional options. Our decision-support system can be integrated into any intelligent agent that utilizes some form of deliberation with separate option generation and selection phases.

Agent decision process

The first step is the identification of potential risks associated with each option. Each identified risk must be evaluated to assess the magnitude of the risk and its probability of occurrence. In the ideal case, the agent has sufficient knowledge to arrive at a meaningful risk assessment. Upon completion, the result of the assessment is returned to the deliberation process which uses the information to aid in the selection of the best possible option. Due to incomplete or uncertain knowledge, risk management may be unable to decide on risk. This triggers knowledge management to acquire the missing information or detailed information on the current situation – including alternative actions. Knowledge acquisition may retrieve knowledge from other agents or directly from external sources/sensors.

A central component of our approach is the representation of decision-support parameters which govern the risk management and knowledge management processes as well as the interactions between them. For example, when RM invokes KM to acquire missing knowledge to help assessment of risk, it communicates the importance of obtaining the missing knowledge to KM. This helps KM selecting the proper strategy. Another parameter used by KM is availability which expresses the probability that an item of knowledge is available from any known source at this time. Availability of knowledge is based on prior experiences and used by KM, for example, in deciding which knowledge items should be acquired (in case there are choices).

As stated before risk is related to uncertainty. Thus the acquisition of facts that can reduce uncertainty is one strategy to handle risk. In this section we present an approach to assess the amount of uncertainty and a strategy to reduce it by invoking knowledge management. Risk management is a continuous process that will trigger further deliberation as soon as a fact is added to the knowledge base, which makes the situation risky. As already mentioned in the introduction, risk arises whenever a subsequent decision must be based on incomplete knowledge and thus might turn out wrong. Our concept of risk management is heavily depending on knowledge. Therefore it can only function in close collaboration with a knowledge management infrastructure. A description of the mechanisms of this of this collaboration and the core task of knowledge-based risk assessment will follow.

Pattern matching for risk identification

The initial task and most important prerequisite for successful risk management is its ability to identify risk and evaluate its potential consequence. Risk identification in an autonomous knowledge-based system can be achieved by matching fractions of the beliefs with patterns.

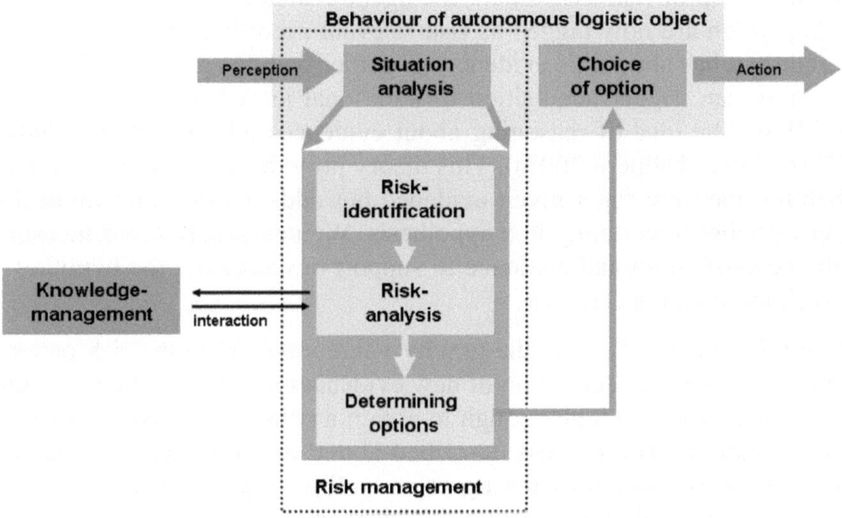

Fig. 3.9 Agent decision process with risk and knowledge magement

In the situation analysis phase of an agent's deliberation cycle (see figure 3.9) incoming perceptions are integrated with the current beliefs. Subsequently the agent generates a list of options that are reachable given the current situation (for details and a formal specification of this process we refer to recent work by Timm (Timm 2004). Risk identification will than work on the set of beliefs relevant to one option and the option itself to search for incidents that may impact the options execution.

Following the approach presented by Lattner (Lattner et al. 2005) we define a risk pattern as a formal description of a situation where certain occurrences may be dangerous for the agent. A risk pattern is defined by a set of predicates with unbound variables which can be unified with the beliefs. Each pattern has a gravity value assigned to it which indicates the possible (i.e., worst case) outcome of the incident described by that pattern. A pattern matching module evaluates the beliefs and substitutes the matching variables in the pattern. It registers all substitutions of variables with matches in a risk pattern. Additionally every substitution is annotated with the gravity value of the pattern.

Risk assessment

In the next step of risk assessment the agent evaluates the evidences (i.e., beliefs), which are now tagged as risk relevant according to the degree of uncertainty it has about this evidence. Together with the gravity value high uncertainty can trigger acquisition of additional knowledge. This evaluation follows the idea of reasoning about evidences introduced by (Shafer 1976) (see also (Halpern 2003)). This theory provides us not only with one probability measure for a given evidence but adds a value indicating the degree of belief or certainty in a hypothesis. We interpret this as a measure for the need of additional evidence to support or contradict the hypothesis and such increase the certainty.

A threshold depending on the gravity value assigned to the risk pattern determines when the acquisition of new evidences will be finished, i.e., the certainty is considered high enough to assign a value to the risk emanating from this pattern. The process described above is continuously evaluated against the world model of the agent as well as every anticipated future world state such enabling proactive risk identification.

3.6.5 Conclusions

New possibilities in reducing damage, lateness and other aberrations to given goals for autonomous logistic objects through the usage of a suitable risk management concept are described in this chapter. Risk Management with its containing parts of risk identification and risk assessment can be a solution to reduce risk in transportation or production for the autonomous objects and is also needed to make the autonomous logistic objects robust against suddenly appearing events which were not considered during the planning phase of the logistic processes. The chapter gives an overview about different levels of risk and Risk Management for planning and controlling the logistic processes by agent based autonomous objects. The handling of information from the real world with implemented methods of risk management to realize risk oriented decisions is a challenging task for an agent based autonomous logistic object. In this chapter the basic risk management concept and a technical realization of a local RM system were introduced and discussed regarding the requirements for agent based logistic objects. To complete the risk management system a component of planning has to be integrated. This is still an open task because until now the risk management can assess risk only on the actual situation and has the ability to evaluate the current knowledge but is not able to predict future world states. To reduce the uncertainty for planning the risk management interacts with the knowledge management. But the complexity in determining the uncertainty and modeling the risk for the complete autonomous process has strong influence on the model (hidden markov or bayes net) to be chosen and on the further development for that it is an important task for the near future.

References

d'Inverno M, Luck M, Georgeff MP, Kinny D, Wooldridge M (2004) Autonomous Agents and Multi-Agent Systems 9(1–2): 5–53

Fagin R, Halpern JY (2003) Reasoning about knowledge. The MIT Press

Finke R (2005) Grundlagen des Risikomanagements. Wiley, Weinheim

Haindl A (1995) Risk Management von Lieferrisiken. VVW Verlag, Karlsruhe

Halpern JY (2003) Reasoning about uncertainty. The MIT Press

Härterich S (1987) Risk Management von industriellen Produktions- und Produktrisiken. Vol. 37, VVW Karlsruhe

Hess HJ, Werk H (1995) Qualitätsmanagement, Risk Management, Produkthaftung. Luchterhand Verlag, Berlin

Kenney WF (1993) Process Risk Management Systems. VCH Publishers, New York (USA), Weinheim (D), Cambridge (UK)

Langer H, Gehrke JD, Hammer J, Lorenz M, Timm IJ, Herzog O (2005) International Journal of Knowledge-Based and Intelligent Engineering Systems, Accepted for publication

Langer H, Gehrke JD, Herzog O (2006) Distributed knowledge management in dynamic environments, in this volume, ed by M. Hülsmann, K. Windt, Springer, Berlin, 2006

Lattner AD, Timm IJ, Lorenz M, Herzog O (2005) Knowledge-based risk assessment for intelligent vehicles, in Proceedings of the IEEE International Conference on Integration of Knowledge Intensive Multi-Agent Systems KIMAS '05Waltham, Massachusetts, USA , pp 191–196
URL http://www.tzi.de/~mlo/download/1000_lattner.pdf

Luhmann N (2003) Soziologie des Risikos. Walter de Gruyter, unveränderter Nachdruck der Ausgabe von 1991

Mikus B (2001) Risikomanagement. Physica Verlag, Berlin

Motsch A (1995) Entscheidung bei partieller Information: Vergleich entscheidungstheoretischer Modellkonzeptionen. Gabler Verlag, Wiesbaden

Nash R (2003) The three pillars of operational risk, Operational Risk - Regulation, Analysis and Management. Pearson Education, London

Pfohl HC (2002): Risiken und Chancen: Strategische Analyse der Supply Chain, Risiko- und Chancenmanagement in der Supply Chain. Erich Schmidt Verlag, Berlin

Rao AS, Georgeff MP (1998) Journal of Logic and Computation 8(3), 293–342

Schmidt B (2000) The Modeling of Human Behavior. SCS Publications, Erlangen

Schwarz R (1996) Ökonomische Ansätze zur Risikoproblematik, Risikoforschung zwischen Disziplinarität und Interdisziplinarität. Rainer Bohn Verlag, Berlin

Seidel UM (2005) Risikomanagement. Weka Media GmbH, Kissing

Shafer G (1976) A Mathematical Theory of Evidence. Princeton University Press, Princeton, NJ

Thrun S, Burgard W, Fox D (2005) Probabilistic Robotics. The MIT Press, Cambridge, MA

Timm IJ (2004) Dynamisches Konfliktmanagement als Verhaltenssteuerung intelligenter agenten. Ph.D. thesis, Universität Bremen, Bremen, Germany

Tumer K, Wolpert D (2004) A survey of collectives. In: Tumer K, Wolpert D (eds) Collectives and the design of complex systems,Springer, pp 1–42

Urban C (2004) Das Referenzmodell PECS. Agentenbasierte Modellierung menschlichen Handelns, Entscheidens und Verhaltens. Ph.D. thesis, Fakultät für Mathematik und Informatik, Universität Passau

Weigand H, Dignum V (2004) "I am autonomous, you are autonomous" Springer-Verlag, Heidelberg, Vol. 2969 of Lecture Notes in Computer Science, pp 227–236

3.7 Autonomy in Software Systems

Ingo J. Timm[1], Peter Knirsch[2], Hans-Jörg Kreowski[2], Andreas Timm-Giel[3]

[1] Faculty of Computer Science and Mathematics, Johann Wolfgang Goethe-Universität Frankfurt, Frankfurt am Main, Germany

[2] Faculty for Mathematics and Computer Science, University of Bremen, Germany

[3] Faculty for Physics and Electrical Engineering/Information Technology, University of Bremen, Germany

3.7.1 Introduction

Looking at the whole logistic network, the structure of logistic processes becomes increasingly complex. Especially in transport logistics, atomisation of transportation processes, multimodal transport chains, international competition, changing ecological and legal constraints along with congestion of traffic infrastructure lead to highly dynamic and complex logistic processes that are difficult to plan (in advance).

The same situation can be found in production logistics. Modern production processes allow highly customized products. But the need for the reduction of costs, emerging virtual enterprises with distributed production plants, and just-in-time production leads to complex and highly dynamic production processes again being difficult to plan.

The described complexity and the arising difficulty in planning is a great challenge for enterprises. Having means to overcome these problems can constitute a significant competitive advantage.

The vision of the Collaborative Research Centre CRC 637 Cooperating Logistic Processes is to equip logistic processes and logistic objects with the capability to take decisions autonomously based on local and partially incomplete information. In consequence, the necessity to plan on a high level of details should be reduced significantly.

Considering transport logistics, this means that transport goods, transport vehicles etc. take decisions, like using a different route because of traffic congestion, locally without reinitiating a new overall planning and optimisation process. Similarly in production logistics, intelligent goods can select different suitable tools for the next production step.

Within the CRC 637 the autonomy in logistic systems and its benefits are investigated from different disciplines. To support autonomy, logistic entities need to have a minimum intelligence. Transport goods need to have some means of interaction, communication, and processing capabilities to take decisions, act, interact and communicate, autonomously. Logistic Systems are distributed and integrate physically mobile entities like transport vehicles or transport goods.

The CRC 637 is developing integrated solutions and management strategies using recent technological advances, on the hardware as well as on the software side, e.g., RFID, WLAN, agent technology.

This chapter presents autonomy as a core property of innovative software systems like agents and autonomous units. In the first section, ideas of agency are introduced. In the following section, autonomous units as a graph transformation-based approach to handling autonomous decision makers in a formal framework are compared with agents (for a detailed introduction of autonomous units the reader is referred to Chapter 2.6). Finally, advanced concepts of agency are discussed.

3.7.2 Ideas of Agency

Since the early 1970s there are various approaches in computer science to design and develop distributed systems to overcome limited computational capacity of single processing units and solving larger problems. Accompanying analysis of other research fields especially, in biology seem to content promising approaches of simple distributed decision entities leading to an emergent somehow intelligent behavior like a human brain. But even simpler contexts, like ant colonies show emergent behavior including "intelligent" solutions resp. global optimization by local interaction.

Let us consider a simple logistics task, which is performed efficiently and reliable by real-world ant colonies: finding the shortest path from the nest to the food source. After random walks in the environment, the ants will identify a new food source. Shortly after the identification, the ants will travel between nest and food on the direct and shortest path (cf. figure 3.10 (a)). This solution is reliable with respect to environmental changes.

If an obstacle is preventing the use of the shortest path (cf. figure 3.10 (b)), the ants will travel around obstacle randomly (cf. figure 3.10 (c)). Again, after short time, the ants will take the shortest path. The underlying algorithm is simple with respect to requirements for coordination effort between the ants and computational complexity within an ant. The ants are using pheromones to mark their path. The intensity of the pheromones decreases continuously over time. Ants traveling to the food source follow the path with the strongest pheromone concentration.

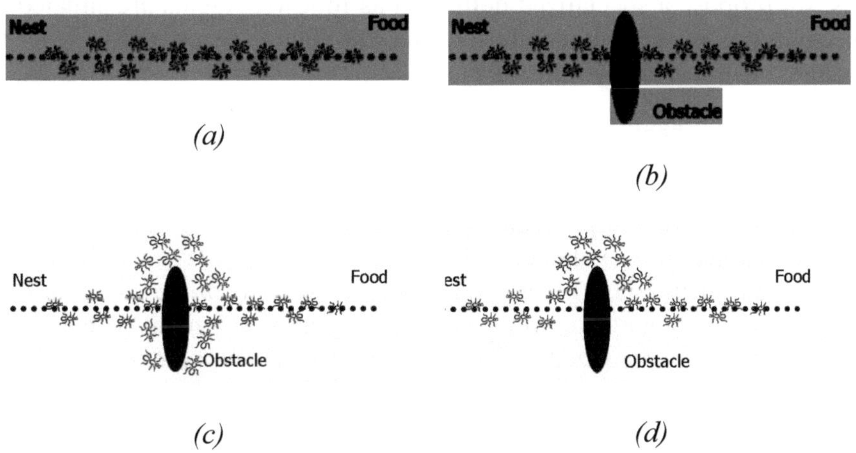

Fig. 3.10 Ant colony is finding the shortest path.

There are several similar examples in natures, where simple decision entities, like ants, bees, birds, termites, are performing complex problem solving by local interaction.

However, logistics tasks in real-world applications are far more complex due to various partially conflicting objectives, competitive behavior of the entities, etc. Thus, central planning in advance causes exponential computational complexity. Additionally, central planning is often prohibited by competing organizational substructures. Consequently significant research is focusing on emergent systems, where global optimization on a macro-level emerges from local interaction on a micro-level. The underlying idea is, to design autonomous entities, which implement simple decision behavior, which gain complexity by interaction with other autonomous entities.

The concept of autonomies entities interacting on a local level has been researched in computer science since the early 1980s. Smith (Smith 1980) invented the contract net approach to negotiate distributed solution in a system consisting of multiple autonomous decision makers with heterogeneous capabilities resp. skills (Smith and Davis, 1988). The actor theory developed important theoretical models for message-based communication of autonomous entities (actors) (Agha 1986). Following developments constituted the research field on autonomous agents and multiagent systems.

There are several classes of agent technology. A widely accepted definition of agents is provided by Pattie Maes: "Agents are software entities that assist people and act on their behalf" (Maes 1994). For agents in logistics, we propose a specialized definition as follows: Agents are situated in an environment, act autonomously, and are able to sense and to react to changes (Knirsch and Timm 1999).

Autonomous agents are modeled as completely free to negotiate and establish any sort of commitment with any other agent (Müller 1996). Following Castelfranchi and Conte (Castelfranchi and Conte 1992), preexisting norms, habits, and procedures are not relevant for the agents' actions. Thus social action is explained only in terms of the agents' mental states as beliefs and intentions. This approach describes the extreme situation of a totally autonomous agent, while in practice partial autonomy is common. This leads to a generalized definition: An agent is autonomous to the extent that its action choices depend on its own experience, rather than on knowledge of the environment that has been built-in by the designer (Russell and Norvig 1994).

From an external view, a system may be defined as autonomous, if it is acting non-deterministically, i.e., the system may function differently in identical situations. However, this does not mean, that an autonomous system has to be non-deterministic. The appearance of non-determinism arises from the limited view on the environmental state (situation). If the internal state of the system is included, an autonomous system might also be deterministic.

A more sophisticated approach to define autonomy resp. autonomous systems is the consideration of properties as introduced in (Timm 2006). In this context, autonomy resp. autonomous agents are best described by the three properties: pro-activity, interaction, and emergence. Pro-activity means, that the agent activates goals resp. initiates actions without specific external events. Therefore, the agent requires the ability to reason about its goals and the current situation, i.e., an explicit representation of goals and environment is required. A main feature of an autonomous agent is the capability of interaction with its environment and other agents. Pro-activity and interaction of agents in multiagent systems cause emerging properties which are not explicitly modeled in advance. The naive formulation of this fundamental assumption is that the system is more than the sum of its parts.

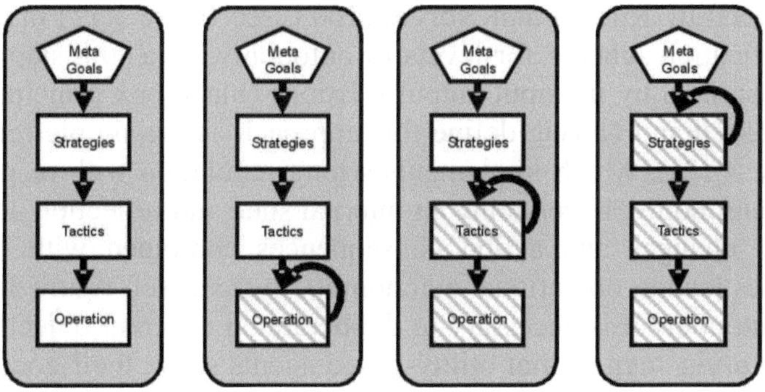

Fig. 3.11 Levels of autonomy

While Caselfranchi and Conte discuss a very high degree of autonomy, different levels of autonomy are introduced, e.g., (Rovatsos and Weiss 2005), (Müller 1997), (Timm 2006). Russell and Norvig classify the environment to differentiate AI approaches (Russell and Norvig 1994, 2003) following the criteria of observable, deterministic, episodic, static, discrete, and agent-oriented environments. In (Timm 2006), a classification scheme for levels of autonomy is introduced (cf. figure 3.11): strong regulation (no autonomy), operational autonomy (reactive systems), tactical autonomy (classical deliberative approaches), and strategic autonomy (complex intelligent systems). Table 3.1 yields the mapping of levels of autonomy to the environmental properties of Russell and Norvig.

Table 3.1. Classification scheme for levels of autonomy

Level of Autonomy	Observable	Deterministic	Episodic	Static	Agents
Strong Regulation	Fully	Deterministic	Episodic	Static	Single
Operational Autonomy	Partial	Deterministic	Episodic	Static	Multi
Tactical Autonomy	Partial	Stochastic	Episodic	Semi	Multi
Strategic Autonomy	Partial	Stochastic	Sequential	Dynamic	Multi

For practical applications or theoretical research a specific architecture has to be developed. The following paragraphs discuss agent architectures as introduced by Russell and Norvig (2003, see figure 3.12) in context of the classification scheme for levels of autonomy. In a first step, an agent can be described by its input/output relations (black-box principle, Müller 1996). Russell and Norvig define this approach as the simple reflex agent (cf. figure 3.12a), which implements strong regulation with respect to the levels of autonomy. Introducing an internal state and reflection about environmental changes and action consequences combined with condition-action rules lead to operational autonomous systems (cf. figure 3.12b). For tactical autonomy it is necessary to deliberate on different objectives; Russell and Norvig suggest that utility-based agents select their goals with respect to the greatest happiness specified by a utility function (cf. figure 3.12c). Finally, the strategic autonomy includes deliberation capabilities on goals, plans, and actions (cf. Figure 3.12d).

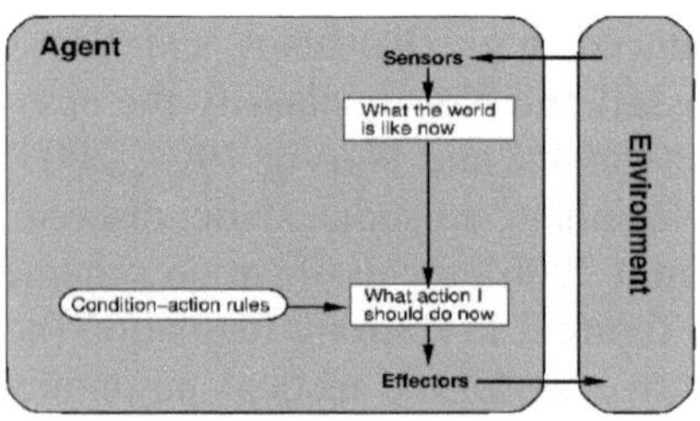

(a) simple reflex agent

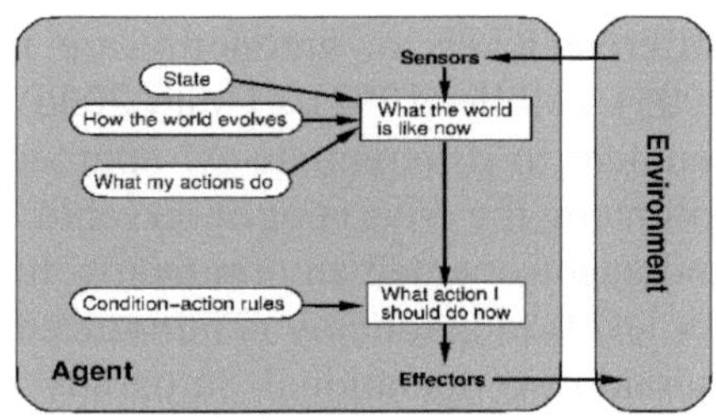

(b) model-based reflex agent

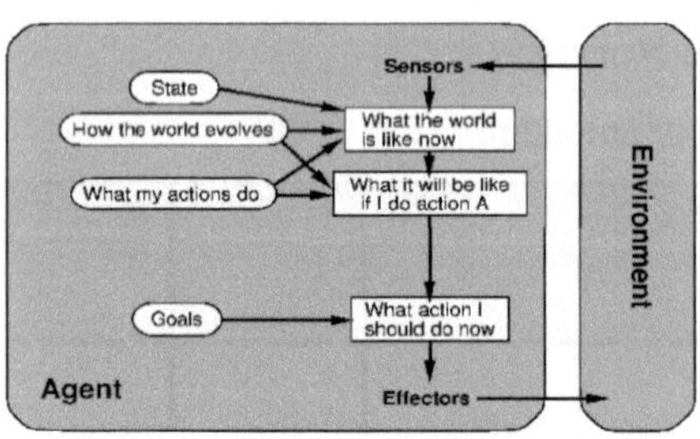

(c) utility-based agent

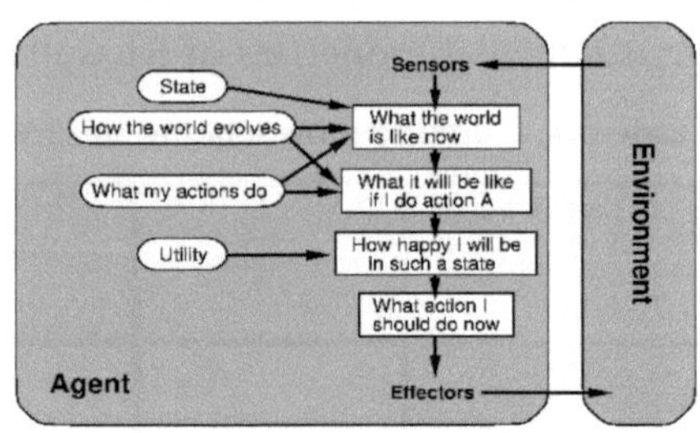

(d) goal-based agent

Fig. 3.12 Agent architectures (Russell and Norvig, 2003, p. 47, p. 49, p. 52, p. 50)

A unified approach to specify architectures in agent technology is the formal specification with (multi-)modal logics. Wooldridge and Lomuscio invented a general framework for the definition of agents as well as multi-agent systems, which may be outlined as follows (cf. figure 3.13,

Wooldridge and Lomuscio 2000). The agent behavior may be based on three phases: perceive, next, do.

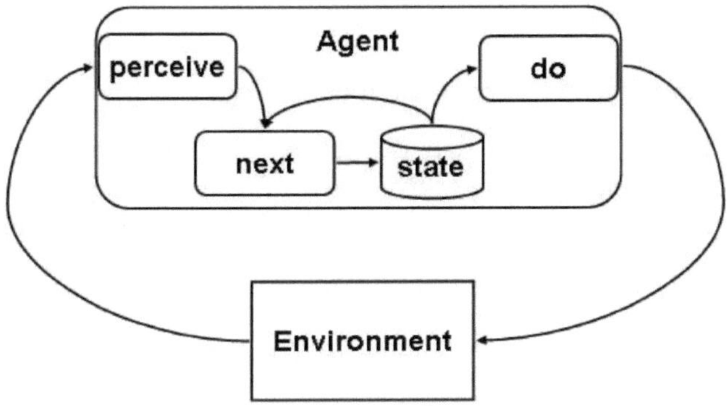

Fig. 3.13 Agent architecture in VSK (Lomuscio and Wooldridge, p. 3)

1. perceive

For each agent *agt*, there exists a unique environment, which is principally visible for it. Agents observe their environment via sensors in order to identify the relevant information constituting its perceptions:

$$perceive: E \rightarrow P \quad (3.1)$$

where E is the set of environmental states and P is the set of perceptible information.

2. next

Depending on the agent's internal architecture and state design, it is able to deliberate, plan, or select appropriate actions for execution. Let L denote the set of local states of agent *agt*. The reasoning process may be formalized by:

$$next: L \times P \rightarrow L \quad (3.2)$$

The local state of an agent may be constituted by highly complex structures. There are several aspects, which have been discussed in order to specify this structure.

3. do

In the third step the agent is selecting an action according to its internal state, which is performed in its environment:

$$do: L \rightarrow Act \tag{3.3}$$

where *Act* is the set of possible local actions.

Summarizing an agent can be defined as a system, consisting of the three decision functions *perceive, next, do*, as well as the accompanying concepts environment, perceptions, local states (including an initial state), and actions.

Definition 1

Given a set E of environmental states, an agent is a system $agt = \langle L, P, Act, perceive, next, do, l_0 \rangle$ where

- L is the set of possible agent's states,
- P is a set of perceptible information,
- *Act* is a set of actions,
- $perceive: E \rightarrow P$ is a function for perceiving environmental states,
- $next: L \times P \rightarrow L$ a local state transformer function,
- $do: L \rightarrow Act$ an action selection function, and
- $l_0 \in L$ the initial state of the agent.

This is a slight modification of the notion of agents of the VSK model. In the original definition, *perceive* is the sequential composition of a visibility function and a see function. The agent environment provides the visibility function for each agent, specifying which parts of the environment are generally perceivable to the agent. The see function belongs to the agent instead of the perceive function. To motivate this separation, let us consider an agent in the Internet as an example. The agent may perceive any web page which is generally accessible, but none which is restricted. Thus the environment "hides" the restricted information to the agent. However, the sensors of the agents may further restrict perceivable information. If an agent does not support a specific protocol for connecting to a web service, the agent may not perceive data provided by the web service, even if the information source is not hidden from the environment.

The visibility function of the environment implements general accessibility to the environment while the see function of an agent maps from environmental state to internal perception representation. However, the separation of the perceive function into two component is not needed in this paper. Hence we have integrated the visibility component into the notion of agents. Therefore, the agent environment is described in a formal sense by Definition 2 as follows.

Definition 2

An environnent is a system $Env = \langle E, Act_1, ..., Act_n, \tau, e_0 \rangle$ where

- E is the set of all possible environmental states;
- Act_i is the set of actions for each $i=1, ..., n$;
- $\tau: E \times Act_1 \times ... \times Act_n \rightarrow E$ is a state transformer function;
- $e_0 \in E$ is the initial state of the environment.

The formal notion of multiagent systems is given in Definition 3. It combines a group of agents $agt_1, ..., agt_n$ with an environment in such a way that the set of actions of the environment coincides with the agents' sets of actions and all agents perceive the environmental states of the environment.

Definition 3

A multiagent system $MAS = \langle Env, agt_1, ..., agt_n \rangle$ consists of an environment $Env = \langle E, Act_1, ..., Act_n, \tau, e_0 \rangle$ and a sequence of agents $agt_i = \langle L_i, P_i, Act_i, perceive^i, next^i, do^i, l_0^i \rangle$ with, $perceive^i: E \rightarrow P_i$ for each $i=1, ..., n$.

The previous definitions are introducing a static structure of the multiagent systems. During runtime, the agents as well as the environment changes with respect to their states. In the approach of VSK, there is a specified starting point in the systems, constituted by the initial states of the agents as well as the initial state of the environment. Essential property of the multiagent system is the function which gains the current state out of prior states. This system dynamics is outlined in Definition 4.

Definition 4

The *system dynamics* of a multiagent system $MAS = \langle Env, agt_1, ..., agt_n \rangle$ is given by sequences of global environmental states $g_0, g_1, \dots, g_t, g_{t+1}, \dots$ where the initial global state is defined as

$$g_0 = \langle e_0, next^1(l_0^1, perceive^1(e_0)), \dots, next^n(l_0^n, perceive^n(e_0)) \rangle \quad (3.4)$$

and, given the global environmental state $g_t = \langle e_t, l^1_t, \dots, l^n_t \rangle$ with $t \in \mathbb{N}$, the next global environmental state is defined as

$$\begin{aligned} &g_{t+1} = \langle \tau(e_t, do^1(l^1_t), \dots, do^n(l^n_t)), l^1_{t+1}, \dots, l^n_{t+1} \rangle \text{ with} \\ &l^i_{t+1} = next^i(l^i_t, perceive^i(e_t)) \text{ for } i \in \{1, \dots, n\}. \end{aligned} \quad (3.5)$$

3.7.3 Ideas of autonomous units

In the following sections, the concepts of autonomous units and communities of autonomous units as they are introduced in the Chapter "Autonomous Units: Basic Concepts and Semantic Foundation" in this book are compared to the framework of agents and multiagent systems. As a short repetition, an autonomous unit (see also (Hölscher et al. 2006)) is a new, formal, and general modelling concept especially designed for the modelling of autonomous behaviors. An autonomous unit has a goal, a certain set of capabilities, and an internal and therefore autonomous control.

Up to now existing modelling approaches do not cover the topic of autonomous control that explicitly while preserving a level of formality of the description that allows defining a precise semantics and proving certain properties of the system. Autonomous units are an extension of the well-studied and proven to be useful transformation units (see, e.g. (Kuske 2000)) which provide a general structuring methodology for rule-based graph transformation systems but only with a sequential semantics. That means that actors could only perform actions one after the other which is not suitable for logistic processes that are characterised by independent actors performing their tasks independently in a not predefined order and even concurrently. The framework of graph transformation as for instance described in (Ehrig et al. 1999) or in (Janssens et al. 2005) allows to model different kinds of semantics ranging from strictly sequential to concurrent behaviour. Autonomous units -which are still under development - constitute an adequate means for modelling complex networks of independent actors in a structured and rule-based way with an explicit representation of autonomy.

Several similarities with multiagent systems make it worth to have a closer look at the used concepts and their relations.

3.7.4 Relationship between autonomous units and agents

The relationship between autonomous units and agents is discussed with respect to the environmental states, the transformation steps, the perception, and the decision making.

Environmental states

Both approaches assume environments in which agents and autonomous units, resp., act and interact. While the environmental states of multiagent systems are not restricted in any way, the information structures underlying communities of autonomous units are assumed to be graphs. If one chooses a particular kind of graphs, it provides some explicit knowledge about the environmental states how they may be manipulated and how they may be visualized for example. One may say that graphs are particular models of environmental states of multiagent systems. But one should notice that graphs are very generic and flexible structures and that many data structures and system states are easily and adequately represented by graphs. Hence the choice of graphs is not much of a restriction.

Transformation steps

With respect to the notion of transformation steps, the relation between multiagent systems and communities of autonomous units is similar. A multiagent system assumes some state transformer function of environmental states dependent on an action performed by each agent, i.e. a function

$$\tau : E \times Act_1 \times \ldots \times Act_n \rightarrow E. \qquad (3.6)$$

It is not specified how an environmental state changes under which actions in the general framework, but must be instantiated in each case of application.

In contrast to this, a transformation step in a community of autonomous units is defined explicitly by a direct derivation, i.e. by an application of a rule to an environment graph yielding another environment graph.

$$G \Rightarrow_r H. \tag{3.7}$$

This provides a particular choice of the environment transformation τ if r is a parallel rule composed of one rule for each autonomous unit. And graph transformation turns out to a model of multiagent systems in this respect.

In another respect, communities of autonomous units have a more general environment transformation than multiagent systems. In addition to the synchronized parallelism of a rule per unit, any kind of sequential or parallel rule may be applied. An autonomous unit can act alone, some – but not necessarily all – of the units may act together, and each of the acting units may apply several rules in parallel in each step. Moreover, there is a concurrent semantics of communities of autonomous units in which synchronized parallelism does not appear explicitly and actions are only ordered in time if they are causally dependent.

Perception

Each agent of a multiagent system has got its individual perception of the environmental states given by the function $vis\colon E \to 2^E$ and $see\colon 2^E \to P$. As they are always applied together, first *vis* then *see*, they may be replaced by a single function $perceive\colon E \to P$ given by $perceive(e) = see(vis(e))$ for all $e \in E$.

An autonomous unit is not equipped with an explicit perception. Nevertheless, there is a counterpart implicit in the approach. Considering the rules of an autonomous unit, they can access an environment graph G by all possible rule applications. In this sense, the set of all direct derivations $G \Rightarrow_r G'$ with rules of the unit is the perception of G.

Depending on the control condition of the unit, the perception of an environment may contain further information. If the control mechanism is based on an evaluation function, for example, then the perception is enlarged by the view the evaluation provides of the environment. But quite often control conditions are used that check only the possible rule applications. In these cases, the control component of a unit does not add anything to the perception.

Decision making

Based on the perception of the actual environmental state *e*, an agent updates its actual local state *l* by applying the function *next*: $L \times P \rightarrow L$ yielding the next local state, i.e. $l' = next\ (l, perceive\ (e))$.

Then it decides about the next action to be performed by applying the function *do*: $L \rightarrow Act$ yielding $do(l')$.

In the framework of autonomous units, this task is done by the control condition of a unit. The control condition checks all possible rule applications of a unit to each actual environment graph and divides them into admitted and forbidden ones. Then one of the admitted rule applications is picked for the next action of the unit. Therefore, the decision of the next action is based on the perception of the environment graph and its restriction to the admitted part, which may be seen as the local state.

3.7.5 Advanced concepts of agency

In the section on ideas of agency, fundamental concepts for agency have been discussed following the VSK specification of Wooldridge and Lomuscio (2000). The basic model of the agents is quite simple, defining a perception (*perceive*), state transformation (*next*), and action (*do*) function. The formal representation of architectures and decision behavior has a strong history in the agent community. Agent's formalization mainly depend the constitution of a suitable formal language. The choice and development of the language depends on the use for internal specification used by agents for reasoning about behavior and actions, external specification used by agents for communicating with other agents, i.e., exchanging pieces of knowledge, or external use on a meta-level by developers for specifying, implementing, validating, and verifying properties of agents' behavior.

Internal specification languages are mainly applied to agents, which implement reasoning capabilities for advanced decision behavior. The formal language is used for the representation of the environment or internal state of the agent. Agents using formal languages for reasoning about knowledge to identify an action or action sequences are referred to as intelligent, deliberative, cognitive or rational. Interaction between agents uses communication languages. These languages specify the process of communication and a mandatory syntax of messages. However, message content is not specified there. The specification of content uses an external language, e.g., OWL (Patel-Schneider et al. 2007). Important aspects of external specification languages are that content can be interpreted in the same way by sender and receiver. The complexity of these languages as well as the underlying coordination mechanism vary, e.g., in market-based coordination models, content languages will consist of simple concepts for price and objects, while in negotiation-based coordination models, logical expressions are exchanged and used explicitly for internal reasoning.

Formal specification with meta-language should enable the design of multiagent systems as well as verification and validation of agents' or multiagents' behavior (Dunne et al. 2003). The language VSK is designed for these purposes. However, the individual agent's should be allowed to use varying formal languages for internal reasoning (Singh et al. 1999). The distinction between specification and implementation languages is not only useful for flexibility but also for expressiveness and efficiency. Designers tend to use a formal language with high expressive power for describing an intended system's behavior. In contrast to this, an implementation is in need of computationally efficient realizations, which – at least – rule out those formal approaches which are not decidable.

Modeling heterogeneous multiagent systems requires the abstraction of individual agents' behavior. The model of the system should only include those actions, which are perceivable to other agents or which change the environment. (Wooldridge and Lomuscio 2000) introduce VSK as a formal model for multiagent system based on multimodal logic. VSK integrates an environment depending visibility function (visibility) and an agent depending perception function (see). These concepts realized as modalities enable varying virtual environments for specific agents. A third modality is used for representation of the local state of agents (knowledge). However, the interaction of desires, beliefs, and intentions is not handled explicitly. Semantically, VSK is based on multimodal sorted first order logic (Wooldridge and Lomuscio 2000) and for temporal aspects it includes the possible worlds semantics, i.e., beliefs resp. propositions about knowledge follow weakS5 (KD-45) modal system (Meyer et al. 1991). In spite of the convincing concept of VSK, the underlying multimodal first-order logic suffers from the well-known logical omniscience problem of weakS5 as well as semi-decidability of first order logic.

In the context of autonomous logistics, a key characteristic of the agents is their physical or virtual mobility. Due to the physical movement of logistics objects, a formal approach has to consider ad hoc networks or restricted visibility of the environment. Furthermore, the mobility of agents within the virtual community has to be considered. The VSK model allows for dynamic manipulation of the accessibility of the environment through the visibility function. Petsch introduced an approach for modeling open agent societies with explicit migration in the formal model including representations of real-world organizations on the formal basis of VSK (Petsch 2006).

The internal decision behavior of agents is in focus of the distributed artificial intelligence community. This also reflects the majority of formal approaches which are focused on enabling intelligent behavior within agents (van der Hoek and Wooldridge 2003), (Wooldridge and Jennings 1995), (Rao and Georgeff 1998), (Fisher and Ghidini 2002), (Nide and Takata 2002), and (Timm 2001). Design of intelligent agents is often based on an explicit, cognitive model of beliefs, desires, and intentions, which are based on (Bratman 1987). BDI-agents use a formal semantics and implement a cognitive model of beliefs, desires, and intentions. The underlying idea is that an agent is creating an explicit world model (beliefs) on the basis of observations and its actions. Additionally, it contains a set of objectives (desires or persistent goals) and a set of goals which are currently pursued (intentions). The agent pursues its goals by autonomously created plans. This decision behavior is outlined in Table 3.2. BDI-agents are "the dominant force" in formal approaches (d'Inverno et al. 2004). Following (Wooldridge 2000) this is caused by their foundation on a widely accepted theory of rational actions of humans, the "great" number of successful complex applications and the availability of a large family of well-understood, sophisticated, and formalized approaches.

Table 3.2. Simplified decision behavior of BDI agents (Wooldridge 2000)

```
beliefs := beliefs'
while (true) do
      get next perception p1;
      beliefs := belief-revision-function(beliefs, p1);
      intentions := deliberate(beliefs);
      plan := planning(beliefs, intentions);
      execute(plan);
end while
```

In 2006, Henesey performed a survey on agent approaches in logistics (Henesey 2006). One of the main conclusions of this survey is that the majority of the agent approaches focus on operation decision support and only rare approaches have been applied in praxis. With respect to the levels of autonomy, the tactical and strategic level as implemented by BDI agents seem to be beneficial to autonomous logistics, as complex internal decision behavior can be modeled explicitly. However, BDI approaches do not focus on system behavior but on agent internal knowledge representation and decision making. In autonomous logistics, there are organizational structures as well as a centralized management defining the boundaries for individual agents' behavior. Here the BDI approach lacks explicit modeling of utility function or mechanisms for reliable behavior of a group of agents. In (Timm 2004) as well as (Scholz et al. 2006) the formal models of BDI especially the logical framework Lora (Wooldridge 2000) and VSK have been integrated as a unified formal basis for systems of intelligent agents.

In the research of the priority research program on "Intelligent Agents and Business Applications" from 2000 to 2006, it has been stated, that flexibility is the key benefit of intelligent agents (Kirn et al. 2006). However the question arises, if the optimization of individual performance within an agent also leads to a global optimization for a group of agents. In current approaches especially in the context of the CRC on autonomous logistics, we are investigating strategic management in multiagent systems. The strategic management is based on autonomous adjustment of the agent's autonomy. The underlying model is based on a social mechanism for reflection within social systems and has been transferred to multiagent theory (Timm and Hillebrandt, 2006).

3.7.6 Conclusions

In this chapter, we have discussed two approaches to modelling autonomy in software systems: multiagent systems and communities of autonomous units. The former is a well-known and widely used logical framework in artificial intelligence. The latter is a rule-based and graph-oriented method recently introduced in the context of the Collaborative Research Centre Autonomous Cooperating Logistic Processes.

As the very first observation in comparison of the two approaches, it has turned out that communities of autonomous units form executable structural models of the axiomatic notion of multiagent systems so that the former provide platform-independent realizations of the latter.

To shed more light on the significance of these observations, future studies will have to work out the relationship in more detail. This will include on one hand to prove that communities of autonomous units do not only follow the structure of multiagent systems, but satisfy also the requested properties. On the other hand, one may employ the well-working decision-making procedures of agents as control mechanism of autonomous units to widen the spectrum of possibilities with respect to the self-control.

References

Agha GA (1986) ACTORS: A Model of Concurrent Computation in Distributed Systems. The MIT Press: Cambridge, Massachusetts

Bratman ME (1987) Intentions, Plans and Practical Reason. Harvard University Press: Cambridge, Massachusetts

Castelfranchi C, Conte R (1992) Emergent functionality among intelligent systems: Cooperation within and without minds. Journal on Artificial Intelligence and Society, 6 (1): 78–87

d'Inverno M, Luck M, Georgeff M, Kinny D; Wooldridge M (2004) The dMARS Architecture: A Specification of the Distributed Multi-Agent Reasoning System. Autonomous Agents and Multi-Agent Systems 9 (1-2): 5-53

Dunne PE; Laurence M; Wooldridge M (2003) Complexity Results for Agent Design Problems. Annals of Mathematics, Computing and Teleinformatics 1 (1):19-36

Ehrig H, Kreowski H-J, Montanari U, Rozenberg G (1999) Handbook of Graph Grammars and Computing by Graph Transformation, vol 3: Concurrency, Parallelism, and Distribution. World Scientific, Singapore

Fisher M; Ghidini C (2002) The ABC of Rational Agent Modelling. In: Castelfranchi C, Johnson WL (eds) Proceedings of the First International Joint Conference on Autonomous Agents and Multiagent Systems (AAMAS 2002). Bologna, Italy, July 15-19, pp. 849-856

Hölscher K, Kreowski H-J, Kuske S (2006) Autonomous Units and their Semantics – the Sequential Case. In: Corradini, A, Ehrig H, Montanari U, Ribeiro L, Rozenberg G (eds) Proc. 3rd International Conference on Graph Transformations (ICGT 2006), Lecture Notes in Computer Science vol 4178, Springer, Berlin Heidelberg New York, pp 245-259

Janssens D, Kreowski H-J, Rozenberg G (2005) Main Concepts of Networks of Transformation Units with Interlinking Semantics. In: Kreowski H-J, Montanari U, Orejas F, Rozenberg G, Taentzer G (eds) Formal Methods in Software and System Modeling, Lecture Notes in Computer Science vol 3393. Springer, Berlin Heidelberg New York, pp 325-342

Knirsch P, Timm IJ (1999) Adaptive Multiagent Systems Applied on Temporal Logsitics Networks. In: Muffato M, Pawar KS (eds) Logistics in the Information Age. SGE Ditoriali: Padova Italy, pp. 213–218

Kuske S (2000) Transformation Units-A Structuring Principle for Graph Transformation Systems. Ph.D. thesis, Bremen

Maes P (1994) Agent that Reduce Work and Information Overload. In: Communications of the ACM 37, Nr.7, pp. 31–40

Meyer J-JC, van der Hoek W, Vreeswijk GAW (1991) Epistemic Logic for Computer Science: A Tutorial (Part One). In: Bulletin of the EATCS 44. European Association for Theoretical Computer Science, pp. 242–270

Müller JP (1996) The Design of Intelligent Agents: A Layered Approach. Lecture Notes in Artificial Intelligence 1177. Springer: Berlin

Müller H-J (1997) Towards agent systems engineering. Data and Knowledge Engineering, 23(3):217–245

Nide N, Takata S (2002) Deduction Systems for BDI Logics Using Sequent Calculus. In: Castelfranchi C, Johnson WL (eds) Proceedings of the First International Joint Conference on Autonomous Agents and Multiagent Systems (AAMAS 2002). Bologna, Italy, July 15-19, pp 928-935

Patel-Schneider PF, Hayes P, Horrocks I (2007) OWL Web Ontology Language Semantics and Abstract Syntax. W3C-Recommendation. http://www.w3.org/TR/owl-semantics/ (last visited Jan. 2007)

Petsch M (2006) Integrative Betrachtung der Offenheit von Multiagentensystemen unter technischen, systemischen, sozialen, und organisatorischen Aspekten. Dissertation. Technische Universität Ilmenau, Fachbereich Wirtschaftswissenschaften, Institut für Wirtschaftsinformatik, September

Rao A, Georgeff M (1998) Decision procedures for BDI logics. Journal of Logic and Computation 8 (3): 293–342

Rovatsos M, Weiss G (2005) Autonomous Software. In: Chang SK (ed): Handbook of Software Engineering and Knowledge Engineering, volume 3: Recent Advances. World Scientific Publishing: River Edge, New Jersey

Russel SJ, Norvig P (1994) Artificial Intelligence: A Modern Approach. Prentice Hall: Englewood Cliffs

Russel SJ, Norvig, P (2003) Artificial Intelligence: A Modern Approach. 2nd Edition. Prentice Hall: Englewood Cliffs, New Jersey

Singh, MP, Rao AS, Georgeff M P (1999) Formal Methods in DAI: Logic-Based Representation and Reasoning. In: Weiss G (ed) Multiagent Systems - A Modern Approach to Distributed Artificial Intelligence. The MIT Press: Cambridge, Massachusetts

Smith RG (1988) The contract net protocol: High-level communication and control in a distributed problem solver. In Bond A H, Gasser L (eds) Readings in Distributed Artificial Intelligence. Morgan Kaufmann, San Mateo

Smith RG, Davis R (1980) Frameworks or cooperation in distributed problem solving. IEEE Transactions on Systems, Man and Cybernetics, 11(1)

Timm IJ (2001): Enterprise Agents Solving Problems: The cobac-Approach. In: Bauknecht K, Brauer W, Mueck Th (eds): Informatik 2001 - Tagungsband der GI/OCG Jahrestagung, 25.-28. September 20011, Universität Wien, pp 952-958

Timm IJ (2004) Dynamisches Konfliktmanagement als Verhaltenssteuerung Intelligenter Agenten. DISKI 283, infix, Köln

Timm IJ (2006) Strategic Management of Autonomous Software Systems. Technical Report. TZI Bericht Nr. 35, Universität Bremen, Bremen

Timm IJ, Hillebrandt F (2006) Reflexion als sozialer Mechanismus zum strategischen Management autonomer Softwaresysteme. In: Schmitt M., Michael F, Hillebrandt F (eds) Reflexive soziale Mechanismen. Von soziologischen Erklärungen zu sozionischen Modellen, VS Verlag: Wiesbaden, pp 255-288

Timm IJ, Scholz T, Herzog O (2006) Capability-Based Emerging Organization of Autonomous Agents for Flexible Production Control. Advanced Engineering Informatics Journal 20: 247-259

van der Hoek W, Wooldridge M (2003) Towards a Logic of Rational Agency. International Logic Journal of the IGPL, 11 (2): 133-157

Wooldridge MJ (2000) Reasoning about Rational Agents. The MIT Press: Cambridge, Massachusetts

Wooldridge M, Jennings N (1995) Intelligent Agents: Theory and Practice. The Knowledge Engineering Review 10 (2): 115–152

Wooldridge M, Lomuscio, A (2000) A Logic of Visibility, Perception, and Knowledge: Completeness and Correspondence Results. In: Proceedings of the Third International Conference on Pure and Applied Practical Reasoning (FAPR-2000). London, UK

3.8 Specifying Adaptive Business Processes within the Production Logistics Domain – A new Modelling Concept and its Challenges

Bernd Scholz-Reiter, Jan Kolditz, Torsten Hildebrandt

Department of Planning and Control of Production Systems, University of Bremen, Germany

3.8.1 Introduction

Today enterprises are exposed to an increasingly dynamic environment. Last but not least increasing competition caused by globalisation more and more requires gaining competitive advantages by improved process control, within and beyond the borders of producing enterprises. One possibility to face increasing dynamics is autonomous control of logistic processes. This shall allow more robust processes in spite of growing environmental as well as internal complexity.

This paper presents the idea of autonomous logistic processes and focuses on a concept for modelling such processes. It is structure as follows: the next section gives a short overview of the concept of autonomous logistic processes. Subsequently section Development of a logistics system based on autonomous cooperating processes presents the overall system development cycle. The main section Modelling autonomous control discusses process modelling under the paradigm of autonomy and starting with requirements introduces our modelling method, important aspects of which are thereafter presented in more detail. The paper is concluded by a short conclusion and an outlook of future work.

3.8.2 Autonomous control of logistic processes

Autonomous control in the context of SFB 637, the research project this work is based on, means processes of decentralized decision making in heterarchical structures. It requires the ability and possibility of interacting system elements to autonomously make goal-oriented decisions. The use of autonomous control aims at achieving a higher robustness of systems and simplified processes achieved by distributed handling of dynamics and complexity due to greater flexibility and autonomy of decision making. Focus of the SFB lies in the areas of production and transport logistics, so the system elements, making their decisions autonomously, are the logistic objects themselves (Scholz-Reiter et al. 2004).

In order to enable logistic objects to be intelligent they have to be provided with smart labels. While today's RFID (radio frequency identification)-chips have very limited capabilities in respect to energy, range, storage capacity and especially information processing (Finkenzeller 2003), near future shall bring highly evolved smart labels that can provide resources alike micro computers to logistic objects. Nowadays RFID is already widely used in industry for identification matters and several visions for future applications exist (Fleisch and Mattern 2005; Heinrich 2005).

With respect to shades of autonomous control, different scenarios are possible, depending on which logistic objects are provided with smart labels and the functionalities they offer. This determines to what extend the logistic objects are able to make decisions. Considering the kind of decision-making by autonomous and therefore potentially intelligent logistic objects, transferring control decisions to goods, machines, storages and conveyors is obvious. Besides scenarios, where only one of the kinds of logistic objects has the ability to autonomously make decisions, arbitrary combinations are possible, depending on whether objects of the respective group are rather autonomously controlled or not.

Different logistic objectives can be assigned to the different groups of objects. For instance the objective of a high utilization can best be assigned to machines, while the objective of low due date deviation can best be assigned to a good. Concrete goal values are only achieved by the interaction of many logistic objects. Often conflicting goals of different objects have to be balanced, e.g. by negotiation. This leads to an increased coordination and communication effort compared to hierarchic forms of finding a decision. The more objects and groups of objects are involved in such a communication and make their decisions autonomously, the more important this point becomes. The number of possible communication relationships roughly grows quadratic in the number of participating objects. With 10 communicating objects there are 45 possible relationships, having 100 objects already leads to 4950. These numbers make clear that communication has to be limited to objects in the immediate spatial and/or logic neighborhood as otherwise control strategies can only hardly be scaled to problems of a realistic size. All these points have to be considered designing a control strategy and for modelling such a system.

3.8.3 Development of a logistics system based on autonomous cooperating processes

The development of an autonomous logistics system can be described on the basis of the Systems Engineering (Haberfellner et al. 2002) procedure model, as shown in figure 3.10. The methodical core is the iteration loop during main study and detail study phase. In the following the single phases of the procedure guiding through the development of an autonomous system and the connection with the more general procedure model of Systems Engineering are described.

1. Initiation

The rather unstructured initiation phase is triggered by sensing a problem and is completed by the decision to start a preliminary study. In our context this might be a problem associated with production planning and control or the assumption of a chance to improve the system's performance by adoption of autonomous cooperating logistic processes.

2. Preliminary Study

During preliminary study the objectives of adopting autonomous cooperating logistic processes have to be defined. Usually you will aim at concrete improvements in the fulfilment of logistic goals. In this regard the considered system and the scope of work have to be stated, for example a certain area of a production system.

Part of the preliminary study is also a situational analysis, which provides an overall understanding of the scope of work, of the existing problems and of the control processes. If required this system analysis can be detailed for certain aspects in later phases.

An important basis for the decision whether to continue or to abandon the project is an estimation of the impact of the solution principle. Therefore it is estimated to what extend an application of autonomous cooperating logistic processes is reasonable and promising. This shall allow a decision whether to start the main study phase or to cancel the project in respect to the development of an autonomous system.

The comparison of autonomous control and alternative methods and therewith the rating of such a solution principle under certain conditions is an issue of ongoing research that will not be discussed here any further (Scholz-Reiter et al. 2006a).

3. Analysis and Design Phase

In this paragraph the single steps of the iteration loop as the methodical core of the development process shown in figure 3.14 are described, followed by a discussion of the connections with the general Systems Engineering procedure.

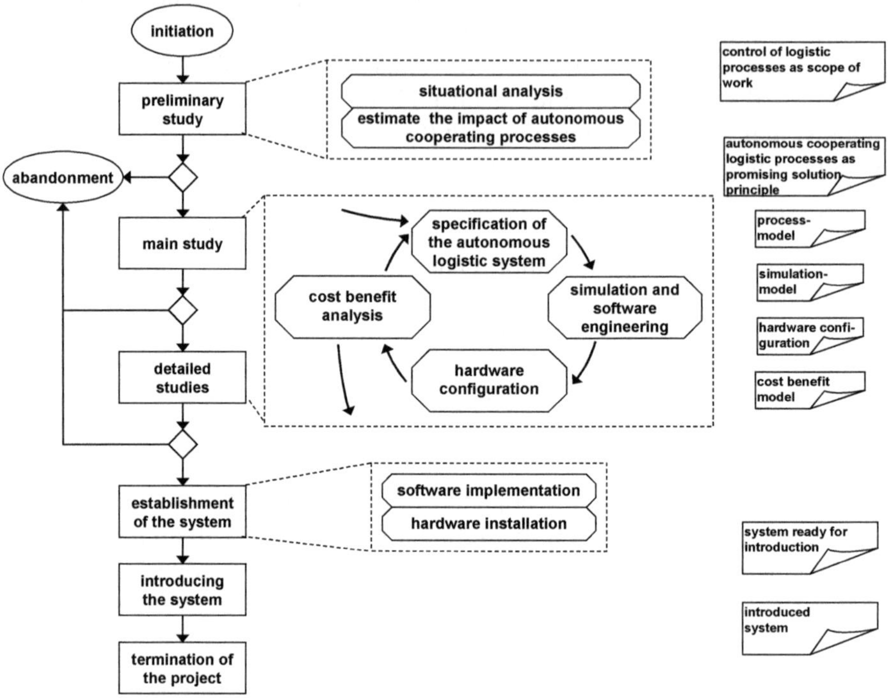

Fig. 3.14 Concretised systems engineering procedure model for autonomous logistics systems

- The first step of the iteration loop consists of the specification of the system. There a semi-formal specification of the proactive elements in an autonomous system as well as identification, design and allocation of decision processes are performed. It has to be clarified which elements are part of the system and which of them intelligent respectively autonomous entities are. To ensure the operability of the system all elements and processes have to be aligned with each other, making this step the basis of the development procedure.

- During the step of simulation and software engineering the design realised before is tested in a simulation first. Especially operability and impact on logistics performance of the whole system are focused here. A central task is the verification of required system behaviour because this is a necessary precondition for industrial application of emergent systems like autonomous logistics processes. The simulation code may already be part of the engineering process of the planned control software if the code is reusable. Otherwise the core software engineering process starts in the implementation phase.
- On the basis of the ideas gained before an estimation of needed hardware equipment for the autonomous system (for example what kind of communication infrastructure) can be made, getting more detailed with every iteration loop. Conclusions may be drawn from the process model as well as from the simulation. For example from allocation of control processes and data packets to entities of the logistic system necessary memory and computing capacity can be derived. Another example is the prediction of the capacity and equipment of the communication infrastructure on the basis of the expected communication volume between logistic system entities resulting from the simulation and the physical distribution of the objects to be arranged during hardware configuration. Attention has to be paid to the fact that although several agreements have been done during the steps before, this step strongly impacts implementation costs.
- Every iteration loop is concluded by a cost benefit analysis. On the basis of the rating and subsequent decision the original process model can be adjusted according to the new conclusions. In case of repeating negative results in this step an application of autonomous logistics processes has to be abandoned for this scenario.

The main study and design studies phases are not separated in two different ones, but combined in one phase. This phase is about an iteration loop, which on one hand serves the generation of variants and on the other hand produces more detailed and concretised solutions with proceeding iterations. A drawback of this abandonment of a phase separation is a lack of a clearly defined main phase, concluded by a decision about the cancellation of the project. But the cost benefit analysis concluding every iteration loop allows a decision about a cancellation on the basis of an economical rating in every cycle. This approach does not conflict with the Systems Engineering procedure because there is a close linkage between main study phase and detailed studies phase intended anyway. It is explicitly recommended to bring forward parts of detailed studies to the main phase if necessary.

Referring to the problem solving cycle as the micro-logic of Systems Engineering (Haberfellner et al. 2002), the focus of the main and detailed study phase lies in the search for solution and the selection (figure 3.15). The search for objectives primarily consists of an analysis of overall concepts and detailed concepts chosen before and the formulation of according objectives for the beginning iteration loop. The specification of an autonomous logistic system represents the synthesis of solutions, the constructive and creative activity. The simulation allows verifying the different solutions developed during the specification step concerning their functionality and capability and therefore represents the analysis of solutions. Afterwards on the basis of logistics as well as complementing objectives an evaluation of the solutions that have been rated as basically suitable is done. In the hardware configuration step the different possible solutions are evaluated in respect of their feasibility concerning hardware-oriented aspects as well as of the anticipated implementation input. The cost benefit analysis provides a basis to economically evaluate and compare the solution variants. Thereafter the decision is made whether to detail and concretise a variant respectively to start the establishment of the system or to cancel the project.

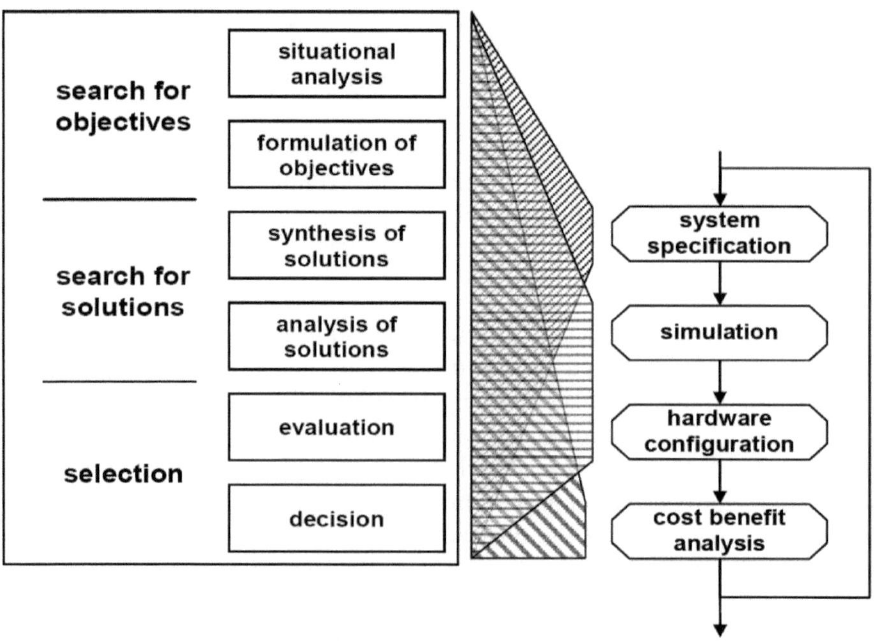

Fig. 3.15 Relevance of the steps of the problem solving cycle for the single steps of the iteration loop

4. Establishment of the system

During the establishment step the autonomous logistics system is realised. The main topics are the software implementation and the creation and integration of facilities and instruments. Ideally the software should be implemented using parts of the program code created during the simulation step.

5. Introduction of the system and termination of the project

Normally the introduction will involve huge and complex systems resulting in hard or even not calculable side effects. Therefore the introduction of the autonomous logistics system should be done stepwise if possible. After verifying the fulfilment of the objectives the system is handed over from the originating project team to the operating institution and the project is terminated.

3.8.4 Modelling autonomous control

Requirements to the modelling method

In this section requirements to the modelling method are formulated and structured following the distinction from requirements engineering of software sytems. Therefore the requirements and necessary characteristics of the modelling method resulting from the definition of Autonomous Logistic Processes are presented first. Subsequently general guidelines towards any modelling method are explained in the form of the Guidelines of Modelling (GoM, (Schütte 1998)), which also serve as general constraints for the modelling method presented here.

Primary requirements for the modelling method result from the fact that analysis and design of autonomous logistic processes has to be made possible for a logistic expert. Using the modelling method it therefore has to be possible to depict the constituent characteristics resulting from the definition of autonomous control given earlier in this book. The general definition results in the requirement that it has to be possible to model autonomous decision making of interacting system elements, i.e. a decentralised decision making in heterarchical structures. More specific requirements result from the specialisation of the general definition towards logistic processes which is relevant here. According to this concretised definition autonomous control of logistic processes is "[...] characterised by the ability of logistic objects to process information, to render and to execute decisions on their own" (Windt et al. 2007). A logistic object fulfilling this definition is called an intelligent logistic object; to support its design implicates an approach focused on these objects. The autonomous control characteristic of information processing requires a possibility to model information processing processes and that they can be assigned to the objects on which they are executed. Rendering of a decision entails possibilities to model the location of a decision, available decision alternatives and if necessary the knowledge needed by the intelligent logistic object for its decision. The characteristic of execution of its decisions finally requires an intelligent logistic object not only to render decisions autonomously, but also to initiate its execution and monitor its execution progress.

Furthermore the models created are the basis for subsequent software implementation. As a requirement this leads to the need to make this transition as frictionless as possible and already consider this during the design of the modelling method.

Following the distinction from the field of requirements engineering for software systems (Kotonya and Sommerville 1998) into functional and non-functional requirements, the requirements presented so far are comparable to functional requirements, which specify, what a system is supposed to do. Contrasting those non-functional requirements represent constraints, how these functional requirements are to be realised. As such non-functional requirements the Guidelines of Modelling (Schütte 1998) can be identified: Relevance, Correctness, Economic Efficiency, Systematic Design, Clarity and Comparability.

Relevance: The guideline of relevance considers the problem adequacy and tractability of model construction that are highly dependent on the constructing engineer's perspective.

Correctness: The guideline of correctness addresses the syntactic and semantic correctness of a model.

Economic Efficiency: The guideline of economic efficiency points out the necessity of economic advantage for modelling projects.

Systematic Design: In order to reduce complexity the guideline of systematic design provides a description of different views of the domain and availability of a view spanning meta-model. A common practice differentiates between static and dynamic views.

Clarity: The Guideline of clarity bears on clearness of models for different users.

Comparability: The possibility of comparing different models has to be guaranteed, which is of particular importance in target/actual comparisons.

These guidelines of model creation, which have to be followed during the modelling process also build the frame of the modelling method. As further non-functional requirements and further general conditions a focus on the logistic expert as the modeller (and user of the method) and a focus on the domain of production logistics can be identified.

It becomes obvious, that these requirements altogether result in partly conflicting requirements to a model or a modelling method. These conflicts have to be identified and balanced.

Overview of the modelling method

The modelling method consists of the components illustrated in figure 3.15. The "Principles", shown in the center of figure 3.16, define the basic structuring of the method. They consist of a view concept, each emphasizing certain aspects of the system to be modeled, as well as elementary guidelines of modelling. The "Meta Model" specifies the modelling elements usable by the modeler in a view-spanning manner. "Diagrams" defines the graphical notation representing these elements and the contexts where they can occur. It defines different diagrams each focusing on different facets of the system and visualizing them. Some examples of these diagrams are discussed later on in conjunction with discussing the view concept.

On the basis of the defined elements a reference model for autonomous cooperating processes is established. This reference model is available to the modeler as a set of building blocks easing model construction. The business process specialist will also get a modelling tool and the procedure model sketched in steps 1 and 2 of the system development process described in the previous section that is intended to guide the user through analysis and specification of autonomous cooperating logistic processes in the surveyed system.

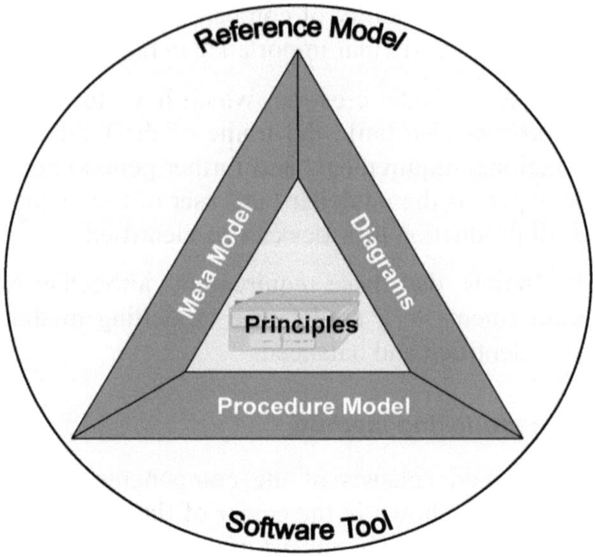

Fig. 3.16 Elements of the proposed modelling method

Modelling concept

Before the next part gives further details of the modelling method and the view concept behind it, we will give an overview of the modelling concept on an abstract level, i.e. shown in the context of different modelling levels (see figure 3.17). The figure shows different modelling levels, from the mapping of the real system at the bottom to the model level as well as from the modelling layers to their respective meta-levels. The distinction between model and meta-model is the same as between the real system level and the model level: the higher level contains explicitly the elements that can be used to model the level below. This means the meta-model-level specifies the elements that can be used to model the system on the model level. Speaking of "elements" this refers only to one aspect of the level transition, the specification of the modelling language. This aspect is called "language-based metaisation" in contrast to "process-based metaisation" which shows the modelling procedure to be used on the level below.

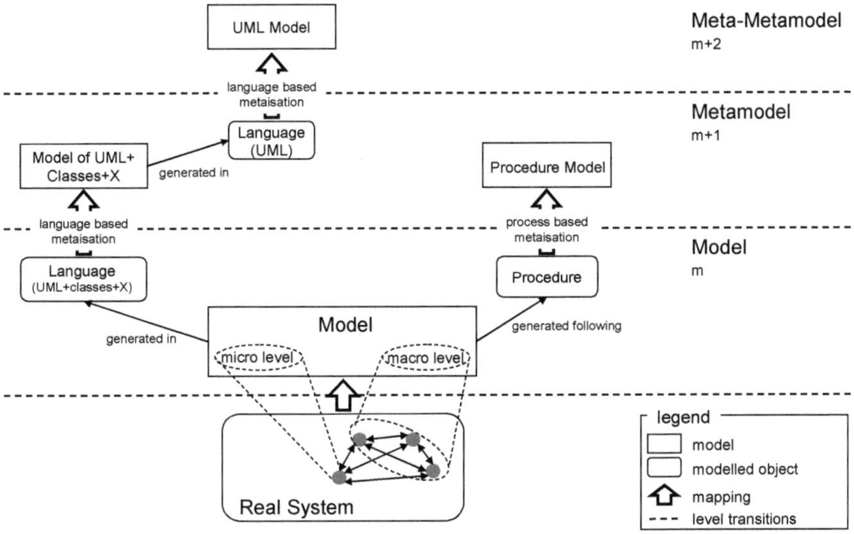

Fig. 3.17 Modelling method in the context of different modelling levels

On the lowest layer of figure 3.17 the (real or thought) system can be found. This is the system to be modeled; the modelling process itself is indicated by the lowest layer transition. Additionally the distinction between a macro- and a micro-level in modelling is indicated. Details regarding this point can be found in the next section. The model on one hand was created in a certain modelling language and on the other hand created following a certain modelling process. Therefore the layer transition from the model to the meta-model-layer distinguishes between language-based and process-based metaisation (for more information on metaisation refer to (Strahringer 1999)). Explicit representation of the creation process leads to the depiction of a procedure model for modelling. The procedure model will be represented using natural language and the process of its creation is not of particular interest to us thus nothing is shown in the figure on the meta-meta-model layer regarding the language- or process-based metaisation of the procedure model.

Concerning the branch of language-based metaisation and the transition from model- to meta-model-layer, the modelling language respectively modelling notation is explicicated. Our modelling notation is based on version 2.0 of the Unified Modelling Language (UML). In addition to that the modeller will be supported by pre-defined domain-specific classes and logistic-specific process-parts and process-templates. The UML notation is extended to better show certain aspects of the logistic system, for example by elements taken from software agent modelling. These extensions of the modelling language are indicated in the figure by the "X".

This (language-based) meta-model again is depicted in a certain way. At this point the distinction between language- and process-based metaisation could be made again, but only the first is of interest here. To represent the modelling notation, as a means of semi-formal modelling, UML will be used. To depict the fact that also this modelling language has to be specified somewhere, the top-layer shows the "model of UML", being the UML specification (see (OMG 2006)). Relative to the modelling we aim at, this specification is on the layer of a meta-meta-model, strictly following language-based metaisation.

Concept of views and notational elements

Creating process models usually leads to a high degree of complexity. A view concept serves as a means to reduce the complexity constructing a model (Scheer 1994) which is also reflected in the guideline of systematic design (see subsection Requirements to modelling). Based on the requirements mentioned above a view concept for modelling of autonomous logistic processes is proposed, whose views are depicted in figure 3.18. A fundamental distinction can be made between a static and dynamic model. The static model describes the structure, the dynamic model the behaviour of the modelled system, according to the basic classification in UML (OMG 2006) that is also appropriate here.

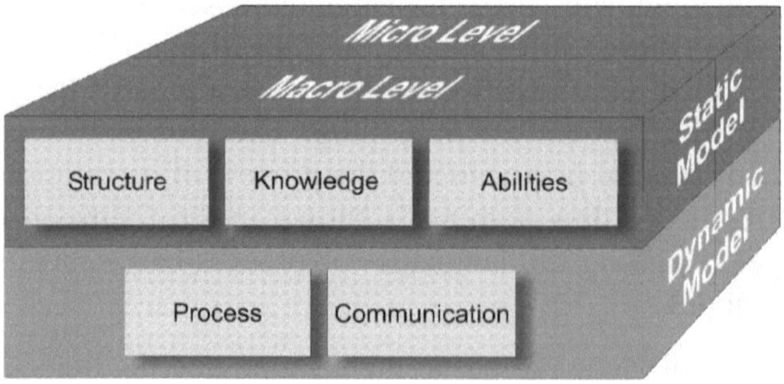

Fig. 3.18 View concept

The structure view that shows the relevant logistic objects is the starting point. The basic elements for this view are UML class diagrams. Besides objects and classes the structure view can show relationships between them, for instance in the form of associations or inheritance relationships.

The knowledge view describes the knowledge, which has to be present in the logistic objects to allow a decentralized decision making. This view focuses on composition and static distribution of the knowledge while not addressing temporal aspects. For this purpose UML-class diagrams are sufficient, while for the just mentioned temporal aspects, a dedicated knowledge representation language would have to be used (Sowa 2000). However it is doubtful how far the additional complexity in using it is compensated by the increased expressiveness. This is especially more important with respect to the intended use of the modelling method by a process expert.

The ability view depicts the abilities of the individual logistic objects. Processes of a logistic system need certain abilities, which have to be provided by the logistic objects. These abilities are supposed to be seen as abstractions of problem types occurring in reality.

The process view depicts the logic-temporal sequence of activities and states of the logistic objects. Here the objects' decision processes can be modelled. The process view plays a central role connecting the views of the static model and depicting the behaviour of logistic objects, so far only viewed statically. The notation elements used for this are activity diagrams as well as state diagrams. These two diagrams are also proposed in business process modelling using the UML (Oestereich 2003).

The communication view presents the contents and temporal sequence of information exchange between logistic objects. Depicting the communication is especially necessary to depict the interaction of autonomously deciding, otherwise only loosely coupled objects to model their interaction (Weiß and Jakob 2005). To display the communication UML-sequence diagrams showing the interacting partners, the messages and their temporal progression as well as class diagrams to display communication contents are supposed to be used.

In addition to the dynamic and static model just described we distinguish a macro and micro perspective. This distinction is also used in methods for software agent development (Weiß 2000). The macro view describes the interaction between the autonomous logistic objects. To some extend, it shows an external view onto the system, its elements and their relations and interactions. On the contrary the micro view describes the actions within and composition of the autonomous logistic objects. For the micro-level especially the process, knowledge and ability view are relevant, while all views proposed are relevant for the macro-level. This means that the micro-macro perspective is orthogonal to the views shown in figure 3.18. Nevertheless not all views use both perspectives to the same extend.

As an example for the static model and to clarify the described modelling concept figure 3.19 shows a part of the classes available to the modeller. He can create instances of the existing classes as well as adapt and/or expand the class model. This means that the diagram is a basis that can be adapted for applications of the modelling method if necessary and furthermore be used to model a concrete scenario by creating instances of these classes, e.g. to model actual machines or work plans. The figure shows some relevant classes and the most important relations between them. For clarity reasons there are no multiplicities included in the diagram and most role names as well as attributes of the classes are omitted. To create the collection of domain specific classes (Scheer 1994), (Loos 1992) and (Schönsleben 2001) were used as references. The models presented there were used in context of information system development and are now adapted to our requirements of modelling autonomous logistic processes.

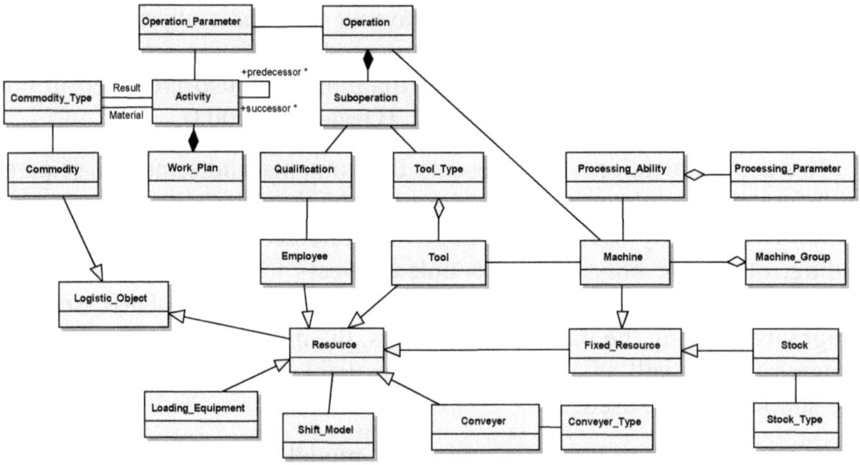

Fig. 3.19 Class diagram showing a part of the taxonomy supporting the user and selected relationships between the classes shown

As central classes "Logistic Object" and "Resource" (itself being a logistic object as indicated by the inheritance relationship) can be identified. Logistic objects are in principle able to be the autonomous objects of autonomous logistic processes. Kinds of logistic objects are commodity, all types of resources and orders (not shown in the selected classes above). Commodity represents a concrete logistic object in a material flow, e.g. an individual end-product, while commodity type is used when a commodity shall be referred to anonymously. A commodity type might be a type of end product, intermediate product or raw material. Work plans, which are an aggregation of "Activities" specify how a commodity can be manufactured, i.e. which work steps to perform and what the required material(s) are and what the result of such a processing or assembly step is. This work plan is specified anonymously, i.e. for "Commodity type"s. "Resource" represents a common base class for physical and rather permanent components of a production system, each of them can be associated with a "Shift Model", which determines resource availability and therefore is an important factor for its capacity. Specialisations of the resource class are machine, tool or stock as well as conveyer, tool, loading equipment and employees, the latter being a software representation or an interface of/to workers on the shop floor.

In order to facilitate a loose coupling of the components of our logistics system there is no static mapping between the activities within a work plan and the machines or other resources to perform them. This is advantageous to achieve a more adaptive behaviour of the system. If new machines are added to the shop floor, they can start processing in a "plug-and-play"-like manner without the necessity to change all existing work plans. Work plans only specify which activity to perform and their parameters, as a simplified example drilling, 5mm wide, 7mm deep. To determine the next machine a commodity asks machines which of them can perform a certain activity. This negotiation process is further specified in the communication and process views. A machine is able to autonomously deduce whether it is able to perform an activity on the basis of its processing abilities stored within it (e.g. able to perform "drilling" in the range of 2-10mm wide, 1-20mm deep). Furthermore it is able to create operations on the basis of activities and processing abilities, which in detail specify which and how long tools and personnel are required to perform such an activity.

As an example for the dynamic model figure 3.20 shows an exemplary sequence diagram as part of the communication view. The example is rather simplified and concentrates just on commodity-machine communication although availability of conveyers must be considered in a resource selection process. The diagram shows a machine object and a commodity object. The exchanged messages are shown chronologically in vertical direction. The commodity requests a machining process answered by the machine with a quote. After the machine has selected a quote (the selection itself with its criteria and algorithms is modelled in the micro level process view) the chosen machine is booked by the commodity, the others are informed about the quote cancellation. In figure 3.20 this is modelled by a combined fragment of the type "alternative".

The presented example also shows some deficits of the UML 2.0 standard with respect to modelling autonomous logistic processes. It is not one commodity communicating with one machine, but one commodity communicating with multiple machines. On the other hand the "maschine selected"-part of the alternative fragment is only executed with one machine. For increased clearness this should be modelled explicitly. One possibility to assure clearness could be an extended notation similar to cardinality which is proposed for software agent modelling with UML using specific extensions (Bauer and Odell 2005).

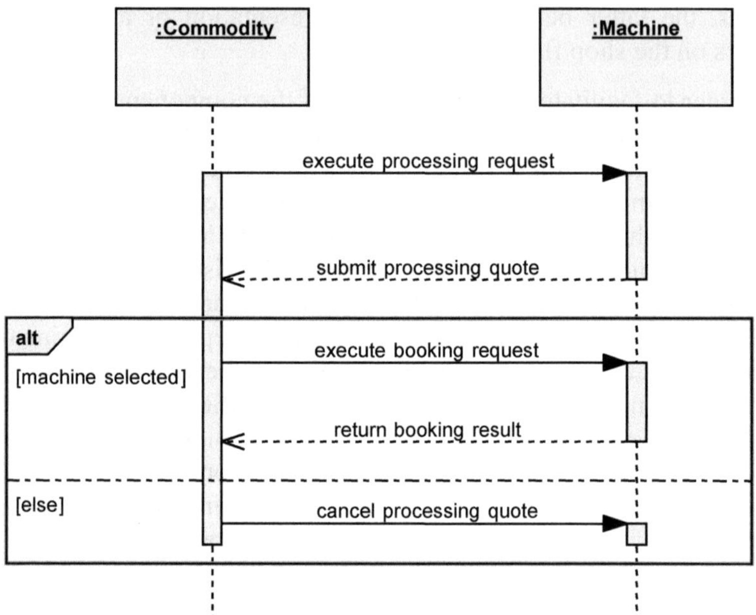

Fig. 3.20. UML sequence diagram machine selection

3.8.5 Fulfilment of requirements

After presenting requirements to the modelling method earlier in this paper, this section will investigate in how far the requirements are fulfilled by the designed modelling method as presented in the previous part of this paper.

First of all the fulfilment of the two general, i.e. non-functional, requirements of a focus on the domain of production logistics and the logistic expert as a modeller will be investigated. The latter requirement can for instance be found in the use of UML as the basis of the modelling notation used. As a graphical, semiformal notation it is broadly used – besides software development (especially agent-oriented approaches are of particular interest here, see for instance AUML (Odell et al. 2001; Bauer et al. 2001)) it is also used for knowledge modelling (Schreiber et al. 2002) or business process modelling (Oestereich 2003). Its broad use makes it likely that the logistic expert assigned to the system design already came in touch with this notation earlier in one context or the other. As it is furthermore an intuitive graphical notation, with its expressiveness reduced to only the sub-set necessary here, the learning effort is accordingly low. The extensions by logistics-specific notational elements and the production logistic reference process also make the method easier accessible for the logistics expert. Both of these points, the extension of the notation with logistics specifics (e.g. a layout diagram) and the reference process consisting of an ontology of production logistic concepts and an exemplary definition of intelligent objects' processes express the requirement "focus on the domain of production logistics". Additionally the use of UML also fulfils the requirement of considering the later phase of software implementation. A language continuously used from the process model to the detailed analysis of the of the software system to be implemented, avoids a break in the development process, as the different fields involved all use the same semantic constructs (Oestereich 2003).

Regarding the primary requirements, supporting the design of intelligent logistic objects implies an approach focussed in these objects and opposes a strict top-down-design approach. This will be accounted for in the procedure model by its use of a mixture of a bottom-up and top-down-approach.

The interacting system elements (especially the intelligent logistic objects, but also other system components) can be shown in the Structure View. Here also intelligent logistic objects can be marked as such and their life-cycle described by an associated state-chart in the Process View. A description of the information-processing processes respectively decision processes also takes place in the Process View using Activity Diagrams. Not only an assignment of processes to the logistic objects they are located on (location of decision) is conducted here, but also the knowledge required for a decision can be modelled explicitly using object nodes. The structure of this knowledge and its initial distribution can in turn be shown in the Knowledge View, using Class Diagrams for the structure of the knowledge objects and Knowledge Maps to show its distribution. The interaction of the system elements among each other and their environment respectively is primarily described in the Communication View. UML Sequence Diagrams can be used here to specify interaction protocols. Event mechanisms (in Activity Diagrams and State Charts) can also be used to depict interaction with the environment and other system elements. They can also be used to initiate decision execution and monitor their execution progress. To be able to not only model direct communication between the intelligent logistic objects but also to allow to specify communication with the environment as a means of interaction is important to model stigmergy-based coordination (for a discussion of a stigmergy-based approach in the context of autonomous logistic processes see (Scholz-Reiter et al. 2006)).

The heterarchical decision structure is not a characteristic of the meta-model respectively the notation, but a property of the processes in their entirety. The reference model created has this property; there is no central entity that renders a decision which is then delegated to executing instances.

3.8.6 Conclusions

This paper addressed the topic of modelling autonomous logistic processes. Therefore after a short definition of autonomous control in the context of logistics, the overall system development process was sketched. After that requirements to a suitable modelling method were derived. The concept of our modelling method was presented subsequently, first giving a rough overview, then detailing selected aspects of it such as the view concept. The last section investigated in how far the designed modelling method fulfils the requirements derived in the beginning of the paper.

Further research will be concerned with the elaboration of the procedure model. The meta-model and graphical notation will be specified formally in a manner suitable to be used in later software implementation. This is important as our work is aimed at the development of a software tool, specifically tailored to support our modelling method comprised of the notation and procedure model as far as possible. With the help of this tool a process expert (e.g. a logistics expert with only little background in computer science) will be supported in modelling and designing autonomous logistic processes.

References

Bauer B, Müller JP, Odell J (2001) Agent UML: A formalism for specifying multiagent software systems. International Journal of Software Engineering and Knowledge Engineering 11(3):207–230

Bauer B, Odell J (2005) UML 2.0 and Agents: How to Build Agent-based Systems with the new UML Standard. Journal of Engineering Applications of Artificial Intelligence 18(2):141–157

Finkenzeller K (2003) RFID-Handbook – Fundamentals and Applications in Contactless Smart Cards Identification, 2nd edn. Wiley and Sons LTD, Swadlincote, UK

Fleisch E, Mattern F (2005) Das Internet der Dinge – Ubiquitous Computing und RFID in der Praxis: Visionen, Technologien, Anwendungen, Handlungsanleitungen. Springer, Berlin

Haberfellner R, Daenzer W F, Huber F (2002) Systems Engineering. Methodik und Praxis, 11. edn. Verlag Industrielle Organisation, Zürich

Heinrich C (2005) RFID and Beyond: Growing Your Business through Real World Awareness. Wiley Publishing, Indianapolis, Indiana

Kotonya G, Sommerville I (1998) Requirements Engineering. John Wiley and Sons Ltd, Chichester et al.

Loos P (1992) Datenstrukturierung in der Fertigung. Oldenbourg, München

Oestereich B (2003) Objektorientierte Geschäftsprozessmodellierung mit der UML. dpunkt Verlag, Heidelberg

Odell J, Parunak HVD, Bauer B (2001) Representing agent interaction protocols in UML. In: Ciancarini P, Wooldridge M (eds) AOSE 2000, LNCS 1957, Springer-Verlag, Berlin Heidelberg, pp 121–140

OMG Unified Modelling Language Specification (version 2.0), Retrieved January 27 2006, http://www.uml.org

Scheer AW (1994) Business Process Engineering – Reference Models for Industrial Companies, 2. edn. Springer, Berlin et al

Scholz-Reiter B, Freitag M, de Beer C, Jagalski T (2005) Modelling dynamics of autonomous logistic processes: Discrete-event versus continuous approaches. CIRP Annals 1(55):413–417

Scholz-Reiter B, Freitag M, de Beer C, Jagalski T (2006) Modeling and simulation of pheromone based shop floor control. In: Cunha PF, Maropoulos P (eds): Proceedings of the 3rd CIRP Conference Digital Enterprise Technology, Setubal, Portugal 18-20 Sept 2006

Scholz-Reiter B, Philipp T, de Beer C, Windt K, Freitag M (2006) Einfluss der strukturellen Komplexität auf den Einsatz von selbststeuernden logistischen Prozessen. In: Pfohl H-C, Wimmer T (eds): Steuerung von Logistiksystemen - auf dem Weg zur Selbststeuerung. Konferenzband zum 3. BVL-Wissenschaftssymposium Logistik. Hamburg: Deutscher Verkehrs-Verlag, pp 11–25

Scholz-Reiter B, Windt K, Freitag, M (2004) Autonomous Logistic Processes - New Demands and First Approaches. In: Monostori L (ed) Proc. 37th CIRP

International Seminar on Manufacturing Systems. Hungarian Academy of Science, Budapest, Hungaria, pp 357–362
Schönsleben P (2001) Integrales Informationsmanagement: Informationssysteme für Geschäftsprozesse – Management, Modellierung, Lebenszyklus und Technologie, 2. edn. Springer, Berlin et al
Schreiber G, Akkermans H, Anjwerden, A, de Hoog R, Shadbolt, N, van de Velde W, Wielinga B (2002) Knowledge Engineering and Management - The CommonKADS Methodology. MIT Press, Cambridge
Schütte R (1998) Grundsätze ordnungsmäßiger Referenzmodellierung: Konstruktion konfigurations- und anpassungsorientierter Modelle. Gabler, Wiesbaden
Sowa J (2000) Knowledge representation: logical, philosophical, and computational foundations. Brooks Cole Publishing Co., Pacific Grove, USA
Strahringer S (1999) Probleme und Gefahren im Umgang mit Meta-Begriffen: ein Plädoyer für eine sorgfältige Begriffsbildung. In: Proceedings of the International Knowledge Technology Forum (KnowTechForum) '99. Potsdam, Germany
v. Uthmann C, Becker J (1999) Guidelines of Modeling (GoM) for Business Process Simulation. In: Scholz-Reiter B, Stahlmann HD, Nethe E (eds) Process Modelling. Springer, Berlin, pp 100-116
Weiß G (2000) (ed) Multiagent Systems – A modern approach to distributed artificial intelligence. The MIT Press, Cambridge, USA
Weiß G, Jakob R (2005) Agentenorientierte Softwareentwicklung. Springer, Berlin Heidelberg
Windt K, Böse F, Philipp T (2007) Autonomy in Logistics - Identification, Characterisation and Application. International Journal of Robotics and CIM, forthcoming 2007

4 Autonomous Control Methods and Examples for the Material Flow Layer

4.1 Approaches to Methods of Autonomous Cooperation and Control and Examples for the Material Flow Layer

Katja Windt[1], Michael Hülsmann[2]

[1] Department of Planning and Control of Production Systems, BIBA, University of Bremen,Germany

[2] Management of Sustainable System Development, Institute for Strategic Competence-Management, Faculty of Business Studies and Economics, University of Bremen, Germany

One aim of the working circle "Autonomous Cooperation and Control" within the Cooperative Research Center 637 was to ensure that the developed definition (chapter 1 Introduction) will be on the one hand precise and valid for an interdisciplinary view, and on the other hand realistically based on production and logistic scenarios. A production scenario is represented by machines, orders, transportation systems, storage areas and the interconnecting material flow. The material flow in conventional production planning and control (PPC)-systems has to be planned, controlled and to be under surveillance. Material flow means the execution of the physical movement of parts and products through production and logistic systems. In order to cope with a wide range of logistic applications, transport and production processes are considered within selected scenarios. High relevance–and therefore logistic potentials–especially for the application of autonomous cooperation and control is offered by the job shop production principle. This is characterized by a functional organization whose departments or work centres are organised around particular types of equipment or operations. Work pieces are transported through the departments in batches (Hernández 2002).

The main challenges of the implementation of autonomous control methods for the material flow layer lies in:

- The ability to model autonomously controlled logistic processes;
- The realization of communication between intelligent items;
- The development of decision algorithms and therefore evaluation systems for the relevant intelligent items;
- The identification of important information for every intelligent item,
- The determination of the anticipation horizon for the selection of important information;
- The feasibility of divisibility of orders or of mergence of intelligent items (e.g. assembly stage);
- The derivation of requirements on production and transportation systems suited for autonomous cooperation and control.

Another challenge concerns the necessity to regard the environment at any one time as a dynamic one. Even during the decision-making process of single logistic objects the environment may be changed. This means that decisions of single logistic objects need to be executed swiftly, especially the decision influences other logistic objects (which will be the case for the majority of logistic objects) (Windt and Freitag 2004).

Finally, the process flow of order processing will change. The identification of these changes is one assumption in order to adapt the appropriate planning and control systems to the new requirements resulting from autonomous cooperation and control.

The aim of this chapter is to show the applicability of the theoretical structures for autonomous cooperation and control. Different logistic scenarios are described: One of the logistic scenarios integrates production and transportation processes on the example of an automobile distribution logistic provider. The transportation processes of cars in automobile terminals as well as the delivery processes are combined with production processes at the terminal, for example car washing and unwaxing, or the assembly of navigation systems (Böse et al. 2005).

The second logistic scenario presented in this edited volume concentrates on the implementation of information and communication systems as one condition for autonomous cooperation and control. In this case the prototype of an 'intelligent container' (Jedermann 2006).

One of the above mentioned challenges of autonomous cooperation and control is addressed in the article from Thorsten Philipp, Christoph de Beer, Katja Windt, Bernd Scholz-Reiter and is entitled **"Evaluation of autonomous logistic processes – Analysis of the influence of structural complexity"**.The coordination of intelligent objects requires advanced planning and control concepts and strategies to realize the autonomous control of logistic processes. In order to prove that the implementation of autonomous control in production systems is more advantageous than conventionally managed systems and to show where the limits are, the development of an adequate evaluation system is essential. This system reflects the degree of achievement of logistic objectives related to the level of autonomous control and the level of complexity. The evaluation system consists of three main components: The correlation between logistic objective achievement and level of autonomous control is heavily dependent on the complexity of the system in question. A measurement and control system for logistic performance was developed for the measurement of the logistic objective achievement. Furthermore, a complexity cube was developed in order to characterize the complexity of production systems and a catalogue of criteria can be used to determine the level of autonomous control. Within this article a vectorial approach to measure the achievement of logistic objectives together with a feedback loop for autonomous processes is introduced. By means of a complexity cube it is possible to operationalize the complexity of production systems with regard to different types of complexity.

Bernd-Ludwig Wenning, Henning Rekersbrink, Andreas Timm-Giel, Carmelita Görg and Bernd Scholz-Reiter concentrate on transportation processes in their article **"Autonomous Control by Means of Distributed Routing"**. To deal with dynamic problems in routing and assignment in logistics the subproject B1 "Reactive Planning and Control" investigates an approach that considers vehicles and packages to be intelligent and autonomous. These logistic items are able to decide about routes and loads by themselves based on local knowledge. This requires replacement of the centralised decision-making approach by a decentralised, distributed autonomous control approach. For this approach, methods and algorithms from other domains of science and technology are evaluated for their suitability for application in transport logistics. One promising technology domain is the wide range of routing algorithms used in communication networks. Distributed routing has already been successful in communication networks for several decades. For a transfer of routing methods from communication networks to logistic networks, it is necessary to identify where these networks are similar and where they exhibit differences. The two kinds of networks are comparable, both involving payloads which have to be transported from a source to a destination. They both have the possibility of resource reservation and are comparable in size and dynamics. But there are also differences in physical existence as well as amount limitations. Handling of loss is completely different in transportation and communication networks. The two networks also use different scales of time. This leads to the conclusion that routing methods from communication networks cannot be transferred directly into logistics. Nevertheless, routing approaches in communication networks can provide inspiration in devising routing approaches for logistic networks. Consequently, a concept for distributed routing in a logistic network is presented. In this concept, vehicles as well as packages are considered as autonomous. They have sufficient intelligence and communication capabilities to obtain their information and to decide on the next steps to be undertaken. The Distributed Logistic Routing Protocol (DLRP) presents a fully distributed routing concept for dynamic logistics. In this concept a vertex is a knowledge broker for the vehicles and packages. Before deciding about a route, a vehicle/package requests current information from the current or next vertex. Each vertex includes relevant information available from its current knowledge-base and forwards the request to neighbour vertices.

The concept has been implemented into a logistic simulation environment to prove its feasibility.

Bernd-Ludwig Wenning, Henning Rekersbrink, Markus Becker, Andreas Timm-Giel, Carmelita Görg and Bernd Scholz-Reiter present a **"Dynamic Transport Reference Scenario"**. Reference scenarios are a common technique in simulations allowing the evaluation and comparison of different algorithms and approaches. For transport logistic processes these approaches can be, for example, different strategies to select the packets to be loaded. Traditional logistic scenarios are not suitable for the investigation of dynamic transport processes. Therefore new reference scenarios are generated which can be used for the evaluation of approaches in these dynamic networks. The components for the modelling of dynamic logistic networks are introduced and evaluation parameters are listed. Based on the definitions of components the scenarios comprise all relevant components, such as location and functionality of vertices, edges, type and initial position of vehicles and distribution of packages. Two selected scenarios, the small 4-vertex scenario and the larger Germany scenario, are described.

When investigating the quality of an approach, there is the need to evaluate its performance levels with respect to the aspired goals. Therefore a set of evaluation criteria is required. Considering transportation logistics, the goal is to achieve a high logistic efficiency, i.e. high performance at low cost. Two sets of possible evaluation measures are introduced: The volume-related measures (consisting of queued packages, inactive vehicles, vehicle utilisation) and the process-related measures (comprising throughput time, punctuality rate per package, trans-shipments per package). The above mentioned combined scenario of transportation and production processes is described by Felix Böse and Katja Windt within the article **"Autonomously Controlled Storage Allocation on an Automobile Terminal"**. In the context of this article a new approach of an autonomously controlled logistics system is introduced, considering as example the storage allocation processes at the E.H.Harms Auto-Terminal Hamburg. The vehicle movement processes at the automobile terminal provide many opportunities for improvement. Based on the described business processes of the conventionally controlled as well as the autonomously controlled storage allocation, two simulation scenarios are developed and evaluated. By establishing autonomous control, vehicles are enabled to render decisions on their own and according to this to determine their way through a logistics network on the basis of an own system of objectives. As result of recent developments in the field of information and communication technologies, the implementation of such an autonomously controlled logistics scenario for an automobile terminal is now feasible.The object of investigation of the simulation study is the transfer times of the vehicles in the automobile terminal. As a main result of the presented simulation study, the new paradigm of autonomous control in logistics provides significant opportunities of time saving in the field of vehicle movement in automobile terminals. Due to the fact that the simulation study was strongly focussed on the storage allocation process as a single part of the vehicle management process chain of automobile terminals, further research is directed to the enlargement of the considered application scenario.

The article **"Intelligent Containers and Sensor Networks – Approaches to Apply Autonomous Cooperation on Systems with Limited Resources"** by Reiner Jedermann, Christian Behrens, Rainer Laur and Walter Lang focuses on RFIDs, sensor networks and low-power microcontrollers are increasingly applied in logistics. They are characterized by restrictions on calculation power, communication range and battery lifetime. The article considers how these new technologies can be utilized for autonomous cooperation and how these processes could be realized in systems with limited resources. Besides tracing of the current freight location by RFID technologies, the monitoring of quality changes that occur during transport is of growing importance. The demand for improved and comprehensive supervision of goods could be best fulfilled by distributed autonomous systems. The prototype of the 'intelligent container' demonstrates how autonomous control could be implemented on a credit-card sized processor module for integration into standard containers or transport vehicles. RFID technologies are used to control the transfer of this mobile freight agent. The implementation of the local data pre-processing and an example quality model for vegetables are described. If the supervision system predicts that the freight quality will fall below an acceptance threshold before arrival, it contacts the transport manager. Furthermore, the extended agent platform for further transport planning is shortly introduced. Sensors that are attached to the freight have to link themselves 'ad hoc' into the communication network of the vehicle. Therefore the text gives an overview of the design, configuration and control of the implementation of a wireless sensor network.

Then architectures, examples and further demands on autonomous cooperative processes running on low-power microcontrollers are discussed. Finally, approaches for future implementations of an autonomous decision system on small battery-powered sensor nodes and logistical freight objects are summarized.

The last article of chapter 4 by Reiner Jedermann, Jan D. Gehrke, Markus Becker, Christian Behrens, Ernesto Morales-Kluge, Otthein Herzog and Walter Lang represents a **"Transport Scenario for the Intelligent Container"**. The article describes how the intelligent container is linked with an agent system for transport coordination including communication gateway and vehicle location. The scenario itself consists out of a traffic network, trucks and loads with their respective positions. It concentrates on the automated monitoring and management of perishable goods. The hardware setting is presented, e.g. sensor configuration, as well as the used controlling methods, e.g. transport coordination and route planning, within the scenario. The article shows how autonomous cooperating and control may improve the processes of supply chain management.

References

Hernández, R, Vollmer, L, Schulze, L: Strukturen. In: Arnold, D, Isermann, H, Kuhn, A, Tempelmeier, H (eds): Handbuch Logistik, VDI-Springer Verlag, Berlin, 2002, B3-14

Windt, K, Freitag, M (2004) „Autonomous Logistic Processes – New Demands and First Approaches“, in: Proceedings of 37th CIRP International Seminar on Manufacturing Systems, May 19-21, Budapest, Hungary

Böse F, Piotrowski J, Windt K (2005) Selbststeuerung in der Automobillogistik. Industrie Management 20(4): 37-41

Jedermann R, Gehrke JD, Lorenz M, Herzog O, Lang W (2006) Realisierung lokaler Selbststeuerung in Echtzeit: Der Übergang zum intelligenten Container. In: Pfohl, HC; Wimmer, T (eds) Wissenschaft und Praxis im Dialog. Steuerung von Logistiksystemen - auf dem Weg zur Selbststeuerung, Deutscher Verkehrs-Verlag, Hamburg, pp 145-166

4.2 Evaluation of Autonomous Logistic Processes – Analysis of the Influence of Structural Complexity

Thorsten Philipp, Christoph de Beer, Katja Windt, Bernd Scholz-Reiter

Department of Planning and Control of Production Systems, BIBA, University of Bremen, Germany

4.2.1 Introduction

The concept of autonomous control requires on one hand logistic objects that are able to receive local information, process these information, and make a decision about their next action. On the other hand, the logistic structure has to provide distributed information about local states and different alternatives to enable decisions generally. These features will be made possible through the development of Ubiquitous Computing technologies (Fleisch et al. 2003).

The application of autonomous control in production logistics can be realized by recent information and communication technologies such as radio frequency identification (RFID), wireless communication networks etc. These technologies enable intelligent and autonomous parts and products to communicate with each other and with their resources such as machines and transportation systems and to process the acquired information. This leads to a coalescence of material flow and information flow and allows every item or product to manage and control its manufacturing process autonomously (Scholz-Reiter et al. 2004). The coordination of these intelligent objects requires advanced planning and control concepts and strategies to realize autonomous control of logistic processes. To develop and analyze such autonomous control strategies dynamic models are required.

In order to prove that the implementation of autonomous control in production systems is more advantageous than conventionally managed systems, it is essential to develop an adequate evaluation system. This system reflects the degree of achievement of logistic objectives related to the level of autonomous control and the level of complexity. Within the Collaborative Research Centre 637 "Autonomous Cooperating Logistic Processes: A

Paradigm Shift and its Limitations" at the university of Bremen (CRC 637) it is investigated in which case the implementation of autonomous control is superior to other approaches and where the limits are (figure 4.1 upper right).

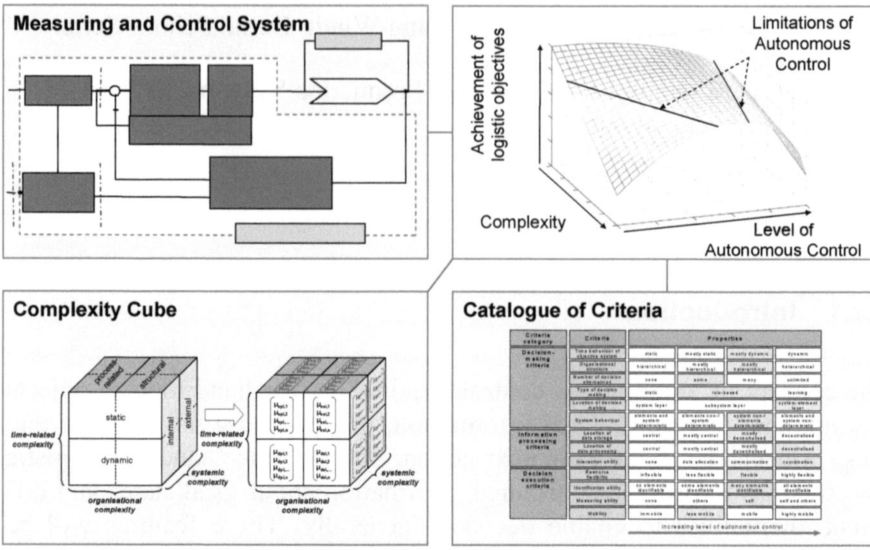

Fig. 4.1 Components of the evaluation system and limitations of autonomous control

In order to determine the limits of autonomy, the axes in the upper right curve in figure 4.1 have to be operationalised. The correlation between logistic objective achievement and level of autonomous control is heavily dependent on the complexity of the considered system. For the measurement of the logistic objective achievement a measure and controlsystem for logistic performance was developed. Furthermore a complexity cube was developed in order to characterize the complexity of production systems and a catalogue of criteria can be used to determine the level of autonomous control. Dynamic models and simulation studies can help to verify the run of the surface build by the single curves and thus the limits of autonomy can be found.

This article will first give a global definition of the term autonomous control in the context of the Collaborative Research Centre as well as a definition in the context of engineering science. Furthermore an approach to measure the complexity of production systems by dint of vectors and a complexity cube is given. In order to measure the achievement of logistic objectives a feedback loop for autonomous processes together with a vectorial approach is introduced. This forms the basis for simulation studies of different autonomous control strategies. Two control methods are analysed in more detail with different levels of complexity of the considered production system in order to verify the hypothesis that autonomous control is a suitable approach to cope with increasing complexity.

4.2.2 Autonomy in production logistic

Based on this global definition of the term autonomous control which is described in chapter 1.1 a definition in the context of engineering science was developed, which is focused on the main tasks of logistic objects in autonomously controlled logistics systems:

"Autonomous control in logistics systems is characterized by the ability of logistic objects to process information, to render and to execute decisions on their own."(Windt et al. 2007)

The paradigm shift expressed in the definition is based on the following assumption: The implementation of autonomous logistic processes provides a better accomplishment of logistic objectives in comparison to conventionally managed processes despite increasing complexity. In order to verify this statement, it is necessary to characterize production systems regarding their level of complexity during the development of an evaluation system.

4.2.3 Complexity of production systems

Existing approaches

The term complexity is widely used. Generally it does not only mean that a system is complicated. Ulrich and Probst understand complexity as a system feature where its degree depends on the number of elements, their interconnectedness and the number of different system states (Ulrich and Probst 1988). An observer judges a system to be complex when it can not be described in a simple manner. In this context Scherer speaks of subjective complexity. Furthermore, he distinguishes between structural complexity which is caused by the number of elements and their interconnectedness and dynamic complexity caused by feedback loops, highly dynamic and nonlinear behavior (Scherer 1998). Moreover, complexity can be understood as interaction between complicatedness and dynamics (Schuh 2005).

An enormous challenge occurs during the operationalization of complexity in the form of a quantifiable complexity level. Some approaches to measure complexity use the measurement of entropy as basis (Deshmuk et al. 1998; Frizelle 1998; Frizelle and Woodcock 1995; Sivadasan et al. 1999; Jones et al. 2002; Karp 1994; Gellmann and Lloyd 1994). In thermodynamic systems entropy can be deemed to be the degree of disorganization of the considered system. Shannon and Weaver developed an equation to measure the amount of information on the basis of the equations for entropy measuring (Shannon and Weaver 1949). This can be used for complexity measurement because the more complex a system is, the more elements and relations are included and the more information is necessary to describe the system. Those considerations were adopted by Frizelle and Woodcock to develop equations to measure complexity in production systems based on the diversity and uncertainty of information within the system (Frizelle and Woodcock 1995). They defined the structural complexity as the expected amount of information necessary to describe the state of a system. In a manufacturing system, the data required calculating the structural complexity can be obtained from the production schedule. Frizelle and Woodcock defined the dynamic or operational complexity as the expected amount of information necessary to describe the state of the system deviating from schedule due to uncertainty.

It is obvious that complexity can not be measured by a single variable. It is necessary to describe complexity by multiple factors which are interdependent but can not be reduced to independent parameters (Schuh 2005). A various number of complexity measurements were developed in the research on complex networks, e.g., the internet (Amara and Ottino 2004) or biological networks (Barabasi and Oltvai 2004). In this context Costa et al. showed that a complex network can be represented by a feature vector (Costa et al. 2005).

This approach is seized for the description of complexity in the following (figure 4.2). By means of this vectorial approach it is possible to measure the complexity of production systems on an ordinal scale. Thus different systems are comparable and measurable concerning their level of complexity.

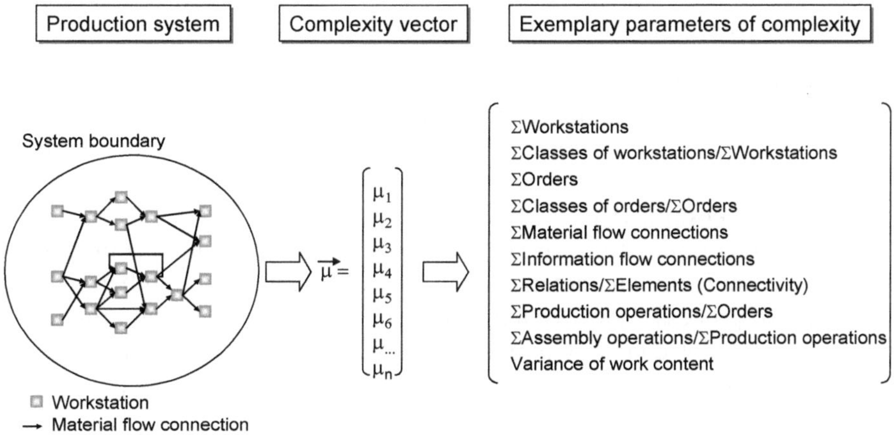

Fig. 4.2 Characterization of a production system's complexity by a vector

The complexity of the total system is accordingly expressed by a complexity vector. In the first instance this vector is an approach to measure the different types of complexity in production systems which has to be specified in further research studies. Several parameters of the systems complexity are exemplarily represented in figure 4.2. By means of this approach it is possible to detect a $\Delta\mu$, which describes the complexity difference of two considered systems. In this manner the production system's complexity can be measured and consequently the effects of changing complexity levels can be analysed.

Complexity in the context of autonomous processes

As described in the chapter before there is a wide range of approaches to describe complexity of systems. Due to the fact that these approaches only refer to single aspects of complexity, as for instance the structure of a considered system, they seem insufficient for an entire understanding of the term complexity in the context of logistic systems, in particular production systems. As shown in (Philipp et al. 2006), it is essential to define different categories of complexity and to refer themselves to each other, to obtain a comprehensive description of the complexity of a production system. In consequence, three categories of complexity *time-related complexity*, *organisational complexity* and *systemic complexity* are derived and referred to each other in a complexity cube. They are defined as follows (Philipp et al. 2006).

Organizational complexity

Organizational complexity consists of process-oriented and structural complexity. Process-oriented complexity defines the number and diverseness of process flows whereas structural complexity describes the number and diverseness of systems elements, their relations and properties.

Time-related complexity

Time-related complexity is divided into a static and a dynamic component. Dynamic complexity characterizes changes with respect to number and diverseness of process flows, systems elements, their relations and properties in time dependent course. Compared to this, static complexity refers to a fix system status at a concrete point in time or in a concrete time period.

Systemic complexity

Systemic complexity deals with internal and external complexity and is determined by the system boundary. Process flows, system elements and their relations and properties which are assigned to the system are part of the internal complexity. Process flows, system elements, their relations and properties outside the system boundary belong to the external complexity.

These three categories of complexity, their characteristics and interdependences are illustrated in figure 4.3 in form of a complexity cube. As explained in chapter 3.1, each area of the complexity cube can be determined by a complexity vector. By defining each area of the cube, the complexity of any production system can be determined. Consequently, the complexity cube provides the opportunity to define and compare different levels of organisational, time-related and systemic complexity of several production systems.

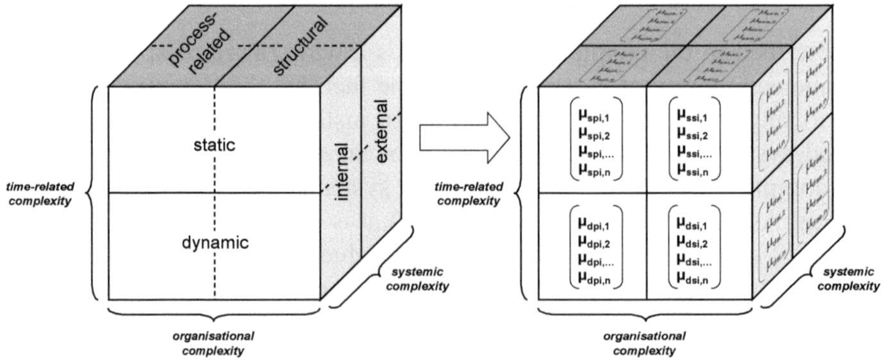

Fig. 4.3 Complexity cube for production systems (Philipp et al. 2006)

In order to get an idea how a specific vector for the different types of complexity looks like, an example for the structural static internal complexity is represented in the following:

$$\vec{\mu}_{ssi} = \begin{pmatrix} \Sigma\text{Human actors} \\ \Sigma\text{Workstations} \\ \Sigma\text{Classes of workstations}/\Sigma\text{Workstations} \\ \Sigma\text{Orders} \\ \Sigma\text{Classes of orders}/\Sigma\text{Orders} \\ \Sigma\text{Material flow connections} \\ \Sigma\text{Classes of material flow connections}/\Sigma\text{Material flow connections} \\ \Sigma\text{Material backflows}/\Sigma\text{Material flows} \\ \Sigma\text{Information flow connections} \\ \Sigma\text{Classes of Information flow connections}/\Sigma\text{Information flow connections} \\ \Sigma\text{Relations}/\Sigma\text{Elements (Connectivity)} \end{pmatrix}.$$

All parameters of this exemplary complexity vector are assigned to the production system (internal), can be determined at a concrete point of time or time period (static) and are referred to the systems elements, relations and properties (structural). According to Wiendahl et al. the human actors play an important role in mastering complex production systems. In this context we focus on human actors as resources and not on their specific individual behaviour (Wiendahl et al. 2005). There are basic parameters like the number of machines or the number of orders which must be included in the complexity vector but generally the choice of measurement parameters to determine the complexity difference of diverse production systems may vary and is highly dependent on the considered system.

4.2.4 Measurement and evaluation of logistic objektives

This chapter will focus on the measurement of the logistic performance of autonomous production logistic systems (e.g. a manufacturing system). Together with the measurement of the level of complexity explained in the previous chapter it allows an investigation of the coherence between the complexity and the performance of production systems.

Feedback loop of autonomous control

The basis for the measurement and evaluation of autonomously controlled logistic processes is a feedback control approach for individual logistic objects as shown in figure 4.4. Former approaches of control loops for production control are for example the works of Petermann and Breithaupt

(Petermann 1996; Breithaupt 2001). The difference of this approach is that the controlled system is the production process while in the works of Petermann and Breithaupt the controlled system was the work system.

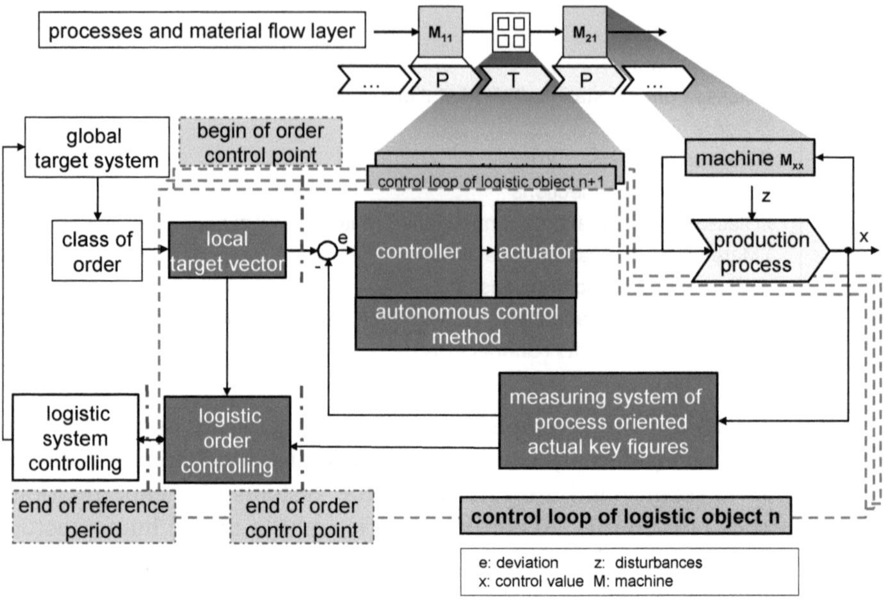

Fig. 4.4 Feedback loop of autonomous control (order view)

In this case the controlled process is a production process. Two logistic objects (an order as well as a resource) are involved in this process. Starting from a global system of objectives (the objectives of the considered production system), target values for varying object classes are deduced. This enables for example from an order's point of view a differentiation between customer orders and storage orders with different target weights for delivery reliability and throughput time of an individual order. Local objectives for individual logistic objects arise based on the object classes' objectives. These local objectives act as reference value for the feedback control approach for autonomously controlled processes. Eventual changes during the production process can immediately be realized through a fast feedback loop by measuring and calculating simultaneously the relevant logistic performance figures. Based on this feedback loop suitable solutions to react on process changes can be found by the evaluation of possible alternatives.

Within the controller (figure 4.4) the deviations of the production process from the local desired values are analysed. All possible alternatives to react on the process deviation will be taken into consideration and are evaluated regarding its forecasted logistic performance. This first evaluation step provides the basis for the following operation procedures of a logistic object through the production floor.

The evaluation-based decision will subsequently be executed by the actuator. For example such a decision might be the change to a different machine if the object decides to change the manufacturing system because of a higher potential of the degree of logistic objective achievement. At the completion of a production order the actual logistic performance figures are immediately compared with the target performance figures (normative-actual value comparison). On this basis the degree of logistic objective achievement of an individual object is calculated. This represents the second step of the evaluation system.

By taking all objects within the entire system into account and in combination with weights of different objects it is possible to determine the degree of logistic objective achievement for the overall system at the end of a reference period for example. The weighting of individual objects or object classes allows to emphasize the importance e.g. of bottleneck machines or specific customer orders. The consideration of the overall system represents the third step of the evaluation system. Through the decentralized feedback control of individual objects an opportunity is given to react on eventual changes or disturbances near real time and thus to increase the logistic performance of the overall system while measuring the individual degree of logistic objective achievement.

Vectorial approach to measure the achievement of logistic objectives

The concrete measuring of the degree of logistic objective achievement and the evaluation of alternatives will be done by means of a vectorial approach. Basis for this approach is the logistic objective vector z as shown in the following form:

$$z = \begin{bmatrix} \text{Due date reliability} \\ \text{Throughput time} \\ \text{Utilization} \\ \text{Work in process} \end{bmatrix} \tag{4.1}$$

This format of the vector applies for target vectors as well as for vectors with the actual values, which are used to determine the logistic performance figures to evaluate logistic objects and to evaluate decision alternatives. In order to consider different weights of the logistic objectives a weighting vector γ is introduced. The target value vectors of logistic objects contain the desired values for the individual logistic objectives. By comparison of the target value z_{target} with the actual value vector z_{actual} it is possible to convert the thereby originated vector $\Delta z_{\text{target-actual}}$ in a vector e with the degrees of individual logistic achievement objective:

$$\Delta z_{\text{target-actual}} \Rightarrow e = \begin{bmatrix} e_{\text{Due date reliability}}\,[\%] \\ e_{\text{Throughput time}}\,[\%] \\ e_{\text{Utilization}}\,[\%] \\ e_{\text{Work in process}}\,[\%] \end{bmatrix} \tag{4.2}$$

with $e_{\text{Due date reliability}}$, $e_{\text{Throughput time}}$, $e_{\text{Utilization}}$ and $e_{\text{Work in process}}$ as degree of logistic objective achievement for each individual objective in [%].
The determination of the degree of logistic objective achievement takes place by normative-actual value comparison of the respective objective considering a given distribution, as shown in figure 4.5 using the example of due date variation.

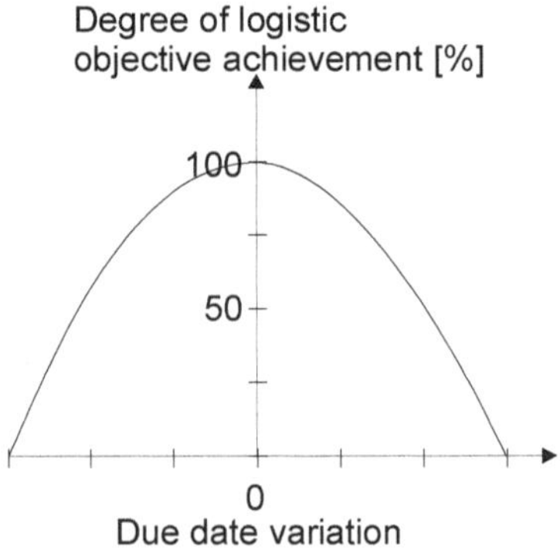

Fig. 4.5 Determination of degree of objective achievement

In this example a due date variation of zero days would lead to 100% objective achievement while a due date variation of two days would approximately lead to only 50% objective achievement. By means of distributions of this type it is possible to determine the logistic objective achievement through reading the difference of target value vector and actual value vector in this diagram. In a next step the achievements of all objectives are aggregated in one degree of logistic objective achievement for the individual object. This is done by introduction of the upper mentioned weighting vector for an individual object. Thus a possibility is given to determine the degree of logistic objective achievement e_{obj} in [%] for an object by calculating the scalar product of weighting vector $\boldsymbol{\gamma}$ and the vector $\boldsymbol{e}$ with the individual degrees of objective achievement:

$$\mathrm{e} \cdot \gamma = \begin{bmatrix} e_{\text{Due date reliability}}\left[\%\right] \\ e_{\text{Throughput time}}\left[\%\right] \\ e_{\text{Utilization}}\left[\%\right] \\ e_{\text{Work in process}}\left[\%\right] \end{bmatrix} \cdot \begin{bmatrix} \gamma_{\text{Due date reliability}} \\ \gamma_{\text{Throughput time}} \\ \gamma_{\text{Utilization}} \\ \gamma_{\text{Work in process}} \end{bmatrix} = e_{\text{obj}}[\%] \tag{4.3}$$

In this case it is very important that the sum of all γ_i within the weighting vector is exactly one to get a proper result in a percentage rate. Consequently, this equation describes the second step of the evaluation system. For the third step of the evaluation system it is essential to aggregate the objects achievement of objectives in one degree of logistic objective achievement for the total system. For this reason it is necessary to implement weights for individual objects, which describe the effects of single objects on the total system. That means that all objects can provide different contributions for the logistic performance of the total system. In this manner it is furthermore possible to consider separately resource classes or order classes. The degree of logistic objective achievement for the total system e_{total} is accordingly determined by:

$$e_{\text{total}} = \frac{\sum_{i=1}^{n} \chi_i \cdot e_{\text{obj}}}{\sum_{i=1}^{n} \chi_i} \tag{4.4}$$

with n as the number of all logistic objects within the system and χ as weighting factor of the logistic object. Through this calculation the degree of logistic objective achievement for production system is ascertainable.

4.2.5 Shop floor scenario

In the following the hypothesis made at the beginning will be verified through simulation studies. In a first step the achievement of logistic objectives, using the example of throughput time, at increasing structural static internal complexity for different autonomous control methods is investigated. For this purpose the previously introduced vectorial approach is implemented with the following weighting vector:

$$\gamma = \begin{bmatrix} \gamma_{\text{Due date reliability}} \\ \gamma_{\text{Throughput time}} \\ \gamma_{\text{Utilization}} \\ \gamma_{\text{Work in process}} \end{bmatrix} = \begin{bmatrix} 0 \\ 1 \\ 0 \\ 0 \end{bmatrix} \tag{4.5}$$

To analyse the ability of an autonomous control to cope with rising complexity a simulation scenario is needed that allows to model different but comparable degrees of complexity and allows for the application of autonomous control methods. Furthermore it should be general enough to be valid for different classes of shop floor types. For these reasons a shop floor model in matrix format has been chosen, see figure 4.5. Subsequent productions steps are modelled horizontally while parallel stations are able to perform resembling processing steps.

At the source the raw materials for each product enter the system. Each product class has a different process plan i.e. a list of operations that have to be fulfilled on the related machine. In case of overload the part can decide autonomously to change the plan and to use a parallel machine instead. The final products leave the system via a drain.

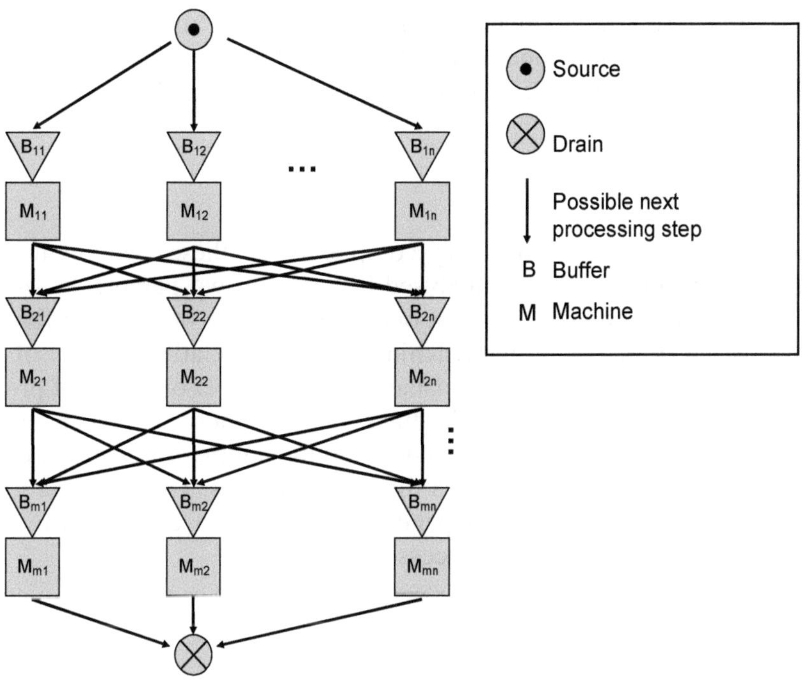

Fig. 4.6 Matrix model of a shop floor.

Autonomous control methods

Two different control methods will be compared. The first method compares the actual buffer states at all the parallel machines that are able to perform the next production steps. Therefore the buffer content is not counted in number of parts but in estimated processing time and the current buffer levels are calculated as the sum of the estimated processing time on the respective machine. When a part has to render the decision about its next processing step it compares the current buffer levels i.e. the estimated waiting time until processing and chooses the buffer with the shortest waiting time. This method will be called “queue length estimator” (QL).

The second method uses data from past events. Every time a processing step is accomplished and a part leaves a machine, the parts generate information’s about the duration of processing and waiting time at the respective machine. The following parts use these data about past events to render the decision about the next production step. The parts compare the mean throughput times from parts of the same class and choose the machine with the lowest mean duration of waiting and processing. This method will be called "pheromone method" (PHE) as it is inspired by the behaviour of social insects which use pheromone trails to find shortest paths.

Simulation model

The ability to cope with rising complexity of these two methods for autonomous control will be analysed by varying two parameters of static structural internal complexity. On one hand, the size of the shop floor will be increased from 3x3 to 9x9 machines while the relative number of product/order classes will be kept constant i.e. the number of different products is equal to the number of parallel lines. On the other hand, the size of the shop floor will be held constant at 4x4 and the number of different product classes will be varied from 4 to 8 different products. The processing plans of the products differ i.e. it depends on the product class on which machines the product should be processed.

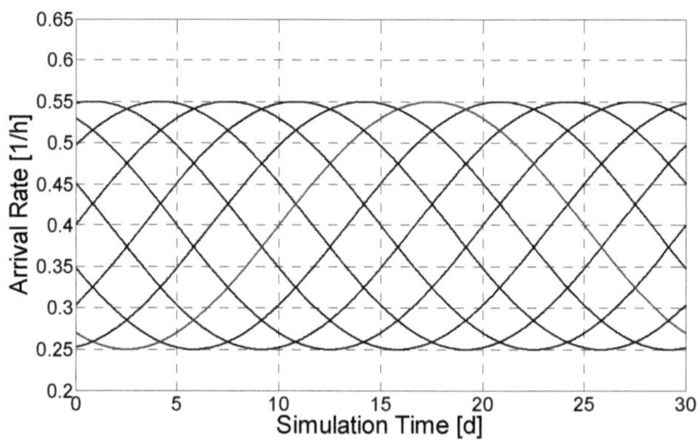

Fig. 4.7 Arrival rate during one simulation period for eight different products

To model a highly dynamic market situation the demand for the different products is set as an oscillating curve with situations of over and under load. The resulting arrival rates of parts that enter the shop floor are shown in figure 4.7.

As simulation period 30 days are chosen. After a phase of two month (with 30 days each) for avoiding transient effects the third month is used to measure the throughput times of every single part that is finished.

For balancing conditions the minimal processing time per manufacturing step is equally 2 hours. This minimal processing time can only be reached if the parts follow exactly the pre-planned processing plan without taking into account the current situation on the shop floor. If the parts decide to use parallel machines instead the throughput time will rise because of transport processes and set up times and higher processing times on parallel machines. This additional time depends on the number of parallel machines that are available for a production step. The additional time t_b is calculated by the distribution of one hour over the number of parallel machines:

$$t_b = {}^{1h}\!/_{N} \tag{4.6}$$

Simulation results

For the simulation experiments a discrete event simulator is used. Figure 4.8 shows the influence of the rising network size on the mean throughput time of the whole orders. This time is measured as the time difference between job release i.e. the appearance of a part at the source and job completion i.e. leaving the shop floor at the drain. The figure shows the mean throughput time for all parts and all different product classes for the two different autonomous control methods. Additionally the minimal throughput time is shown which is a linear rising function of the network size because more production steps have to be undertaken as the shop floor size is increased. It appears that the rising system size has no effect on the mean throughput time applying the Queue Length Estimator as the curve is nearly parallel to the minimal throughput time. The Pheromome Method on the other hand shows a more and more worse performance as the mean throughput time rises exponentially with increasing network size.

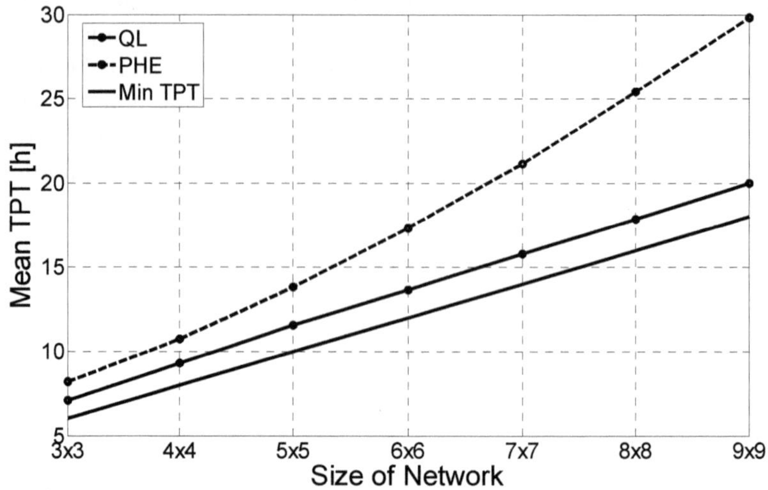

Fig. 4.8 Mean throughput time for different network sizes

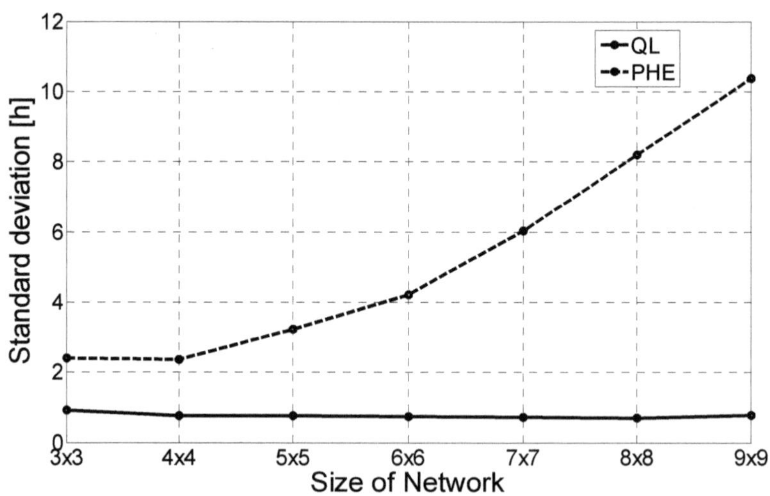

Fig. 4.9 Standard deviation of the throughput time for different network sizes

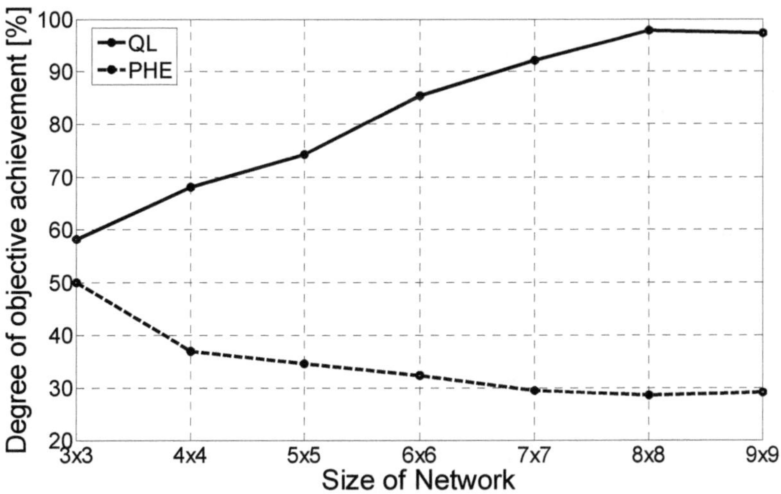

Fig. 4.10 Fraction of parts that are finished within 120% of the minimal throughput time for rising network size.

One realizes the same effect in the standard deviation of the throughput times which is displayed in figure 4.9. With rising network size the standard deviation is even decreasing for the QL method. For the PHE method also the standard deviation of the through put time is rising with higher network size.

The mean and the standard deviation are important measurements for the predictability of the throughput time and therefore essential for the due date reliability. Figure 4.10 shows the fraction of parts (called degree of job achievement) that are finished within 120% of the minimal throughput time. For the QL method this fraction rises with larger network size while for the PHE method this fraction decreases. This follows directly from the data for mean and variance. For the QL method mean and variance have a constant run. Therefore more and more parts are within the tolerance limit of 120% whose absolute value is rising analogue to the minimal throughput time. Accordingly the decreasing run of the curve for the PHE method follows from the data about mean and variance.

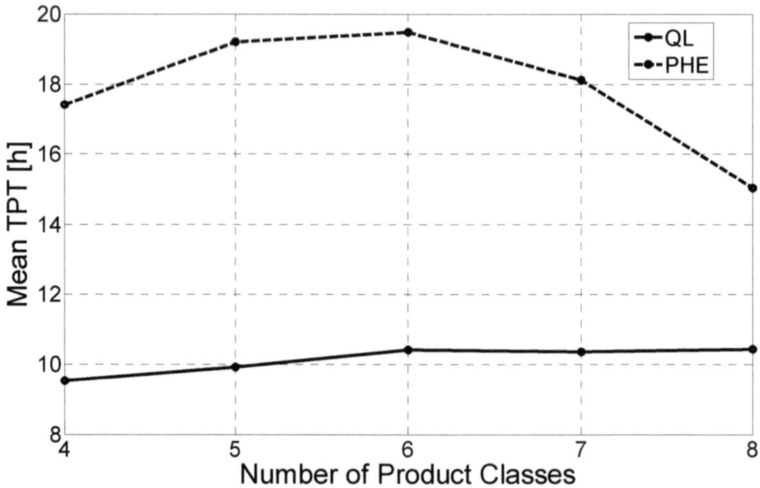

Fig. 4.11 Mean throughput time for different number of product classes

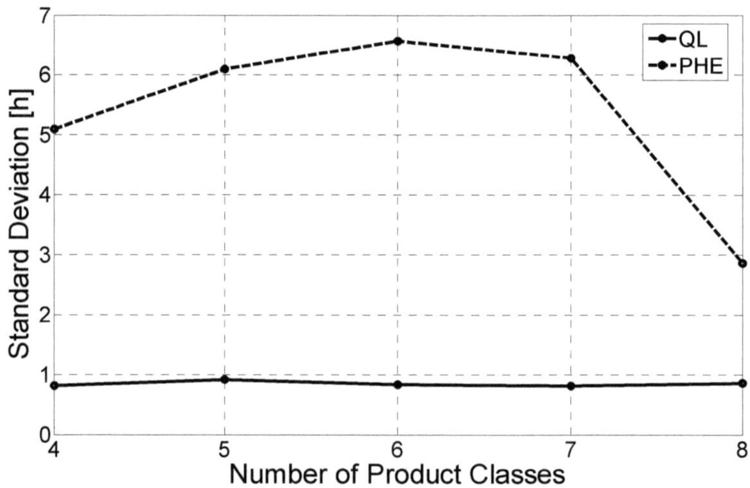

Fig. 4.12 Standard deviation of the throughput time for different number of product classes

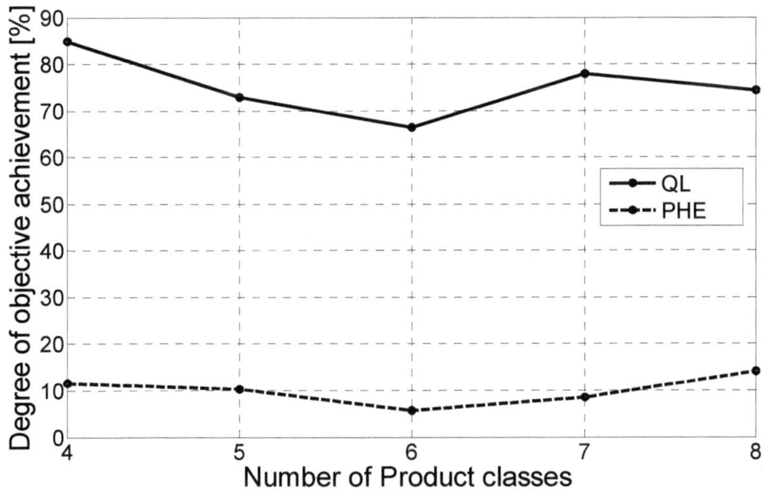

Fig. 4.13 Fraction of parts that are finalised within 120% of the minimal throughput time for different number of product classes

In a second step the number of different product classes is varied. Figure 4.11 shows the mean throughput time within a 4x4 shop floor for four to eight different products. Again the QL method shows a better performance than the PHE method but a trend is observed that for a rising number of product classes the performance of the PHE method is getting better. The same effect can be seen in figure 4.12 where the standard deviation of the throughput time is shown and for seven and eight product classes the PHE method is showing a decreasing standard deviation. Figure 4.13 underlines this effect in showing the fraction of parts that are finished within 120% of the minimal throughput time and which are rising for the PHE method from six to eight different products.

Interpretation

The appliance of the QL method shows a constant performance in face of rising static structural internal complexity i.e. a higher number of machines on the shop floor while the PHE method is not able to maintain a sufficient performance. An exponential increase in mean and standard deviation of the throughput times is observed. This is also caused by the fact that with a rising number of machines the number of possible parallel machines is increased and therefore the switching onto other less utilised machines is facilitated. Because the PHE method shows in general a slower behaviour than the QL method the ability to switch more frequently is not exploited.

In the second case of a higher number of different order or product classes than parallel machines also the order arrival is modified. Because the mean utilization should be comparable the mean arrival rate has to be lowered every time a new product class is added to the model. Therefore the higher number of product classes causes also a more balanced utilisation of the system. This reduces the possibility and the necessity to change the processing plan and to move to a parallel machine. This improves the situation for the slower PHE method and allows for a trend to better results at a higher number of product classes.

The major difference between the two methods is the character of the used information. The QL method uses information about estimated processing times while the PHE method uses information about past events. Because the PHE method calculates a mean value of the past throughput times this method reacts more slowly on highly dynamic situations with fast changing system conditions. This causes fewer switches to parallel machines.

As a result one can state that in situations of a high number of machines that have to be equally utilised the QL method is more advisable because it shows a constant performance despite rising structural complexity.

The PHE method shows here a decreasing performance. In case of a high number of different products the PHE method could be an alternative. In particular when the trend is extrapolated the PHE method could show a better performance than the QL method.

4.2.6 Conclusions and outlook

At the beginning of this paper an assumption has been made that decentralised systems with autonomous control methods could be an approach to cope with rising complexity. A global definition as well as a definition in the context of engineering science was given. To verify in which cases the implementation of autonomous processes is of advantage in relation to conventionally managed processes an evaluation system is necessary. Main tasks regarding the development of this evaluation system are the operationalisation of the logistic objective achievement, the level of autonomy and the production systems complexity.

Within this article a vectorial approach to measure the achievement of logistic objectives together with a feedback loop for autonomous processes was introduced. By means of a complexity cube it is also possible to operationalize the complexity of production systems regarding different types of complexity.

In simulation studies the ability to cope with rising complexity of two different autonomous control methods has been compared. Thereby different trends have been determined. The QL method based on a "look ahead approach" shows a constant performance at rising system complexity. It is obvious that systems of this size can also be controlled by traditional centralised PPC systems. But, if one extrapolates the trend there will be certainly a critical size were the constant performance of the QL method is superior to a centralized PPC method.

The PHE method based on a "look back approach" shows a slowly reacting behaviour and could be an alternative if it is not favourable to have permanent processing plan changes. So far the quality and dependability of data used by the two methods have not been taken into account. It seems to be realistic that information about past events are more reliable than information about future events. The smaller error in the information could further improve the performance of the QL method in comparison to the PHE method.

Further Research has to be done on the development of the evaluation system regarding the operationalization of the level of autonomous control and the definition of complexity parameters for the different vectors in the complexity cube. Furthermore additional simulation studies will help identifying for which types of increasing complexity the implementation of autonomously controlled processes is of advantage.

References

Amara L, Ottino J (2004) Complex networks. European Physical Journal B

Barabasi A., Oltvai Z (2004) Network biology: understanding the cells functional organization. Nature 5: 101-113

Breithaupt J. (2001) Rückstandsorientierte Produktionsregelung von Fertigungsbereichen. VDI Verl, Düsseldorf

Costa L, Rodrigues F, Travieso G, Villas Boas P (2005) Characterization of complex networks: A survey of measurements. cond-mat/0505185

Deshmukh A, Talavage J, Barash M (1998) Complexity in manufacturing systems. Part 1: analysis of static complexity. IIE Trans 30: 645-55

Fleisch E, Kickuth M, Dierks M (2003) Ubiquitous Computing: Auswirkungen auf die Industrie. Industrie Management 19(6): 29-31

Freitag M, Herzog O, Scholz-Reiter B (2004) Selbststeuerung logistischer Prozesse – Ein Paradigmenwechsel und seine Grenzen. Industrie Management, 20(1): 23-27

Frizelle G (1998) The management of complexity in manufacturing. Business Intelligence, London
Frizelle G, Woodcock E (1995) Measuring complexity as an aid to develop operational strategy. International Journal of Operations and Production Management 15(5): 26-39
Gellmann M, Lloyd S (1996) Information measures, effective complexity, and total information. Complexity 1(1): 44-52
Jones A, Reeker L, Deshmukh AV (2002) On information and performance of complex manufacturing systems. Proceedings of the Manufacturing Complexity Network Conference
Petermann D (1996) Modellbasierte Produktionsregelung. VDI-Verl, Düsseldorf
Philipp T, Böse F, Windt K (2006) Autonomously Controlled Processes - Characterisation of Complex Production Networks. In: Cunha, P.; Maropoulos, P. (eds.): Proceedings of 3rd CIRP Conference in Digital Enterprise Technology. Setubal, Portugal
Scherer E (1998) The Reality of Shop Floor Control – Approaches to Systems Innovation. In: Scherer E. (ed): Shop Floor Control – A Systems Perspective. Springer Verlag, Berlin
Scholz-Reiter B, Windt K, Freitag, M (2004) Autonomous Logistic Processes - New Demands and First Approaches. In: Monostori L (ed) Proc. 37th CIRP International Seminar on Manufacturing Systems. Hungarian Academy of Science, Budapest, Hungaria, pp 357-362
Schuh G (2005) Produktkomplexität managen. Carl Hanser Verlag, München
Shannon CE, Weaver W (1949) The mathematical theory of communication. Urbana, IL: University of Illinois Press
Sivadasan S, Efstathiou J, Shirazi R, Alvez J, Frizelle G, Calinescu A (1999) Information complexity as a determining factor in the evolution of supply chain. Proceedings of the International Workshop on Emergent Synthesis
Ulrich H, Probst G (1988) Anleitung zum ganzheitlichen Denken und Handeln. Haupt, Bern, Stuttgart
Wiendahl HH, v Cieminski G, Wiendahl HP (2005) Stumbling blocks of PPC: Towards the holistic configuration of PPC systems. Production Planning and Control 16(7): 634-651
Windt K, Böse F, Philipp T (2007) Autonomy in Logistics – Identification, Characterisation and Application. International Journal of Computer Integrated Manufacturing. Forthcoming

4.3 Autonomous Control by Means of Distributed Routing

Bernd-Ludwig Wenning[1], Henning Rekersbrink[2], Andreas Timm-Giel[1], Carmelita Görg[1], Bernd Scholz-Reiter[2]

[1] Communication Networks, University of Bremen, Germany
[2] Department of Planning and Control of Production Systems, University of Bremen, Germany

4.3.1 Introduction

In current logistic practices, routing and assignment of transport orders to vehicles are done centrally by a dispatching system and/or a human dispatcher. Here, the dispatching problem is generally of static nature and is solved either by the use of heuristics, e.g. evolutionary algorithms or Tabu search, or by applying "rules" that are gained from experience, when done by a human dispatcher.

The modern logistic systems permit incorporation of dynamic features into the dispatching problem. Here, dynamic means that not all orders are known a-priori, and an order can change its attributes with time. In most solution methods, the dynamic problem is broken into a sequence of static problems, so that the same or similar heuristic approaches can be used sequentially. The problem is thus repeatedly solved at the central planning instance whenever some change occurs in the order situation. Such algorithms are known as online algorithms (Fiat and Woeginger 1998, Gutenschwager et al. 2004).

In the subproject B1 "Reactive Planning and Control", a completely different approach for dealing with dynamic problems is introduced and investigated: Vehicles and packages are considered to be intelligent and autonomous. They can decide about routes and loads by themselves based on local knowledge. This requires replacement of the centralised decision-making approach by a decentralised, distributed autonomous control approach. For this approach, methods and algorithms from other domains of science and technology are evaluated for their suitability for application in transport logistics. One promising technology domain is the wide range of routing algorithms used in communication networks.

4.3.2 Routing algorithms in communication networks

Distributed routing as such has already been successful in communication networks for several decades. Therefore, routing methods used in communication networks are identified to be interesting for use in transport networks.

As far as use of routing algorithms is concerned, communication networks can be classified into infrastructure-based networks and ad-hoc networks. These two different types have specific properties that lead to a significant difference in the way routing is done.

Infrastructure-based networks

Currently, most communication networks are infrastructure-based. In this type of networks, there is a hierarchy present where routing is usually done by dedicated nodes, called routers, within the network. Their responsibility is to keep track of the network status and enable attached nodes to communicate with others. Usually, the topology of infrastructure-based networks is not very dynamic, as the routing information there can be valid for a long time.

Large-scale networks often consist of several subnetworks which are interconnected through router to router connections. There can also be several levels of hierarchy there, like for example in Internet - local provider - company level network - department level network and so on. At different levels of the hierarchy, different routing methods may be used.

Basically, routing protocols in infrastructure-based networks are divided into Interior Gateway Protocols (IGP) and Exterior Gateway Protocols (EGP), depending on whether they route within one network or between networks. The most prominent IGPs are Routing Information Protocol (RIP) (Malkin 1998) and Open Shortest Path First (OSPF) (Moy 1998). As EGP, the Border Gateway Protocol (BGP) (Rekhter et al. 2006) is most widely used and can be considered as the "quasi-standard" routing protocol in the Internet.

Ad-hoc networks

In ad-hoc networks, there is no fixed infrastructure and hierarchy. Mostly, the term ad-hoc networks is used for mobile/wireless ad-hoc networks where wireless devices „spontaneously" form a network. In such networks, there are no nodes that are specifically dedicated for routing, but each node may act as a router. Further, due to the node mobility, the network topology is not necessarily fixed once the network is established, and may change very frequently as nodes move or even leave the network. This means that routing in ad-hoc networks has to cope with the dynamic changes in network topology. Several different approaches to solve this problem have led to a vast amount of routing algorithms which can be classified into three categories: Proactive routing, reactive routing and hybrid routing (Perkins 2001).

Proactive routing

When proactive routing is used, each node in the network maintains a routing table for all other nodes in the network. The nodes exchange their route information either on a regular basis or as soon as they detect a change. The advantage of proactive routing is that up-to-date information about the routes and thus the network status is always available. The drawback is that it needs a high signalling overhead to maintain the routing tables, especially in highly dynamic networks.

The most common examples of proactive routing protocols are Destination Sequenced Distance Vector (DSDV) (Perkins and Bhagwat 1994) and Optimized Link State Routing (OLSR) (Clausen and Jacquet 2003).

Reactive or on-demand routing

In contrast to proactive routing, reactive routing, often referred to as on-demand routing, does not constantly maintain routing tables on all nodes. Here, routes are reactively detected when they are needed, i.e. the node that wants to send something starts the route discovery process by sending a route request to its neighbours. This request propagates through the network until a route to the destination is found, then a route reply is sent back to the originator, which then leads to the establishment of the data link.

The obvious advantage is that there is less signalling overhead related to the maintenance of route tables. A drawback is that route discovery takes some time, which results in an initial delay for the sender before it can transmit its data. Further, in large scale ad-hoc networks, frequent route request floods can also produce a high signalling overhead.

Examples for on-demand routing protocols are Dynamic Source Routing (DSR) (Johnson and Maltz 1996), Ad-Hoc On-Demand Distance Vector (AODV) (Perkins et al. 2003) and Dynamic MANET On-Demand Routing (DYMO).

Hybrid routing

Hybrid routing tries to combine the advantages of proactive and reactive protocols. One example is Zone Routing Protocol (ZRP) (Haas 1997), where routing is done proactively for routes to nodes inside a limited zone and on-demand for routes to nodes outside the zone.

Context aware routing

A special class of ad-hoc protocols that is currently emerging covers more than just link quality or hop counts: Context aware routing protocols are designed to include information about the context of a node. This context information can be information about the node's location, energy resources, importance of the transmission and so on. In most cases they consider one context only, e.g. energy of the individual nodes. The context aware routing protocols extend the existing proactive and reactive protocols.

4.3.3 Comparison of logistic and communication networks

For a transfer of routing methods from communication networks to logistic networks, it is necessary to identify where these networks are similar and where they have differences. Obvious similarity between both networks is that in both, payloads have to be transported from a source to a destination. Generally, there are different routes available for such a transport, so that the best route has to be chosen based on some selection criteria. However, the criteria that influence the decision between two or more route options can be very different and specific to the network type.

Another similarity is the possibility for resource reservation in both networks. In both cases, it is related to a Quality of Service (QoS), in case of logistics, this means fulfilling certain transport conditions, in case of communication networks, it means guaranteeing the fulfilment of bandwidth requirements, loss probability limits etc.

Size and dynamics of both network types are also comparable. The autonomous-control approaches for transport logistics are specifically targeted for efficient operation of dynamic large-scale networks, which is achieved in communication networks by using decentralized control.

There are also significant differences between communication and logistic networks. One difference is that there are entities such as vehicles, containers and pallets in a logistic network that are physically existent and limited in their number, whereas there is nothing comparable in communication networks, especially concerning the persistence, but also concerning the hierarchy. This hierarchy of movable objects leads to the possibility of conflicting interests concerning the route choice. If, for example, the load's goal is a fast or just-in-time transport, and the vehicle's goal is maximum utilisation of its cargo space, they might prefer different routes to reach their individual goals.

Furthermore, there is a difference in how to handle losses. In communication networks, a packet loss is not unusual, and the packet can be retransmitted. This is not the same in logistics, as a piece of good can not be duplicated easily, making a retransmission either very expensive or even impossible.

A very significant difference between both networks is the scale of time. In communication networks, both the route formation and the actual data transmission work on time periods in the range of seconds or milliseconds. The time that is required for route selection is generally not negligible in comparison to the transmission time. In logistics, on the other hand, the transport of the payload takes much longer (hours, days). This implies that the time needed to determine a route is far less compared to the transport duration and therefore, it is permissible to do more communication and calculations in order to get the best route for the current conditions.

This leads to the conclusion that routing methods from communication networks cannot be transferred directly into logistics. Nevertheless, routing approaches in communication networks can inspire in devising routing approaches for logistic networks. In doing this, it is desirable to address the special requirements of transport networks while keeping the advantages of the proven communication network methods such as robustness and automatic failure recovery to the maximum possible extent.

For distributed routing of autonomous components, it is necessary that they collect information that influences the routing decisions. This can be information about the current status of edges[7], such as traffic jams and information about other components' plans if they have influence on the route. This information retrieval is a point where aspects from communication networks can be used. Assuming the information is available at the vertices[8], it can be collected similar to a route discovery process in ad-hoc routing algorithms: Route request messages are sent from the entity that needs the information. These requests are propagated through the network from vertex to vertex until they reach their destination, then a route reply message is sent back.

4.3.4 A distributed routing concept

In the following, a concept for distributed routing in a logistic network is presented. In this concept, vehicles as well as packages are considered as autonomous. They have sufficient intelligence and communication capabilities to get their information and to decide on the next steps to be undertaken.

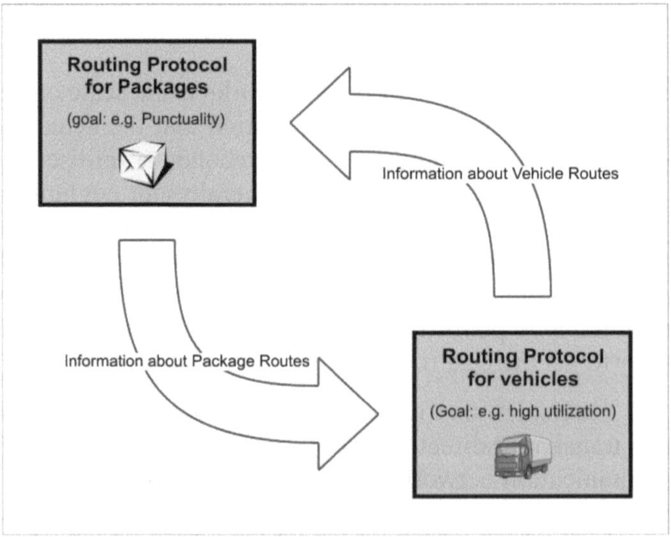

Fig. 4.14 Interdependence of routes

[7] Edges are connections between locations (vertices) in a logistic network, e.g. roads. For details about the definition of logistic network components see chapter 4.2: Dynamic Transport Reference Scenarios.

[8] Vertices are locations in the logistic network where edges meet, e.g. depots. See also footnote 2.

In this concept, next steps mean calculating a route or deciding about being loaded into a vehicle (from the package's view) or picking up a package (from the vehicle's view). If both the vehicles and the packages determine routes based on their individual goals, the dilemma arises that the routes are most probably different. To make it worse, the decisions are interdependent: The package needs knowledge about vehicle routes to find candidate vehicles and the vehicle needs knowledge about the package routes to be able to find an efficient route where its capacity is best utilized. Figure 4.14 illustrates this interdependence.

The interdependence implicitly gives rise to another issue: The knowledge of each other's existence, i.e. how does the package know which vehicles are there, and further: How does the vehicle know about the packages? If there is no way to get to know about each other, they cannot communicate and thus cannot exchange their information.

There are two possibilities to solve this problem:

- Direct communication: An entity, say a package that enters the system, broadcasts some information about itself and collects responses from all other present entities. This is very inefficient and would lead to a high load of communication signalling, and the entities which are currently out of communication range might not get the information.
- Indirect communication: This assumes the presence of some kind of knowledge brokers or repositories in the network. In this way, both the vehicles and the packages know entities to whom they can send their information and where they retrieve other information.

Distributed Logistic Routing Protocol (DLRP)

Due to the drawbacks of the other solution, the indirect communication was chosen as the way to solve the interdependence problem. As it is not intended to introduce an additional central repository, which would in fact foil the idea of a distributed system, the vertices that are present in the logistic network are chosen as the "relays" for indirect communication and therefore as the knowledge brokers. This fits perfectly into the distributed nature of the concept, as each vertex has only a part of the global knowledge, rather than the complete knowledge about all routes and all packages in the system.

In detail, the concept, named "Distributed Logistic Routing Protocol" (DLRP), operates as follows (Scholz-Reiter et al. 2006):

The vertex is a knowledge broker for the vehicles and packages. Before deciding about a route, a vehicle/package requests current information from the current or next vertex. Each vertex includes relevant information available from its current knowledge-base and forwards the request to neighbour vertices. The neighbour vertices do the same and forward it further. This way, the request is propagated through the network until the destination or a predefined hop limit is reached. Then the last vertex creates a reply message that is sent back directly to the originator of the request. This reply contains all the information that has been collected during the propagation of the request message through the network, including the last vertex. In general, an entity can receive more than one route reply as there are multiple paths possible. As it is not known how many replies would get back, a timeout and an upper limit for the number of replies are specified in order to trigger the decision process without long waiting periods.

After receiving the reply messages, the entity is ready to make its route decision based on its individual preferences and the data received. After making the decision, it withdraws its old route if any, and announces its new route to all relevant vertices. This way, the vertices get an information update, which will be used in processing the future requests. Figure 4.15 shows the information flow in DLRP.

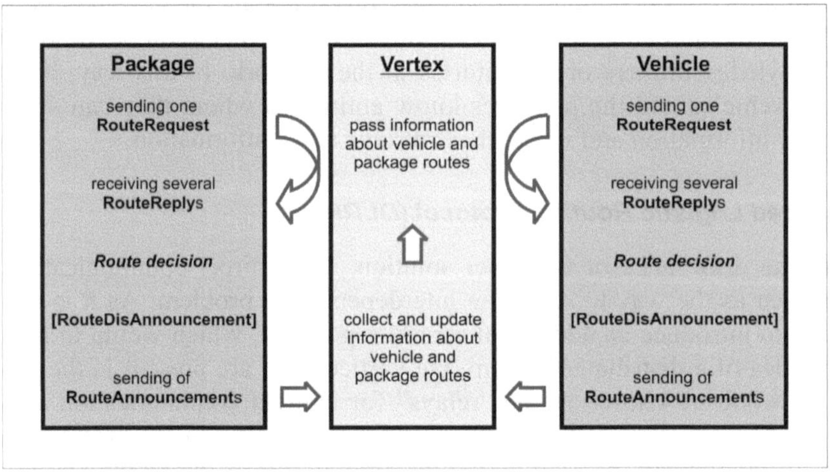

Fig. 4.15 DLRP information flow

This approach also leads to uncertain knowledge: As a package does not know in advance whether a specific vehicle picks it up or not, it looks for a set of alternative routes to increase the probability to reach its destination in time. All these alternative routes are announced to the vertices, so that the announced package routes are just valid with a certain probability. If a package is picked up by a vehicle, unused routes have to be cancelled again. Vehicles on the other hand do not necessarily stick to a single route, so the vehicle routes also are uncertain. The vehicles check the current state of their options whenever they reach a vertex. If they find a route that is better than the original one, they can either change their decision depending on their individual settings, or stick to the old one.

The DLRP itself does not specify the functions that are used by the packages and vehicles to decide about their routes, it just specifies the interaction. Therefore, it should be regarded as an interaction framework which provides a basis for distributed information management and decision making in logistic scenarios. The logistic performance that can be achieved with this framework strongly depends on how the logistic entities utilise the information they can obtain. There are several possibilities for decision making, for example fixed rule sets (e.g. always take the shortest route), heuristic, probabilistic or fuzzy logic approaches etc. Some of these options are under investigation for their use in the DLRP framework.

Currently, the DLRP functionality has been successfully implemented in a logistic simulation environment. Now, suitable decision making approaches are being developed and evaluated by simulation in that simulation environment.

In the first decision-making approaches, the vertices calculate a metric based on the route announcements of vehicles and packages at this vertex, excluding those from the vehicle/package that initiated the current routing process and those from vehicles/packages that are expected to be later than it. This metric m is calculated according to the following formula:

$$m = \sum m_{package} - \sum m_{vehicle} \tag{4.7}$$

Here, $m_{package}$ is the individual metric related to a package route announcement, and $m_{vehicle}$ is the one related to a vehicle route announcement. The package metric is determined as follows:

$$m_{package} = \frac{0.25^{d_1}}{d_2} \tag{4.8}$$

Here, d_1 is the distance between the package's current location and metric-calculating vertex and d_2 is the distance from this vertex to the package's destination. In this way, the closer the package is to the vertex, the more relevant it is for routing. The vehicle-related metric is calculated similarly:

$$m_{vehicle} = C \times 0.5^{d_1} \tag{4.9}$$

The distance d_1 is again the distance between the vehicle's location and the vertex, and C is the vehicle's capacity.

The package's goal is now to find a route with a low metric, as a low metric means it is more likely to find free vehicle capacity there. Vehicles on the other hand try to find high metrics in order to maximize the utilisation of their capacity.

From the simulations, additional constraints were derived that have to be taken into account for the route decision process:

- Vehicles should prefer continuing a route they have started. Therefore, in each re-routing step, routes that continue the current one get a bonus.
- Package routes should not lead the package back to where it came from when a package is recalculating its routes.
- If there are several alternative routes from one package registered at a vertex, only one of them (preferably the one with earliest arrival time) is considered for metric calculation.

These constraints have shown to be crucial for the approach to operate as intended.

The decision-making approach presented here is currently being investigated in detail to evaluate its performance, and in-depth results will be shown in publications in the near future.

4.3.5 Conclusions and outlook

This chapter presents the DLRP, a fully distributed routing concept for dynamic logistics. The concept has been implemented into a logistic simulation environment to prove its feasibility. For performance evaluation, different decision functions are being investigated using simulations within this concept in order to obtain an efficient solution for routing in dynamic logistic environments.

References

Clausen T, Jacquet P (2003) Optimized Link State Protocol (OLSR). Internet Request for Comments 3626. http://tools.ietf.org/html/rfc3626

Fiat A, Woeginger GJ (1998) Online Algorithms: The State of the Art. Springer, Berlin

Gutenschwager K, Niklaus C, Voß S (2004) Dispatching of an Electric Monorail System: Applying Metaheuristics to an Online Pickup and Delivery Problem. Transportation Science 38 (4): 434-446

Haas Z (1997) A New Routing Protocol for the Reconfigurable Wireless Networks. Proceedings of the IEEE International Conference on Universal Personal Communications, pp 562-566

Johnson D, Maltz D (1996) Dynamic Source Routing in Ad Hoc Wireless Networks. In: Imielinski T, Korth H (eds) Mobile Computing. Kluwer Academic Publishers, Dordrecht, pp 153-181

Malkin G (1998) RIP Version 2. Internet Request for Comments 2453. http://tools.ietf.org/html/rfc2453

Moy J (1998) OSPF Version 2. Internet Request for Comments 2328. http://tools.ietf.org/html/rfc2328

Perkins CE (2001) Ad Hoc Networking. Addison-Wesley, Boston

Perkins C, Belding-Royer E, Das S (2003) Ad hoc On-Demand Distance Vector (AODV) Routing. Internet Request for Comments 3561. http://tools.ietf.org/html/rfc3561

Perkins C, Bhagwat P (1994) Highly Dynamic Destination-Sequenced Distance-Vector Routing (DSDV) for Mobile Computers. Proceedings of the Conference on Communications Architectures, Protocols and Applications, pp 234-244

Rekhter Y, Li T, Hares S (2006) A Border Gateway Protocol 4 (BGP-4). Internet Request for Comments 4271. http://tools.ietf.org/html/rfc4271

Scholz-Reiter B, Rekersbrink H, Freitag M (2006) Kooperierende Routingprotokolle zur Selbststeuerung von Transportnetzen. Industrie Management 3: 7-10

4.4 Dynamic Transport Reference Scenarios

Bernd-Ludwig Wenning[1], Henning Rekersbrink[2], Markus Becker[1], Andreas Timm-Giel[1], Carmelita Görg[1], Bernd Scholz-Reiter[2]

[1] Communication Networks, University of Bremen, Germany

[2] Department of Planning and Control of Production Systems, University of Bremen, Germany

4.4.1 Introduction

Reference scenarios are a common technique in simulations allowing the evaluation and comparison of different algorithms and approaches. For transport logistic processes these approaches can be for example different strategies to select the packets to be loaded.

Different reference scenarios are required ranging from simple scenarios for easy understanding the effects up to complex and realistic scenarios comprising all major factors to be considered. As the focus here is on dynamic transport problems, the scenarios should facilitate representation of such dynamics.

4.4.2 Traditional scenarios

There are few scenarios which are commonly used to model logistic transport processes. Well-known examples are the Solomon Instances (Solomon 1987) and scenarios derived from them. The Solomon Instances are scenarios for so-called "vehicle routing and scheduling problems with time windows". They consist of a list of orders, their locations and their time constraints and of a set of vehicles that have to serve the orders. Derived scenarios can also be used for "pickup and delivery problems" when pairs of orders from the original scenarios are combined to orders that have to be picked up in one location and delivered to another. However, these scenarios have major drawbacks for modelling dynamic transport processes as investigated in the CRC:

- They assume direct connections between all locations in the scenario;
- They are not dynamic in the sense that all destinations and transport orders are known in advance;
- No "travelling obstacles" such as traffic jams or road closures are assumed.

This leads to the conclusion that the traditional logistic scenarios are not suitable for the investigation of dynamic transport processes. Therefore, new scenarios have been developed and are presented here. The scenarios describe all relevant elements of the logistic transport process.

4.4.3 Components of dynamic transport logistic scenarios

In the following, the terms for the description of a general model for dynamic multi-modal transport networks are defined. The set of terms described here build the basis for the description of scenarios.

A model of a transport network has to represent on the one hand the infrastructure, i.e. the route network, the trans-shipment points, storage facilities and other locally fixed objects which can be shown on a map or a weighted network graph. For the representation of the route network, directed graphs are used, so parts of the terminology (vertex, edge) originate from graph theory. On the other hand, the model has to represent the movable parts of the transport process, i.e. the goods to be transported (packages) and the carriers for these goods (vehicles). Three elementary information carriers, order, suborder and shipment are introduced, which can be assigned to different packages or groups of packages. These elements permit the representation of data related to the packages including the possibility that packages can be aggregated to larger load units for sections of the transportation route taking into consideration that a given transport order can include goods for several destinations. In the following, the components of the model and their characteristics are briefly described.

Vertices

Vertices in general are static points in the network where two or more edges meet. At vertices, load bundling/unbundling and trans-shipment tasks can take place.

The description of a vertex includes the definition of functional units located inside the vertex, like storage facilities and trans-shipment possibilities. In a multi-modal network, the transition between edges of different types in a vertex is closely linked with trans-shipment processes.

Possible types of vertices:

- Pure furcation point: A pure furcation point is a vertex without storage or trans-shipment facility, load or unload possibility. A vertex of this type, however, permits route continuation in different directions;
- Pure trans-shipment point: This is a location where only trans-shipments can take place, but the direction of travel cannot be changed. For example, this is a port where a (one way) street is terminating. The arriving trucks wait (requiring parking capacity) until a RORO[9] ship with free transport capacity arrives and transports them over water to the next vertex (harbour) where the trucks can leave the ship;
- Multi-modal vertex with limited trans-shipment possibility: This is a type of vertex which generally allows transport mode changes, but might have restrictions concerning mode change directions due to the limitations of available equipment. An example is a train station located at a road which has the capability to transfer loads from trucks to trains, but not from trains to trucks or from one truck to another;
- Pure storage vertex: A vertex which just provides storage functionality. An example for a storage vertex can be a highway car park where trucks can wait for the duration of the weekend driving ban[10]. Trans-shipment possibilities or route forks do not exist in general;

Sources and sinks

Sources and sinks are special vertices or functional units assigned to vertices of a network. A source is the sender of a package and a sink is the receiver of the package. The function of a source is to generate transport orders, suborders or shipments and the packages assigned to these orders. The rules, lists or distributions with which transport orders are generated at a source strongly depend on the logistic scenario considered.

Sinks receive packages and complete the transport orders. Once a transport order is completed, the order and the related packages are removed

---

[9] Roll-On/Roll-Off, a type of ship where vehicles (cars, trucks, sometimes also trains) can directly drive onto the deck

[10] In Germany, heavy trucks are not permitted to drive between 0h and 22h on Sundays and public holidays, except for transports of fresh food like fish, milk, vegetables.

from the network. Sources and sinks have to be able to store packages until a vehicle with adequate space picks them up or an order is completed.

Edges

The physical connections between vertices, like roads, railways or water ways are named edges. All edges are considered to be directed. An edge therefore has an origin and a destination vertex and a fixed length. In addition, it carries information about permitted transport velocity which usually depends on the type of vehicle and the time of the day.

In multi-modal transport networks, different types of edges are possible. This leads to the possibility of having several directed edges of same or different types between two vertices, which can even be absolutely equivalent for certain types of vehicles.

Vehicles

All means of transport carrying packages along edges of a network are called vehicles. Vehicles are limited in number and can not arbitrarily enter or leave the scenario.

Each vehicle is assigned a type, e.g. ship, aircraft or truck. Each type can have further sub-type specifications, e.g. container truck (for a special type of container), hazardous material truck with special trans-shipment equipment, etc. The type of vehicle contains attributes to give vehicle dependant information about the goods the vehicle can carry and the conditions under which these goods can be transported and (un-)loaded. Further, the type implies the ability to use certain edges.

The speed with which the goods can be transported is at least limited by a maximum speed assigned to the vehicle. Further, a vehicle has a defined load capacity. A vehicle in use can have its capacity unused, partially used or fully used, depending on the orders the vehicle is carrying out.

Packages

Each form of transport good in a fixed packing is called a package. This means, in the model, a package is the smallest unit of goods to be transported. The kind of content of a package which can imply special transport conditions and treatment during trans-shipment (e.g. frozen goods, hazardous goods etc.) is described by its type.

A package has volume and mass - or more generally, it occupies load capacity of a vehicle during transport and storage capacity during intermediate storage. Packages undergo processes of load forming in the logistic context. In the presented formalism, this load forming (bundling) is expressed using the concepts of orders, suborders and shipments.

Orders, suborders and shipments

The concept of a transport order as a model component provides information which is mandatory for the description of a logistic network. The transport order contains all the information needed for carrying out the transport of a package or a group of packages. In addition, the order may contain several suborders. There is the possibility also to specify the desired contractor if necessary.

The original order of the transport goods is generated at its source. An order is generated when there is a need to transfer goods from one location to another. For each package, there is a related transport order which contains the information needed for the execution of the transport. An order is generally completed at the destination, which is the relevant sink. The order is completed only after all the packages belonging to the order have reached the sink and have been grouped together.

Shipments are information objects describing the non-interrupted transport of a fixed amount of goods between exactly two nodes and using exactly one vehicle. This means that the shipment is only temporarily existent and it is assigned to the vehicle that is processing this shipment. As a vehicle can transport packages from different orders simultaneously, a shipment can contain packages from several orders and suborders.

4.4.4 Evaluation criteria for transport scenarios

When investigating the quality of an approach, there is the need to evaluate its performance levels with respect to the aspired goals. Therefore a set of evaluation criteria is required. Considering transportation logistics, the goal is to achieve a high logistic efficiency, i.e. high performance at low cost. Two sets of possible evaluation measures are introduced in the following:

Volume-related measures

- Queued packages: This is the number of packages that are located at a vertex and waiting for transport. The higher this number, the more storage is required at a vertex, resulting in increased cost;
- Inactive vehicles: The number of inactive vehicles can be seen as a measure for efficient vehicle usage. If there is a constant number of inactive vehicles in a simulation, this means the proposed approach needs less than the allocated number, indicating potential for cost saving;
- Vehicle utilisation: This indicator gives the capacity utilisation of the active vehicles. High utilisation means the vehicles are well loaded most of the time, and there are only few empty trips.

Process-related measures

- Throughput time: This is the time from the generation of packages up to the completion of the transport order. It is an absolute measure for the completion without considering whether all the requirements given in the order are met or not;
- Punctuality rate: This is the percentage of orders that are completed in time. A high punctuality rate is one of the key measures of an efficient transport process;
- Distance per package: This compares the actual distance taken by a package with the minimum distance between source and sink. This way, it is possible to evaluate how "straight" the transport path is. Longer distances imply higher costs and increased risks for the packages;
- Trans-shipments per package: Every trans-shipment operation means risks and added costs. Therefore the number of trans-shipments should be kept as low as possible.

Most of the measures introduced here need to be used in conjunction. Otherwise, the overall performance could be bad regardless of one or two measures being good. For example, the vehicle utilisation could be kept high by carrying packages around on unnecessarily long trips, leading to bad values in other measures such as the throughput time and distance per package and thus decreasing the overall performance.

Economic measures are not explicitly included in the described set. However, they depend on the aforementioned volume- and process-related measures. To derive an economic evaluation for the investigated logistic scenarios, additional cost models are required that map the described measures to costs and revenues. Such models are beyond the scope of this chapter.

4.4.5 Example scenarios

Based on the definitions of components as described above, reference scenarios have been generated. For modelling of logistic processes, they comprise all relevant components, such as location and functionality of vertices, edges, type and initial position of vehicles and distribution of packages. Two selected scenarios, the small 4-vertex scenario and the larger Germany scenario, are described in the following subsections. The 4-vertex scenario is designed for basic testing and understanding the impact of algorithms and approaches. The Germany scenario is based on cities and motorway connections in Germany, it is needed especially for complex investigations, e.g. routing algorithms requiring the existence of multiple routes. These scenarios are intended to be used as extensible basis for investigations of dynamic logistic processes.

The 4-vertex scenario

The network used as the physical base of this scenario is shown in figure 4.16. This network has only four vertices and the edges are of different types such as Highway, Road, and Railway, representing the multimodality even in this small example scenario. An arbitrary number of vehicles of four different types can exist in the network and carry packages according to their specifications.

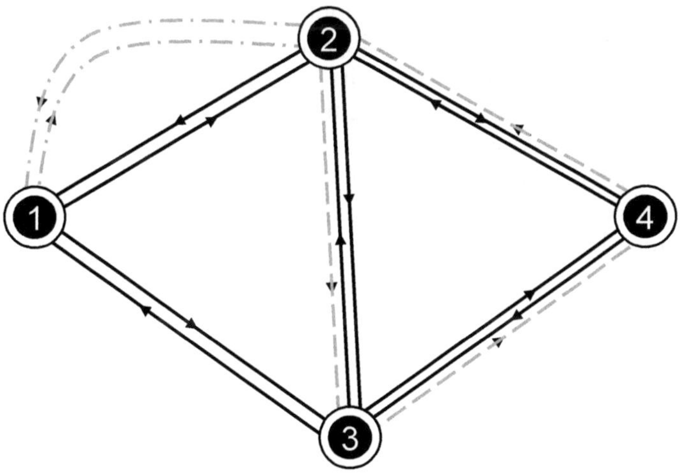

Fig. 4.16 The 4-Vertex Scenario Topology

The network contains four vertices, numbered 1 through 4. These vertices are start or end points of different edges, and represent sources and sinks of transport goods and have various trans-shipment facilities. It is supposed that a vehicle arriving at a vertex can change to any other edge present in that vertex, given the edge accommodates that vehicle. Table 4.7. lists the vertices along with their properties.

Table 4.7. Vertex properties in the 4-vertex scenario

Vertex ID	Type	Trans-shipment type	Trans-shipment capacity [pu/h]	Trans-shipment cost per unit
1	General	Road→Road	23	1
2	General	Road→Road	40	1
		Road→Rail	100	5
3	General	Road→Road	50	2
		Road→Rail	25	3
		Rail→Road	80	3
4	General	Road→Road	120	4
		Road→Rail	42	4
		Rail→Road	70	4

The table also contains the trans-shipment options for each vertex. The capacities in package units per hour [pu/h] apply only to real trans-shipment operations. A fixed loading/unloading time of half an hour is applied at each of the sources/sinks irrespective of the number of packages handled. It should be noted, that the transport mode of a package cannot be changed in all directions in every vertex. It is assumed that all vertices have unrestricted storage capacities for intermediate storage both for vehicles and for packages. Thus vehicles that are not involved in transport operations must idle at a vertex.

Table 4.8. Source properties in the 4-vertex scenario

Source ID	Location (Vertex)	Output rate [pu/h]	Destinations	Requirements (package type)
S1	1	10	40% --> 2	none (A)
			30% --> 3	
			30% --> 4	
S2	1	12.5	30% --> 2	cooling (C)
			20% --> 3	
			50% --> 4	
S3	3	15	40% --> 1	none (A)
			60% --> 4	
S4	4	10	70% --> 1	careful handling (B)
			30% --> 2	
S5	4	2.5	100% --> 1	none (A)
S6	4	25	50% --> 1	cooling (C)
			50% --> 3	

Sources in the sample network are the points where new packages and their "transport orders" are generated. The sources and their properties are given in Table 4.8. All sources are located in already existent vertices of the network and the arrivals of packages are modelled as a Poisson process (a discrete memoryless process (Trivedi 2002)). It is further assumed that a source has an unlimited waiting space where the packages can be stored until a vehicle picks them up and transports them to their destinations. As shown in Table 4.8., the sources are not uniformly distributed over the set of vertices and their output rates are different. This allows the investigation of unbalanced load conditions. In this scenario, all vertices act as sinks, as the source specifications include all vertices in the „Destinations“ column (see Table 4.8.).

For simplicity, it is assumed that there is only one general form of freight that should be transported, namely packages of unified size. Each package belongs to one of three different types, A, B, or C depending on handling requirements and risks involved (see Table 4.11. for definition of the package types).

Three different types of edges are present in the network: Simple road, highway and railway. While simple roads (interrupted and dotted line in the figure) and highways (continuous line) are bidirectional connections between vertices usable for vehicles of class S, the railway (interrupted line) is a ring which is uni-directional and can be used only by vehicles of type R (for vehicle parameters see Table 4.10.). The parameters for edges, especially the path length and the allowed maximum velocity, are given in Table 4.9.

Table 4.9. Edge properties in the 4-vertex scenario

Edge ID	Start Vertex	End Vertex	Type	Length	max. Speed
E1	1	2	Highway	370	100
E2	1	2	Road	300	80
E3	1	3	Highway	250	100
E4	2	1	Highway	380	100
E5	2	1	Road	300	60
E6	2	3	Railway	400	80
E7	2	3	Highway	480	100
E8	2	4	Highway	490	100
E9	3	1	Highway	250	90
E10	3	2	Highway	400	100
E11	3	4	Railway	700	180
E12	3	4	Highway	770	100
E13	4	2	Highway	450	100
E14	4	2	Railway	500	120
E15	4	3	Highway	700	100

For vehicles, a maximum transport capacity and speed is defined. The routes of the vehicles except for the trains and their loading priorities are not predefined. The trains travel only in a closed ring in one direction.

The vehicles available in the scenario are characterized by the attributes given in Table 4.10. The number of vehicles and their capacities are over-dimensioned for the load that is given in the scenario. This means if an approach fails to handle the load with the given vehicles, it can be considered being very inefficient. Efficient approaches can do with far less than the given number of vehicles.

Table 4.10. Vehicle properties in the 4-vertex scenario

Vehicle IDs	# of Vehicles	Type	Capacity [pu]	max. Speed	Allowed Edge Types
V01 .. V20	20	Light Truck	60	120	Road/Highway
V21 .. V25	5	Cooling Truck	100	100	Road/Highway
V26 .. V40	15	Truck	200	80	Road/Highway
V41 .. V44	4	Freight Train	2000	200	Railway

If a vehicle arrives at a vertex, the scenario allows the following actions: It can deliver packages at a sink, load new packages from a source, do trans-shipment operations by unloading a number of packages and loading other ones, wait or continue its route. In trans-shipments the specified rates and restrictions given in Table 4.7. apply.

As mentioned above, a relatively simple concept of packages is used in the scenario, where the only variable relevant for transport is the number of packages. However, some risks and special transport requirements are assigned to packages in the model. Therefore, three types of packages are introduced and defined in Table 4.11.

Table 4.11. Package types in the 4-vertex scenario

Package Type	Required Vehicle Type	Specialties
A	any	no specialties
B	any	5% risk of breaking during trans-shipment,
		0.5% risk per hour of breaking during train transport
C	cooling vehicle	destroyed when transported in a non-cooling vehicle

The Germany scenario

The Germany scenario is based on a network of 18 cities in Germany, as shown in figure 4.17. The edges between the vertices represent highway connections between those cities. This makes the scenario a single-mode scenario limited to highway traffic. The edges are directed. However, in figure 4.17 the directions of the edges are not shown for simplicity, and each link in the figure stand for two edges, i.e. one per direction. Thus, there are a total of 70 edges in this scenario.

In contrast to the small scenario described earlier, this scenario gives more choices for alternative routes, especially between vertices far away from each other. Therefore, it is well suited for investigation of routing algorithms. Some investigations have been completed using this scenario effectively (Wenning et al. 2005, Becker et al. 2006).

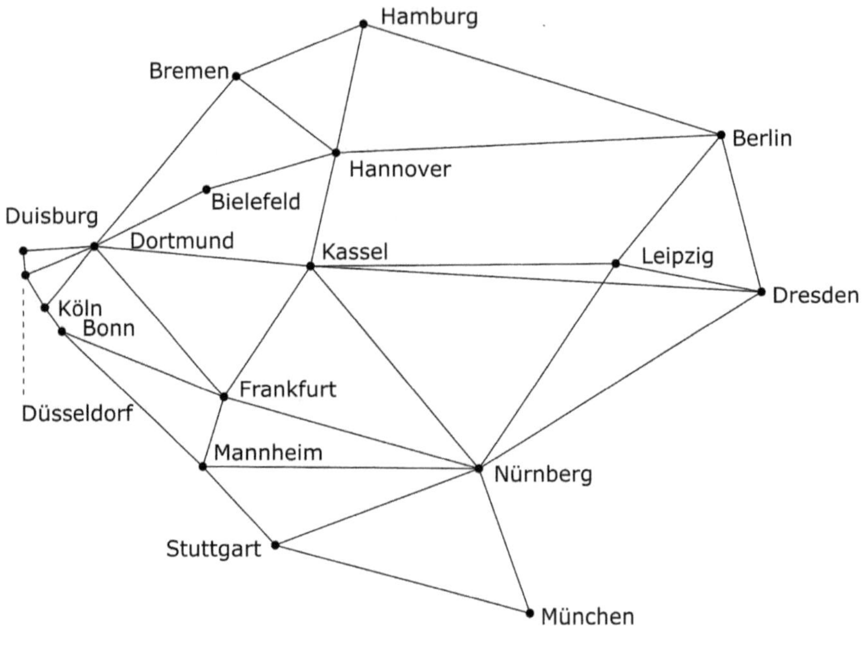

Fig. 4.17 The Germany scenario

Each of the vertices in this scenario is origin for some packages and destination for others, which means that there is a package source at each vertex, and each vertex is acting as a sink. The output rate of the sources depends on the size of the city, ranging from 2 pu/h in Kassel up to 34 pu/h in Berlin. The vehicle distribution also depends on the city size. In total, there are 71 vehicles, each with a capacity of 60 pu and a maximum speed of 120 km/h. The basic version of the scenario assumes a fixed maximum edge speed of 100 km/h, but it provides the opportunity to introduce random occurrence of traffic jams individually for each of the edges, specified by an occurrence probability, an average delay that each vehicle experiences and an average duration of the traffic jam.

In addition to the logistic network, this scenario is overlaid with a definition of the communication capabilities on the edges. All edges are fully covered with GPRS, and partially covered with UMTS. Figure 4.18 shows the GPRS and UMTS coverage. The idea behind the integration of communication capabilities is to simulate also the communication volume that arises from the autonomy and cooperation of the logistic components. This way, the simulations can also be used to study aspects concerning the wireless traffic that is generated.

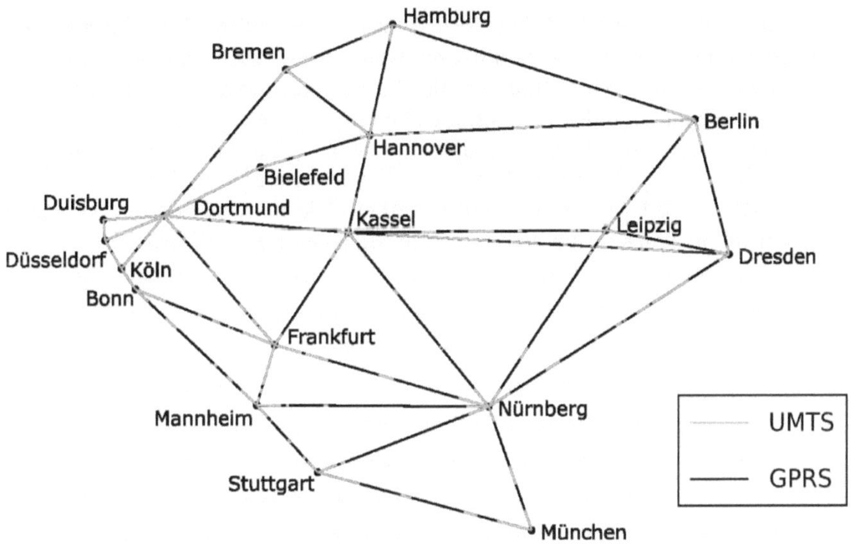

Fig. 4.18 GPRS and UMTS coverage in the Germany scenario

The use of this scenario and its components, with especial emphasis on communication parts, in a discrete-event simulation is presented in detail in (Becker et al. 2005).

4.4.6 Conclusions

In this chapter, components for modelling of dynamic logistic networks have been introduced and evaluation parameters have been listed. Two example scenarios are given which can be used for the evaluation of approaches in these dynamic networks. These scenarios are examples that might not contain all aspects relevant for a specific approach, but they can easily be extended or other scenarios can be created based on the defined components.

References

Becker M, Wenning BL, Görg C (2005) Integrated Simulation of Communication Networks and Logistical Networks – Using Object-oriented Programming Language Features to Enhance Modelling. In: Ince N, Topus E (eds) Modeling and Simulation Tools for Emerging Telecommunication Networks – Needs, Trends, Challenges, Solutions. Springer, pp 279-287

Becker M, Wenning BL, Görg C, Gehrke JD, Lorenz M, Herzog O (2006) Agent-based and Event-discrete Simulation of Autonomous Logistic Processes. In: Borutzky W, Orsoni A, Zobel R (eds) Proceedings of the 20th European Conference on Modelling and Simulation, Bonn, St. Augustin, pp 566-571

Solomon MM (1987) Algorithms for the Vehicle Routing and Scheduling Problems with Time Window Constraints. Operations Research 35(2): 254-265

Trivedi KS (2002) Probability and Statistics with Reliability, Queuing and Computer Science Applications, 2nd edn. Wiley and Sons, New York

Wenning BL, Görg C, Peters K (2005) Ereignisdiskrete Modellierung von Selbststeuerung in Transportnetzen. Industrie Management 5: 53-56

4.5 Autonomously Controlled Storage Allocation on an Automobile Terminal

Felix Böse, Katja Windt

Department of Planning and Control of Production Systems, BIBA, University of Bremen, Germany

4.5.1 Introduction

Today, planning and control of logistic processes on automobile terminals are generally executed by centralised logistics systems, which in many cases cannot cope with the high requirements for flexible order processing due to increasing dynamics and complexity. The main business processes on automobile terminals – notification of vehicles by automobile manufacturer, transport to automobile terminal, storage and technical treatment as well as delivery to automobile dealer – are planned and controlled by a central application software system. By establishing autonomous control, vehicles are enabled to render decisions on their own and according to this determine their way through a logistics network on the basis of an own system of objectives.

The idea of autonomous control is to develop decentralised and heterarchical planning and controlling methods in contrast to existing central and hierarchical aligned planning and controlling approaches (Scholz-Reiter et al. 2006). Decision functions are shifted to logistic objects. In the context of autonomous control logistic objects are defined as material items (e.g. vehicles, storage areas) or immaterial items (e.g. customer orders) of a networked logistic system, which have the ability to interact with other logistic objects of the considered system. Autonomous logistic objects are able to act independently according to their own objectives and navigate through the logistic network themselves (Windt et al. 2006). Figure 4.19 illustrates the described paradigm shift in logistics from conventional control to autonomous control.

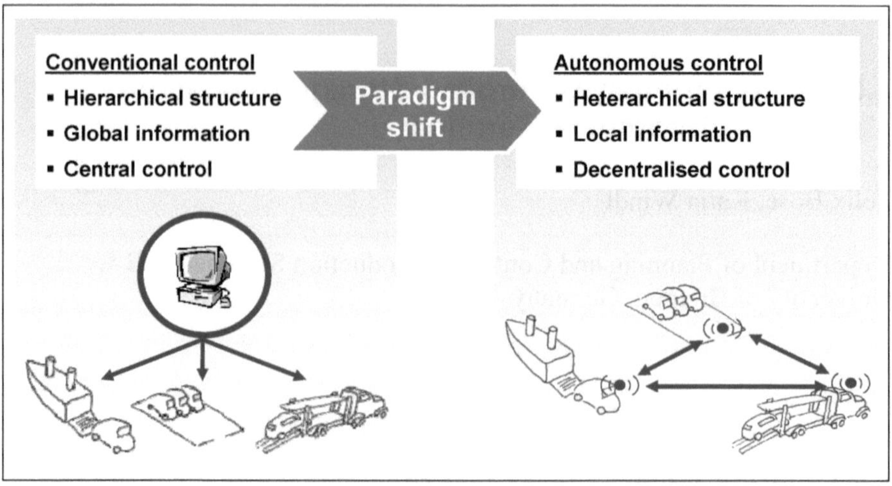

Fig. 4.19 Paradigm shift from conventional control to autonomous control

An essential condition of autonomous control is a high degree of interoperability. Autonomous logistic objects must be able to communicate with other objects and exchange data, in order to ensure the availability of all relevant data needed for decision-making. Recent developments by information and communication technologies (ICT) are of particular importance concerning the fulfilment of this fundamental requirement, including RFID (Radio Frequency Identification) for identification, GPS (Global Positioning System) for positioning or UMTS (Universal Mobile Telecommunications System) and WLAN (Wireless Local Area Network) for communication tasks (Böse et al. 2005; Böse and Lampe 2005).

In the context of this article a new approach of an autonomously controlled logistics system is investigated using as example the vehicle movement processes on the E.H.Harms Auto-Terminal-Hamburg. Several opportunities for improvement by implementing autonomously controlled logistic processes are identified and investigated by means of a simulation study. This case study is a result of the cooperation project "Autonomous Control in Automobile Logistics" between the company E.H.Harms GmbH & Co. KG Automobile-Logistics and the University of Bremen. This research is funded by the German Research Foundation (DFG) as the Collaborative Research Centre 637 "Autonomous Cooperating Logistic Processes - A Paradigm Shift and its Limitations" (SFB 637) at the University of Bremen (Scholz-Reiter et al. 2004).

4.5.2 Initial situation

E.H. Harms develops and provides complex services for new and used vehicles in the range of transport, handling, technical treatment and storage. The group of companies, consisting of E.H.H. Automobile Transports, E.H.H. Auto-Terminals and E.H.H. Car Shipping, has established a Europe-wide logistics network on the basis of automobile terminals at strategically important traffic junctions. Every vehicle passes a set of process steps in the automobile logistics network: collection of vehicles at automobile manufacturer, multi-modal transport to automobile terminal via road, rail or inland waterway/sea, storage and technical treatment as well as delivery to automobile dealer. This article focuses on the logistics order processing of the E.H.Harms Auto-Terminal-Hamburg (EHH Auto-Terminal). The vehicle movement processes of an automobile terminal are illustrated in figure 4.20

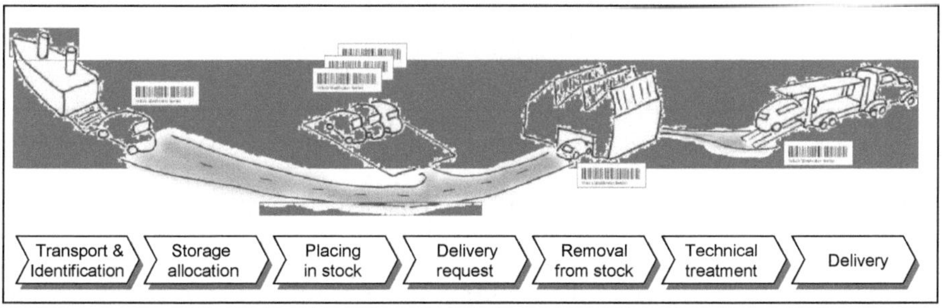

Fig. 4.20 Vehicle movement processes of an automobile terminal

After delivery each vehicle is identified by its vehicle identification number (VIN) from the terminal staff using mobile data entry devices (MDE) which can read barcodes placed inside the vehicle behind the windscreen. The VIN allows an assignment of the vehicle to its storage and technical treatment orders stored in the logistic IT-system. Based on predefined priorities the IT-system allocates a storage location of a storage area to each vehicle. A handling employee moves the vehicle to the assigned storage location. After removal from stock the vehicles possibly run through several technical treatment stations as fuel station or car wash. The sequence of the technical treatment stations is specified in the technical treatment order of the vehicle. Upon completion of all technical treatment tasks the vehicle is provided on the shipment area for transportation to the automobile dealer (Böse et al. 2006).

4.5.3 Opportunities for improvement

The vehicle movement processes on the automobile terminal provide many opportunities for improvement (Böse et al. 2005; Fischer 2004). In particular they result from the centralised storage allocation which is illustrated in figure 4.21 with the Business Process Modelling Notation (BPMN) (Owen and Raj 2003).

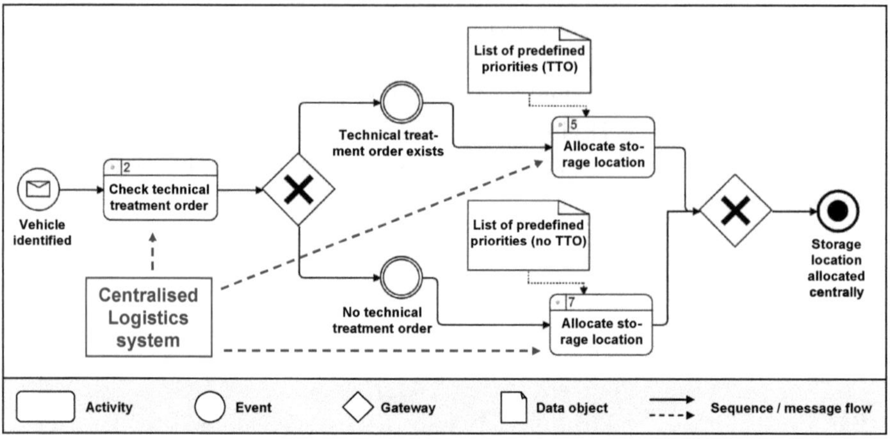

Fig. 4.21 Centralised storage allocation based on predefined priorities

Each vehicle is allocated to a storage location in a storage area on the basis of fixed and predefined priorities. Even though these priorities consider if there are possible technical treatment orders assigned to the vehicles, there is no differentiation regarding the type of technical treatment and therewith the location of the technical treatment stations which are partially a long way away from each other. As a result of the fix prioritization of the storage areas for vehicles with or without technical treatment orders, a flexible selection of storage areas in consideration of future process steps is not possible. Furthermore, the parking time – meaning the time of a vehicle in a storage area to be parked by a handling employee at a designated storage location - is not taken into account in the scope of the storage allocation process. This is of particular importance due to the fact that the needed parking times of storage areas can heavily differ depending on their stock level. As a result time saved due to the short distance between current vehicle location and selected storage area is possibly compensated by a long parking time in the storage area.

4.5.4 Objective target

To realize the opportunities for improvement concerning the storage allocation and the related vehicle movement processes described above, a decentralised decision-making approach for autonomously controlled logistics systems is developed. According to the definition of autonomous control, autonomous logistic objects are enabled to process information, render and execute decisions on their own (Böse and Windt 2007). In consequence, both the vehicles and the storage areas have their own master data and act independently regarding their local objective system (compare figure 4.22).

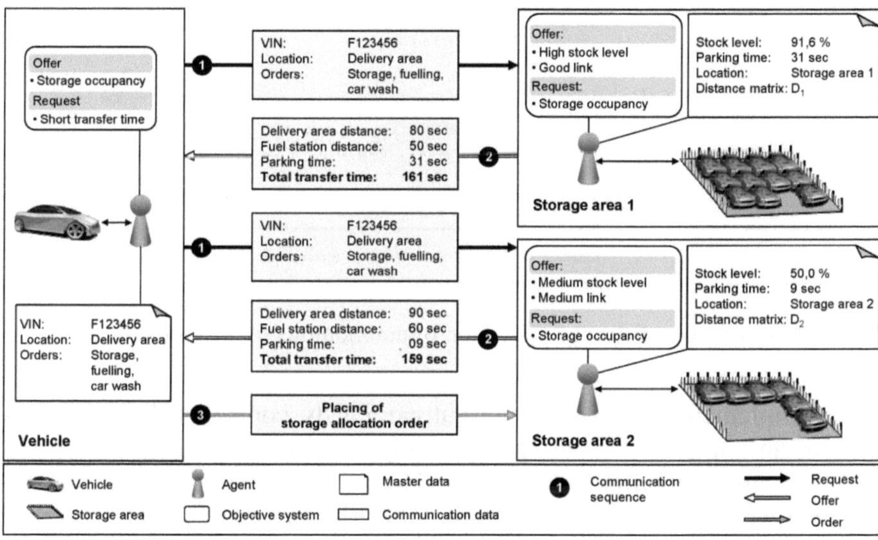

Fig. 4.22 Autonomously controlled decision-making of vehicles and storage areas

Each vehicle has the objective of short transfer times on the terminal area and provides every single storage area the occupancy of a storage location. On the other hand, the objective of the storage areas is high storage occupancy. They offer the inquiring vehicle the total transfer time which consists of the transfer time from the current vehicle location to the storage area, the parking time on the storage area as well as the future transfer time of the vehicle to the first technical treatment station after removal from stock. Depending on the stock level and the position of the storage areas in the automobile terminal, the storage areas can offer a more or less convenient storage time and link to the next technical treatment station. The belonging times described above are added to the total transfer time and transmitted to the inquiring vehicle that compares the received total transfer times of all storage areas and chooses the best-rated. Based on this autonomously controlled decision-making approach, the underlying process chain of the decentralised storage allocation by vehicles and storage areas acting as autonomous logistic objects is illustrated in figure 4.23.

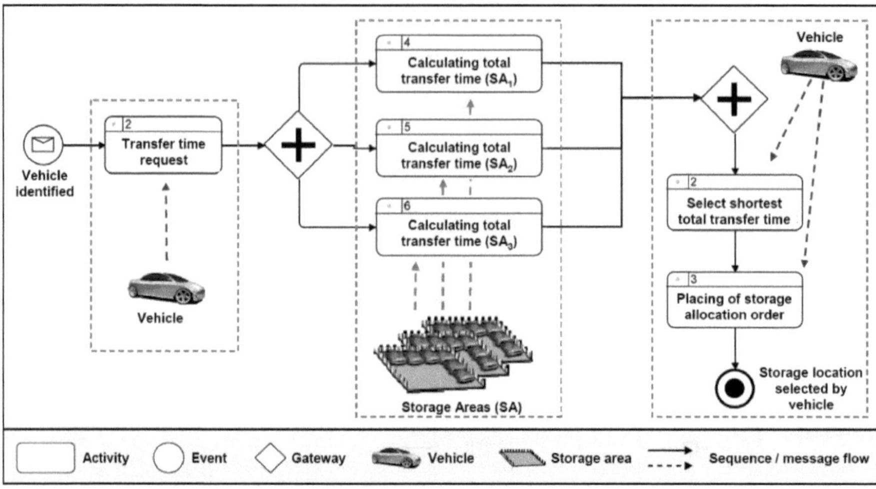

Fig. 4.23 Decentralised storage allocation of autonomous logistic objects

The implementation of such an autonomously controlled logistics scenario of an automobile terminal is already feasible with today's information and communication technologies. The vehicles can be fitted with passive read/write transponders containing the relevant vehicle data as well as the belonging technical treatment orders. The initial data storage on the transponder can be effected by means of a MDE reading an existing barcode and forwarding the information onto the transponder. For this purpose each handling employee is provided with an MDE with integrated transponder reader which enables reading and storing of relevant data on the tags. Furthermore, the MDE contains a communication module based on WLAN that allows the data exchange of the vehicle with other autonomous logistic objects, especially the storage areas, a GPS localisation module for vehicle positioning as well as a user interface. Based on this technological equipment, the process chain of vehicle storage allocation can be described as follows.

After reading the vehicle and technical treatment order data from the transponder placed inside the vehicle a logistic planning and control software system on the MDE determines the best-rated storage area. After that, the handling employee moves the vehicle to the designated storage area and parks the automobile on the fastest reachable storage location (chaotic stock keeping). The MDE in place of the vehicle determines its position on the storage area via satellite using the GPS module and communicates the current storage location to the storage area. Because every vehicle is moved by a handling employee fitted with an MDE, the storage locations of all vehicles on the automobile terminal are always available. As a consequence each storage area has real time information on its stock level at any time. In the following the introduced decentralised decision-making approach for the autonomously controlled storage allocation of an automobile terminal is evaluated by means of a simulation study.

4.5.5 Simulation model

The object of investigation of the simulation study is the transfer times of the vehicles on the automobile terminal. The total transfer time of a vehicle on an automobile terminal TT_{total} consists of the transfer time from the delivery area to the storage location $TT_{storage}$, the transfer time to the technical treatment stations $TT_{technical\ treatment}$ as well as the transfer time from the storage location, respectively the current technical treatment station to the shipment area $TT_{disposition}$ (see Eq. 4.10).

$$TT_{\text{total}} = TT_{\text{storage}} + TT_{\text{technical treatment}} + TT_{\text{disposition}} \tag{4.10}$$

The transfer time from the delivery area to the storage location $TT_{storage}$ is divided into the transfer time from the delivery area to the storage area $TT_{storage\ area}$ and the parking time on the storage area $TT_{storage\ location}$ (see Eq. 4.11).

$$TT_{\text{storage}} = TT_{\text{storage area}} + TT_{\text{storage location}} \tag{4.11}$$

The transfer time to the technical treatment stations $TT_{technical\ treatment}$ is composed of the variable transfer time from the storage area to the first technical treatment station after removal from stock $TT_{technical\ treatment,\ variable}$ and the fixed transfer time between the technical treatment stations $TT_{technical\ treatment,\ fixed}$ (see Eq. 4.12).

$$TT_{\text{technical treatment}} = TT_{\text{technical treatment, variable}} + TT_{\text{technical treatment, fixed}} \tag{4.12}$$

Finally, the transfer time from the storage location, respectively the current technical treatment station to the shipment area $TT_{disposition}$ consists of the variable transfer time from the storage area to the shipment area $TT_{disposition, variable}$ and the fixed transfer time form the last technical treatment station to the shipment area $TT_{disposition, fixed}$ (see Eq. 4.13).

$$TT_{disposition} = TT_{disposition, variable} + TT_{disposition, fixed} \tag{4.13}$$

The transfer times described above show both fixed and variable time slices. For example, the transfer time of a vehicle between technical treatment stations is fixed because of the predetermined handling sequence in technical order processing. For instance, a vehicle is always moved to the car wash after executing technical services or installations in workshops. A variable time slice is the transfer time of a vehicle from the storage area to the shipping area because this time slice depends on the previously made decision regarding the storage area. Recapitulating, only such vehicle movement processes contain opportunities for improvement regarding the total transfer time which have a variable starting or end point. In the considered example these are the vehicle movement processes from or to the selected storage area in the context of placing in or removal from storage. Each vehicle can determine the best possible storage location and minimize its total transfer time on the automobile terminal area considering the distance between delivery area and storage areas, the stock levels of the storage areas as well as the first destination after removal from storage.

The basis of the simulation study is real vehicle and technical treatment order data of 124.000 vehicles of the EHH Auto-Terminal for the time period of one year. In addition to the delivery area the simulation model includes seven storage areas with an average of 1500 storage locations, nine technical treatment stations with belonging buffers as well as the shipping area. The distances between the technical treatment stations and the several areas of the automobile terminal are represented in a transportation time matrix which contains the transfer times of a vehicle between all considered locations. Based on the described business processes of the conventionally controlled as well as the autonomously controlled storage allocation, two simulation scenarios are developed as follows:

Conventionally controlled Scenario S_C

The storage allocation is executed centralised on the basis of fixed and predefined rules which contain an order of priority of all storage areas for both vehicles with and without assigned technical treatment orders. Depending on the existence of a technical treatment order, each vehicle is assigned to the consecutively next available storage location on the currently prioritised storage location.

Autonomously controlled scenario S_A

The storage allocation is executed decentralised by the autonomous logistics objects. Each vehicle chooses that storage area which offers the shortest total transfer time. Placing in storage is accomplished chaotically, i.e., the handling employee moves the vehicle to the designated storage area and parks the automobile on the fastest reachable storage location.

Based on these simulation scenarios two simulation models are developed and investigated by means of the simulation tool eM-Plant. Figure 4.24 illustrates the implementation of the autonomously controlled simulation model in eM-Plant.

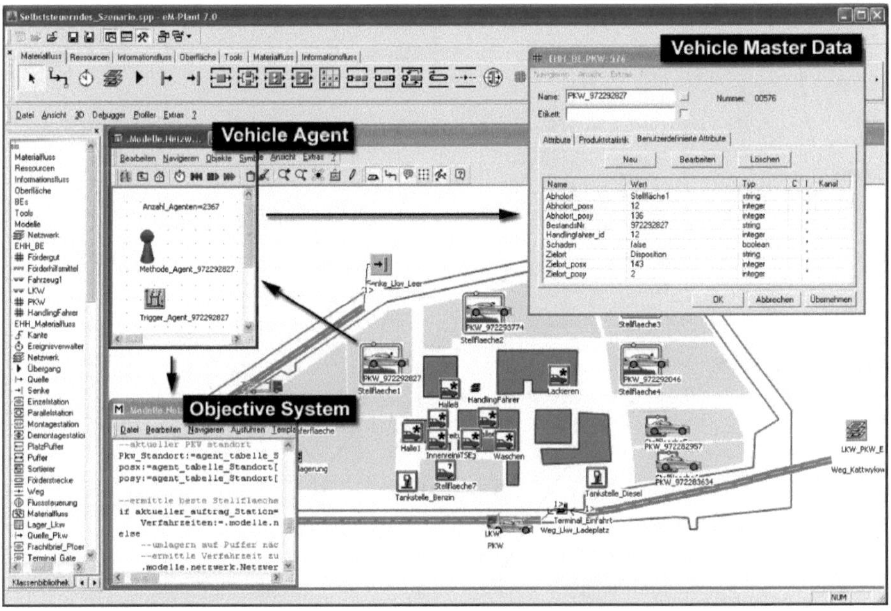

Fig. 4.24 Simulation model of the autonomously controlled scenario S_A

In this simulation model each autonomous logistic object is represented by a virtual agent, for example a vehicle agent. According to multi-agent systems (Ferber 1999) the vehicle agent has its own master data, which are stored in tables. The objective system and the decision functions for planning and control of the vehicle movement processes on the automobile terminal are described in the form of knowledge-based methods.

4.5.6 Results

The main results of the simulation runs are illustrated in figure 4.25 At first the frequencies of the total transfer times per vehicle TT_{total} of the conventionally controlled scenario S_C as well as the autonomously controlled scenario S_A are drawn in respectively one histogram. For the purpose of comparability of these simulation scenarios a continuous frequency distribution is deviated by approximation (compare at the top of figure 4.25).

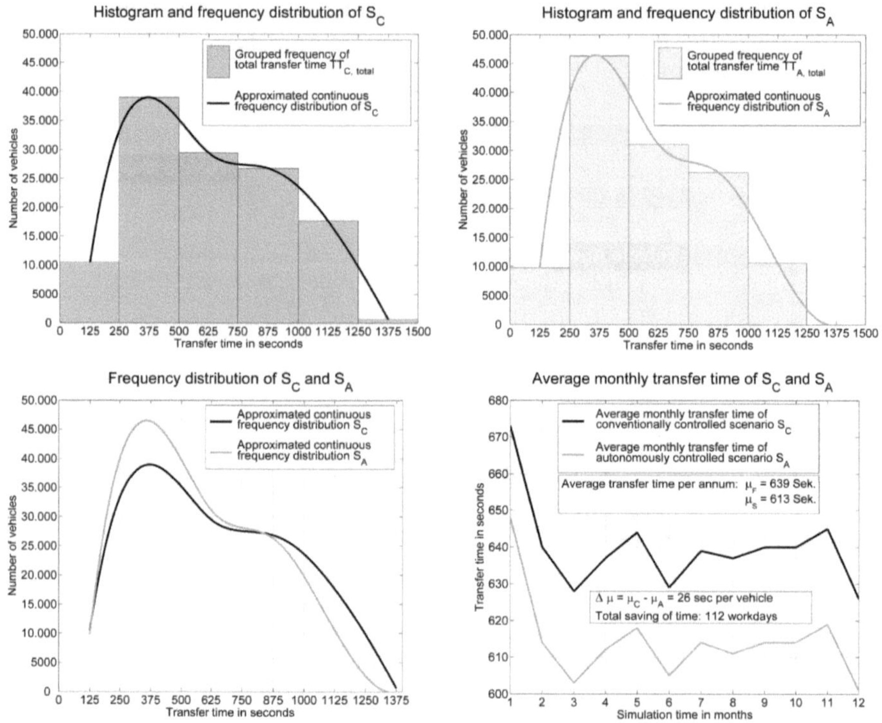

Fig. 4.25 Results of the simulation study

Merging the two frequencies a significant upsetting of the curve of the autonomously controlled scenario becomes apparent compared to the conventionally controlled scenario. In the autonomously controlled scenario S_A more vehicles show a lower total transfer time and fewer vehicles a higher total transfer time than in the conventionally controlled scenario S_C (compare figure 4.25 in the lower left corner). Due to the fact that the data pool of both simulation scenarios is identical regarding the number of considered vehicles, the continuous frequency distributions have an intersection. The average monthly transfer time of the conventionally controlled scenario S_C and the autonomously controlled scenario S_A are represented in the lower right corner of figure 4.25. Both scenarios show a similar behaviour of the curve, but the curve of the autonomously controlled scenario is shifted down. As a consequence, S_A has a lower average monthly transfer time than S_C. The distance between the curves varies depending on the stock level of the storage areas. The higher the stock levels of the storage areas the longer the parking times in the case of chaotic stock keeping and the lower the time saving of the autonomously controlled scenario. The high total transfer times in January result from a large initial stock of the storage areas. The variation of the curve behaviour throughout the year depends on the variable number of moved vehicles and the amount of technical treatment orders. Over the year the average time saving of the autonomously controlled scenario S_A adds up to 26 seconds per vehicle compared to the conventionally controlled scenario S_C. Over all vehicles a total time saving of 112 workdays arises for the vehicle movement processes on the automobile terminal.

4.5.7 Conclusions and outlook

In the context of this article a new approach of an autonomously controlled logistics system was introduced considering as example the storage allocation processes on the E.H.Harms Auto-Terminal Hamburg. As a main result of the presented simulation study the new paradigm of autonomous control in logistics provides significant opportunities of time saving in the field of vehicle movement on automobile terminals.

Due to the fact that the simulation study was strongly focussed on the storage allocation process as a single part of the vehicle management process chain of automobile terminals, further research is directed to the enlargement of the considered application scenario as follows:

- **Consideration of other business processes**
  In addition to the vehicle movement processes other important business processes of logistic order processing of an automobile terminal are included, for example order sequencing of technical treatment stations.
- **Including new autonomous logistic objects**
  In conjunction with additionally considered business processes exemplary mentioned before, new autonomous logistic objects are included

in the simulation model, for example technical treatment stations, shuttle busses or orders.

- **Adding new logistic objectives**
  Like the vehicles and storage areas, the new autonomous logistic objects posses own master data and an own objective system. Therefore it is necessary to add new logistic objectives. Technical treatment stations for instance aim for the goal high utilization while orders have the objective high due date punctuality.
- **Investigation of disturbances**
  To verify the thesis that the allocation of planning and control tasks to autonomously controlled logistic objects effects a higher achievement of logistic objectives because of a better coping with high dynamics and complexity in today's logistics systems, several disturbances are added to the simulation model (e.g. break down of technical treatment stations or rush orders).

The main objectives of these enlargements of the simulation scenario are both to investigate and evaluate other fields of application of autonomous control in the context of logistic order processing of an automobile terminal and to emphasize the significant advantages of autonomous control like better coping with complexity and dynamics as well as higher flexibility and robustness of logistics systems.

References

Böse F, Lampe W (2005) Adoption of RFID in Logistics. In: Proceedings of IBIMA International Business Information Management Association Conference, Cairo, CD-ROM

Böse F, Piotrowski J, Windt K (2005) Selbststeuerung in der Automobil-Logistik. Industriemanagement, 20(4): 37-40

Böse F, Lampe W, Scholz-Reiter B (2006) Netzwerk für Millionen Räder. FasTEr – Eine Transponderlösung macht mobil. In: RFID im Blick, special issue RFID in Bremen, Verlag & Freie Medien, Amelinghausen, pp 20-23

Böse F, Windt K (2007) Catalogue of Criteria for Autonomous Control in Logistics. In: Hülsmann M, Windt K (eds) Understanding Autonomous Cooperation and Control - The Impact of Autonomy on Management, Information, Communication, and Material Flow. Springer, Heidelberg

Ferber J (1999) Multi-agent systems: an introduction to distributed artificial intelligence. Harlow

Fischer T (2004) Multi-Agenten-Systeme im Fahrzeugumschlag: Agentenbasierte Planungsunterstützung für Seehafen-Automobilterminals. Dt. Univ.-Verl., Wiesbaden

Owen M and Raj J (2003) BPMN and Business Process Management - Introduction to the New Business Process Modelling Standard. White Paper, Popkin Software

Scholz-Reiter B, Windt K, Freitag M (2004) Autonomous Logistic Processes – New Demands and First Approaches. In: Proceedings of 37th CIRP International Seminar on Manufacturing Systems, Budapest, pp 357–362

Scholz-Reiter B, Windt K, Kolditz J, Böse F, Hildebrandt T, Philipp T, Höhns H (2006) New Concepts of Modelling and Evaluating Autonomous Logistic Processes. In: Chryssolouris G, Mourtzis D (eds) Manufacturing, Modelling, Management and Control, Elsevier, Oxford

Windt K, Böse F, Philipp T (2007) Autonomy in Logistics – Identification, Characterisation and Application. In: International Journal of Robotics and CIM, Pergamon Press Ltd, forthcoming

4.6 Intelligent Containers and Sensor Networks Approaches to apply Autonomous Cooperation on Systems with limited Resources

Reiner Jedermann[1], Christian Behrens[2], Rainer Laur[2], Walter Lang[1]

[1] Institute for Microsensors, Actuators and Systems (IMSAS), University of Bremen

[2] Institute for Electromagnetic Theory and Microelectronics (ITEM), University of Bremen

4.6.1 Introduction

RFIDs, sensor networks and low-power microcontrollers are increasingly applied in logistics. They are characterized by restrictions on calculation power, communication range and battery lifetime. In this article we consider how these new technologies can be utilized for autonomous cooperation and how these processes could be realized on systems with limited resources.

Besides tracing of the current freight location by RFID technologies, the monitoring of quality changes that occur during transport is of growing importance. The demand for improved and comprehensive supervision of goods could be best fulfilled by distributed autonomous systems.

The 'intelligent container' as autonomous supervision system

The prototype of our 'intelligent container' demonstrates how autonomous control could be implemented on a credit-card sized processor module for integration into standard containers or transport vehicles (figure 4.26). The processor provides a platform for local interpretation and pre-processing of sensor information. The system automatically adapts to the specific requirements of the transport good. An extended electronic consignment note that is implemented as software agent contains individual transport- and monitoring instructions. RFID technologies are used to control the transfer of this mobile freight agent. The implementation of the local data pre-processing and an example quality model for vegetables are described in section 2. If the supervision system predicts that the freight quality will drop below an acceptance threshold before arrival, it contacts the transport manager. The extended agent platform for further transport planning is shortly introduced in section 3.

Fig.4.26 Reduced scale (1:8) prototype of the intelligent container.

Loaded freight items are scanned by the RFID-Reader on left hand side. Sensor nodes supervise the environmental conditions (middle). A processor module on the right hand side executes a software agent containing specific transport instructions and quality modelling. The module for external mobile communication is placed on the right hand side panel.

Autonomous control of wireless sensor networks

Incorrect packing or poor isolation could lead to local temperature maxima or 'hot spots'. Because of the number of required sensors a wireless solution is the most suitable way to monitor spatial deviations of environment parameters. Sensors that are attached to the freight have to link themselves 'ad hoc' into the communication network of the vehicle. Section 4 gives an overview over the design, configuration and control of our implementation of a wireless sensor network. Standard algorithms for self-configuration already exhibit features of autonomous cooperation. Because service intervals should be prolonged as long as possible and there is no practicable solution for recharging, battery lifetime is more crucial as in other common mobile applications like cell phones. Besides improvements on the hardware and communication protocols we focus on energy saving by intelligent control. The energy consumption mainly depends on the number of measurement and communication cycles. An intelligent decision system could reduce their required number. Section 5 discusses architectures, examples and further demands on autonomous cooperative processes running on low-power microcontrollers. Approaches for future implementations of an autonomous decision system on small battery powered sensor nodes and logistical freight objects are summarized in section 6.

Requirements of improved supervision and control systems

The design of improved transport supervision and control systems has to consider limitations of communication bandwidth as well as requirements for just in time decisions and extended sensor monitoring. Special attention has to be paid to the following aspects:

Mobile communication

Communication is a substantial component for the implementation of networks of distributed autonomous processes. Technologies for secure and cost efficient communication have to be provided. During system design, the bandwidths of the communication links have to be considered. To save costs for mobile services, the transferred data volume should be reduced by shifting interpretation and decision processes to the physical origin of the data as close as possible. The effects of moving the scope of communication from the transmission of sensor raw data towards the transfer of conclusions and decision rules are handled in detail by Markus Becker.

Extended sensor monitoring

For a detailed sensor monitoring, it is not sufficient to distinguish between 'intact' and 'damaged' goods. Quality losses depend on the duration and amount of deviations from the optimal transport conditions. Spatial variations of environmental parameters have to be assigned to the affected packages. A concise prediction of quality changes assumes complex data and decision guidelines.

Robustness

Transport monitoring systems have to work in rough environments. Communication links might not be available or some of the involved systems could be damaged. Sensor measurements might be faulty. Solutions for supervision and control systems should be robust enough to continue their work despite system failures in their neighbourhood.

Just in time decisions

Corrections to the supply chain should be carried out as fast as possible, at least before the next part of the production chain is entered. A permanent supervision is necessary to avoid a freight reaching its destination with insufficient quality. Decisions should be made synchronously to the timespan that is required by the related real-world processes. This is, for example, the time that is left before the last turning point for a changed route is passed, or the time needed by the thermal mass of the freight to warm up if the reefer aggregate fails. The decision process assumes that the related information could be transferred from the sensing to the execution unit (actuator) without violating timing restrictions. Additional communication links should be avoided to minimize the risk of delayed decisions due to communication failure.

Networking of embedded systems

The communication budget of each sensor node is very limited due to its battery capacity. For this reason central data collection and evaluation has to be replaced by local processing and data compression. It should be considered whether decision processes could be divided into smaller units and distributed among a network of low-cost microcontrollers. The idea of networking embedded measurement systems is comparable to the approach of ubiquitous computing, that was originally meant to embed miniaturized processors into everyday objects (Mattern 2005).

Application in food logistics

The logistics of food and especially agricultural products is an outstanding example for dynamic demands that are placed on transport planning. Planning has to take into account that market and order position are subject of permanent alterations. Although road transport from Spain to North Europe is about tree days, large customers like retailer chains expect their orders to be fulfilled within 20 hours (Dannenberg 2006). Changing weather conditions affect both sides of the supply chain. If the conditions are too bad for harvest, the purchaser has to fall back to an alternative cultivation area. The consumer behaviour is weather dependent, as well. Certain fruits like melons are not very well sold during rainy periods.

Transport planning has to take in account that product quality can fall below an acceptance limit which leaves the transport without shelf life during retail and thus without economic value. To avoid economic losses supervision of food quality during transport can be applied. The evaluation of the huge amount of data that is produced by detailed supervision assumes concise knowledge of the product.

Supervision devices can de divided into data loggers and telemetric remote monitoring systems. In many cases, several incompatible technologies are used within a single transport. Data loggers are packed together with the freight, the reefer aggregate records temperature and humidity values; the temperature is manually read once per day on sea transports. Sensor protocols are mainly used to settle liability questions after damage has occurred.

Telemetric systems are on their way into food logistics. In contrast to data loggers, they enable early corrections in the transport planning. The System of Cargobull Telematics[11] sends periodically data about position (GPS) temperature, tire pressure and state of the reefer aggregate. Doors are only to be opened in predefined allowed areas. Otherwise an alarm is sent over mobile communication, for which GPRS is currently used. IBM and Maersk announced a similar module called TREC that can be mounted to the door of a container. It measures temperature, altitude and light and transfers data over satellite communication[12]. Standard tariffs for the Cargobull system offer temperature and position information updated every 15 minutes for a monthly rate about 50 Euros. To keep inside an inclusive volume that is negotiated with the network provider, high-level data compression is necessary. The inclusion of additional environmental data and their spatial distribution requires advanced data pre-processing and interpretation to avoid increased communication costs that will not be accepted by transport companies.

4.6.2 Local data pre-processing

Evaluation of quality changes demands not only detailed information about environment parameters but also guidelines on how this data should be interpreted. In this section, we show as an example how deviations from the optimal transport conditions for certain vegetables could be related to quality changes.

---

11 http://www.cargobull.de/en/produkte_und_dienstleistungen/ cargobull_telematics/Produkte/default.jsp

12 IBM press bulletin, see RFID-Journal http://www.rfidjournal.com/article/articleprint/1884/-1/1

Perception systems for intelligent agents

Distributed autonomous control systems are mainly realized by software agents. The perception of the external world is an important feature of intelligent agents according to Bigus (Bigus 2001, p. 235). Agents need an internal representation of their environment for decision-making. An intelligent agent has to avoid to be overwhelmed by the flow of information by filtering or pre-processing the incoming data. The perception system of a fully automated transport planning can be divided in two parts. In the 'inside' of the means of transport, dynamic parameters like the number and kind of loaded goods, as well as the temperature and other environmental conditions have to be supervised and interpreted. On the 'outside', permanent changes in transport orders, cost and the effects of the traffic situation to the expected transport time have to be considered.

Automated interpretation of environmental data in food logistics

The inspection of food quality is, in practise, carried out by visual inspection of only a small part of the total freight. Most of the more scientific ways like measurement of firmness or starch content require opening of the package and destruction of the fruit. Furthermore, although visual inspection or chemical tests provide information about the current quality, they cannot predict future quality changes as function of the transport conditions over time. For real-time transport control quality changes have to be assessed based on parameters that are suitable for continuous monitoring. These are environmental condition like temperature, humidity and the composition of the atmosphere.

In the recent years there has been a lot of research in the modelling of quality (Tijskens 2004). As example for various modelling approaches we consider the keeping quality model. Tijskens and Polderdijk (Tijskens and Polderdijk 1996) found that the time-span that is available for transport and storage before the quality falls below an acceptance threshold depends of the inverse sum of a number of temperature dependent coefficients. These coefficients can be calculated as a function of the environmental temperature by the law of Arrhenius with the reaction specific activation energy as parameter. Parameter sets for 60 different agricultural products are listed (p 178). Figure 4.27 shows the maximum transport and storage time for tomatoes as example. The product lifetime is reduced by senescence (mostly during high temperature transport) and chilling injury (low temperature transport). To account for changing temperature conditions during transport the model was formulated in a dynamical form (p. 182).

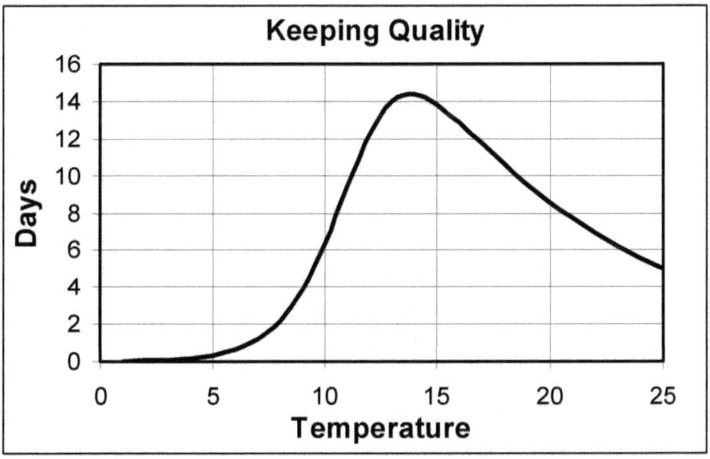

Fig. 4.27 Keeping Quality for tomatoes according to Tijskens

Temperature dependency of the maximum time-span for transport and storage before the product quality falls below an acceptance limit.

To make these models more accurate, they have to be extended to include the initial quality at harvest, which depends on the climate conditions and other influence factors. Especially the gaseous hormone ethylene has an important impact on the ripening of a number of agricultural products. Additional research is necessary to determine the quantitative effects of ethylene as well as for the development of miniaturized cost effective sensors for mobile measurement of ethylene concentrations.

Implementation of a local perception system

To increase the robustness of the system and to reduce the communication volume, assessment of the environmental conditions were implemented as local processes. The means of transport is equipped with a processor module that provides a platform for the perception system. In our technical implementation we shifted the product specific perception processes and the necessary technical investments from the transport packing to the level of the transport vehicles or warehouses for practical reasons: Transport packing rarely returns to the sender. Expensive sensor or processor equipment would be lost after the end of transport. The means of transport has to be furnished with RFID readers to scan for new freight items, sensors for supervision of the transport conditions, external communication and a processor platform. Our prototype in figure 4.26 shows an example implementation of the required hardware. The technical system is described in (Jedermann et al. 2006b).

During transport the freight items enters the dominion of local supervision systems that represent the involved transport vehicles and warehouses. By separation of the perception processes from the physical object the mobility of freight specific instructions becomes another crucial feature of the system. In our solution the perception was realized as mobile software agent. The software and the object are linked by address information stored on a passive RFID-Tag that is attached to the freight. The agent accompanies the physical object along the supply chain as part of an extended electronic consignment note containing the transport and supervision instructions. At transhipment the address of the system that currently holds the consignment note is read from the RFID-Tag. With this information the transfer of the mobile agent through the communication network is initiated. The local supervision systems form an intelligent infrastructure that provides for sensors and processing power to the loaded freight items (Figure 4.28).

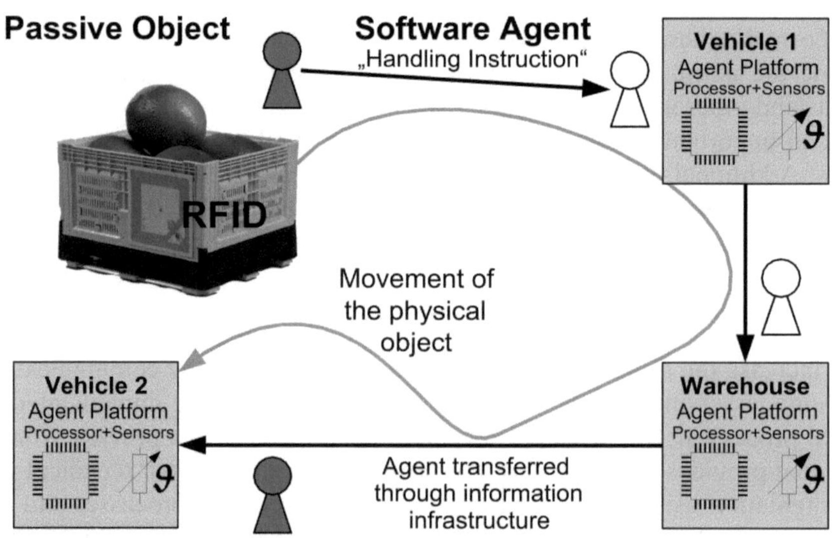

Fig.4.28 The link between physical object and mobile perception system

The freight is handled by different vehicles and warehouses along the supply chain. Arriving items are recognized by an RFID reader. A mobile software agent that contains the individual perception systems is transferred in parallel to the freight object. Required sensor systems and processing power are provided by the local platforms

As first step towards the goal of distributing autonomous processes to miniaturized systems, we examined facilities to run software agents on embedded processors with a computational power comparable to handheld PDAs. We selected an ARM-XScale processor module that provides about 10% of the clock rate and memory of a standard PC. The reduced processor architecture additionally slows down the code execution. Measurements by Jedermann (Jedermann 2006a) showed that the execution of agent systems takes 50 to 100 times longer than on a PC. That article also describes necessary optimizations in the agent framework. A special real-time JAVA virtual machine was required to run JAVA as basis of the framework on the embedded processor.

An implementation of the keeping quality model was worked out in cooperation with Horticultural Production Chains, Wageningen University (Jedermann et al. 2006c). The quality model is executed as software agent on the local system. If it predicts that quality will drop below an acceptance limit before the destination could be reached, the agent contacts the route and transport planning instances to initiate necessary reactions. The external communication is carried out by a unit developed by ComNets, University Bremen that switches between different mobile networks (e.g. WLAN, GPRS or UMTS) depending on availability.

4.6.3 Relation to the definition of autonomous cooperation

The local perception system depicts the essential features of autonomous cooperation according to the definition that was described in chapter 1.

The interpretation of the sensor measurements is organized as a **decentralized concept**. An individual software entity represents each specific transport good; the perception agents are executed close to the current location of the physical object by a distributed network of processor platforms. Each means of transport has own sensor and processor resources at his disposal.

Transhipments are carried out among partners that are on the same **heterarchical** system level without a central operator. Data are transferred peer to peer between vehicles and warehouses. Different communication standards can be used side by side.

The concept of the intelligent container allows for fully **autonomous** supervision, even if external communication links or remote processor platforms fail. The system reacts to unexpected events like sensor failures, temperature rise by sunlight or defects of the reefer aggregate without interference from humans or other systems.

The intelligent container **interacts** with other systems to retrieve freight specific information and the consignment note. The perception system negotiates with the sensors how to distribute the measurement task, which is performed in **cooperation** of several sensor nodes.

To improve the cooperation with the sensor network and the transport planning the autonomous transport supervision system has to go beyond calculation of quality models. Especially situations that allow for **alternative** reactions have to be considered. The **decision** system is still under development. It will be extended to fully cover the following topics:

- **Distribution of the measurement task**: A number of the available sensors are selected according to their tolerance, location and remaining battery lifetime.
- **Plausibility checking**: The system decides whether unusual single sensor values should be handled as measurement error or as an indicator for a spatial or time limited deviation of the environment.
- **Quality assessment:** The perception process decides whether a current deviation of the transport condition leads to an unacceptable quality losses.
- **Reactive planning**: The transport planning selects between different options to react to foreseeable quality losses.
- **Energy reduction**: Intelligent sensors could minimize their energy consumption by reducing the number of measurement and communication cycles. Section 5 discusses the feasibility for equipping miniaturized sensor units with a decision system.

The consequences of these decisions depend on future events like changes in communication quality and network topology and unknown external influences to the environmental conditions. The system behaviour could also depend on internal states that are unknown or not measurable like the harvest conditions of the product for example. The decision-making has therefore to be regarded as **non-deterministic** process with no clear right/wrong decisions. The possible consequences of several possible reactions have to be weighed up instead.

4.6.4 Linking quality information and transport planning

Decision processes of the transport planning system are based on two sources: quality information and external factors like traffic situation and the market of available transport capacities. Transport decisions are made in cooperation of different agents. A freight attendant (FA) acts from the point of view of a single transport item and coordinates its complete transport. The FA negotiates with different agents, which represent a means of transport (MTA). The MTA endeavours to maximize the use of their capacities. The acceptance of a transport orders by the MTA depends on the transport costs, destination and time schedule. The transport request is compared against the sensor equipment and reefer capacities of the vehicle.

Demonstration system for dynamic transport planning

The 'intelligent container' and the agent based transport planning developed by the TZI (Centre for Computing Technologies, University Bremen) were linked to a common demonstrator. FA and MTA verify regularly whether the requirements of the freight could be fulfilled. If a risk is detected the FA searches for alternative plans that could possibly include a change of the means of transport. In this case the FA and MTA start to negotiate about changing the destination of the vehicle.

The freight and vehicle administration agents currently have to run on a standard notebook as separate software platform. The perception agents on the embedded system and the PC based planning agents use both the JADE framework (Bellifemine et al. 2003). The FA and MTA are not tied to a particular location. An optimization of their consumption of processing power and memory would allow executing them on the same embedded platform inside the means of transport as the perception agents. The plans for future development will at least shift the FA to the embedded system. The software approach of the demonstration system is described in (Jedermann et al. 2006d).

Examples for dynamic planning

The described system shows its most advantages if the shelf life and the transport duration have a comparable magnitude, which is the case for most fresh fruits as well as fresh meat and fish. Because of their distinctive ripening behaviour bananas are excellent examples for the use of quality information for dynamic planning. Bananas are harvested in an unripe 'green' state. After their two or three weeks ship transport they are exposed to ethylene in special ripening rooms for up to one week. During this forced ripening process starch is converted to sugar and the colour changes to yellow. The aim of warehouse keeping is to have an even mix of different ripening states in stock. This process can be improved by a system that permanently monitors the ripening state and sends notifications about quality changes one or two weeks ahead of the planned arrival of the vessel. The further distribution of the fruits also demands careful planning. Weekly deliveries to retailers are partly composed of bananas in three different ripening states. This allows them to offer bananas in perfect condition on a daily basis.

Another example might be the road transport of strawberries from south to north Europe. Bad weather conditions at harvest could cause severe quality problems. If the content of some trucks is lost, the remaining vehicles could be redirected to share the remaining undamaged freight evenly among the costumers and fulfil all delivery commitments at least partly.

4.6.5 Measurement of spatial distributed environmental parameters

Deviations of the environmental conditions could affect the means of transport as a whole or only a spatial limited share of the freight. To detect the latter case a multi-point measurement is required. If the difference from the prescribed transport condition rises, it could result in a local quality loss. The detection of these local quality losses is a crucial issue, because already the decay of smaller parts of the freight could endanger the whole transport. To identify such risks in good time it is necessary to distribute sensors over the entire length of the container. The use of wireless sensor networks for data transmission could reduce additional installation costs.

Examples for local parameter deviations

Especially the temperature in reefer containers is subject to severe fluctuations. Measurements with several data loggers showed differences of 5°C over the length of a container (Tanner and Amos 2003; Punt and Huysamer 2005). These or even greater deviations are caused by bad thermal isolation or wrong packing that blocks or short cuts the air stream of the reefer aggregate. Reefer containers and vehicles are not designed to cool down food from harvest conditions to transport temperature. European regulations like the HACCP[13] concept demand that only pre-cooled goods

[13] Hazard Analysis and Critical Control Points (HACCP) is a systematic preventative approach to food safety that addresses physical, chemical and biological hazards as a means of prevention rather than finished product inspection.

are loaded. But violations of the rule happen not only outside Europe. Absorbing the heat of 'warm' goods could take more than a day with large difference between air and freight core temperature as a side effect.

The formation of local 'ripening spots' has to be avoided. This effect is mainly observed at see transports of tropical fruits. The intensified metabolism processes lead to a further temperature rise. For this reason the transport of fruits at a temperature between 10 °C and 15 °C requires more energy for cooling than deep frozen goods. Besides the temperature rise the fruits start to produce ethylene themselves, which stimulates ripening processes in neighbouring fruits. This effect could in the end lead to a total loss of the transport. To estimate the effects of local losses onto the behaviour of the whole freight requires further modelling.

Key features of wireless sensor networks

Wireless sensor networks (WSN) consist of tiny-networked embedded devices, which act as the network nodes. These nodes are formed by a microcontroller, a RF-Interface and sensors and are usually powered using batteries. While designing such systems, usually COTS (Commercial Off-The-Shelf) components are used in order to reduce system prices. The consequences of this concept will be shown later.

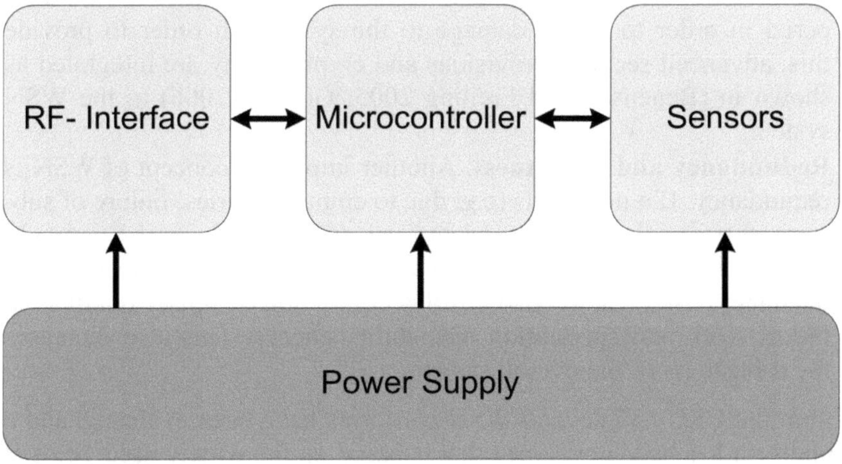

Fig.4.29 Wireless sensor node architecture

In the past five years, a lot of research projects on various aspects of WSN have made extensive advances in this field possible. The major aspects were system design (Handziski et al. 2005), communication protocol design (Woo and Culler 2001) and how to interact with a WSN

(Madden et al. 2003). Some of these research projects brought forth advanced concepts and systems, which are paving the way to WSN applications in industry.

The concept of WSN offers a lot of benefits for the integration of intelligent sensor systems into logistic supply chains. These benefits are wireless ad hoc communication, security and robustness.

- **Wireless ad hoc communication:** All nodes in the networked are linked wirelessly. This gives the opportunity of monitoring the transported assets continuously. If any disturbance is detected, the WSN system may inform autonomous cooperating units, which will evaluate the disturbance and may trigger necessary reactions. No extra effort has to be made for the integration of the additional sensor nodes into the WSN. The systems may either be fitted to the freight items or may be deployed before or after the loading process. After being turned on and detected by the network within the transport medium the nodes autonomously log into the network. As an RF interface an IEEE802.15.4-compliant transceiver was chosen, which is used by many other WSN platforms (IEEE 2003).
- Privacy and Security are important aspects for the integration of WSN into autonomous cooperating logistic processes. The data of the systems may neither be readable to unauthorized parties nor be tampered in order to inflict damage to the system. In order to provide this, advanced security provisions and cryptography are integrated as shown in (Benenson and Freiling 2005; Gorecki 2006) to the WSN system.
- **Redundancy and robustness**: Another important concept of WSN is redundancy. If a node fails (e.g. due to empty batteries, failure of subsystems), its role is simply taken over another node in the network. The application of COTS components is simply a consequence of redundancy. This feature also enables correlation of sensor readings in the network and application plausibility checks. This also increases the robustness of the overall system.

Within the CRC 637 several WSN platforms have been evaluated and a new platform has been developed that is based on the widely used Telos B platform by Moteiv[14]. In contrast to Telos B, our WSN prototype system, offers a modular sensor interface. This enables eased usage of several different sensor types with the proposed platform.

[14] www.moteiv.com

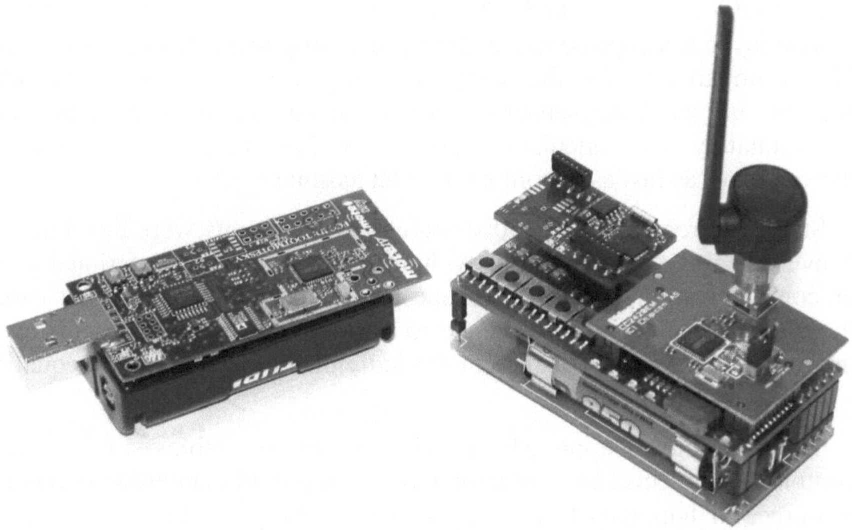

Fig. 4.30 Commercial Moteiv system (left) and SFB637 WSN system (right)

The above description shows that wireless sensor network constitute a promising technology for the integration of sensors in supply chain management.

4.6.6 Applying autonomous cooperation in sensor networks

The previous sections described how the concept of autonomous cooperation was implemented to improve the supervision of quality parameters at the level of transport containers or vehicles. The following chapter will discuss to what extent this concept could be adapted for the coordination of the sensor systems inside the container and whether it is possible to reduce classical agent system architecture for these resource-constrained devices.

The self-configuration of sensor networks can already be regarded as an autonomous cooperating process. Rogers (Rogers et al. 2005) gives an example how the message forwarding is organized by local control. The autonomous approach should be extended to other systems tasks. Especially the question when and how often energy-consuming measurements and communication procedures have to be triggered demands an intelligent selection among different alternatives.

As previously described, the hardware that is used at the transport system layer has approximately 10% processing resources of conventional PCs. Compared to this the computational power of the microcontrollers applied in wireless sensor nodes is even a hundred times less. Unfortunately the calculation power of low-energy microcontrollers increases not as fast as Moore's Law[15] let assume.

Standard architectures for distributed agent systems were developed on conventional PC based systems without restriction to computational power or communication. It has to be questioned whether and to what extend these approaches could be adapted to resource-constrained systems and if the restrictions in performance of the individual elements can be equalised by increasing their number.

The well-known approach of Grid computing enables solving highly complex problems by collaboratively employing unused computing resources of hundreds to thousands of PCs. As given by Walter (Walter 2005), problems that employ the processing of several independent data streams can be solved by microcontroller clusters by application of tools and methods from Grid computing. Especially problems that require a high number of interrupt-triggered tasks can be solved more efficiently by using several coupled embedded systems then by employing single high-performance CPUs. The grid computing approach may not be applied directly for the distribution of autonomous cooperating processes within a network, as Grid computing is based on centralized control and hierarchical structures.

Approaches for reduced hardware

Solutions have to be found for the distribution of tasks in a network of computationally small systems. Some of the approaches that have been applied to this question like swarm intelligence, fuzzy and agent architectures are introduced in the following sections.

Swarm intelligence

The intelligence of swarms observed in nature is often quoted as example solution. Ants search the shortest path by following a pheromone track left by their predecessors (Bonabeau et al. 1999). A swarm of fishes agrees on a swimming direction without requiring a communication intensive voting.

[15] From recent developments Moore's Law extrapolates that the complexity of integrated circuits doubles every 18 month. But this mainly applies to PC components where the huge market volume allows for large technical investments.

By coordination with their imminent neighbours the individuals balance contradictory information about the best direction and avoid break-up of the swarm (Pöppe 2005). The intelligence of the swarm cannot be concluded from an isolated view on the behaviour of a single individual. It is not even necessary that each individual knows the super ordinate aim of the swarm.

Deliberative and reactive agents

Wooldridge and Jennings (Wooldridge and Jennings 1995, p. 24) describe deliberative architectures as the classical or symbolic AI methodology for building agents. A deliberative agent is characterized by

- an explicitly represented symbolic model of the world and
- decision making via logical or at least pseudo-logical reasoning

Unsolved problems in applying the deliberative approach on time-constrained systems have led to the development of reactive architectures. Wooldridge and Jennings (Wooldridge and Jennings 1995 p. 27) define

„a reactive architecture to be one that does not include any kind of central symbolic world model, and does not use complex symbolic reasoning".

Especially the 'subsumption architecture' from Brooks 1986 has gained much attention. Different vertically layered behaviour patterns are continuously computed in parallel. From this 'behaviour set' a single behaviour is chosen to dominate the reaction of the system. The higher layer patterns decide whether they superimpose the lower layers. Brooks employed this approach for the control of robots. For example if a module is activated that cares for returning to the power station for recharging the batteries all lower layers will be blocked. The behaviours for exploration of the surroundings and keeping a minimum distance to obstacles are no longer executed.

Using this layered approach increases the overall robustness of the system. If a single layer fails the whole system keeps its capacity of acting. The essential difference to conventional systems is that the robot does not employ a view of its world. No symbolic world model needs to be developed, as the reactions depend only on current observations of the environment. The robot responds to changes in its surrounding in a form that corresponds to reflexes. This approach may be also combined with a symbolic representation of the world, E.g. the robot generates a map in order to reach distant destinations (Bergmann 1998). Using this extremely simple architecture concerning its computational complexity Brooks achieved astounding results:

> But despite this simplicity, Brooks has demonstrated the robots doing tasks that would be impressive if they were accomplished by symbolic AI systems. (Wooldridge and Jennings 1995, p. 28)

Learning

The ability to learn from previous actions is seen as another prerequisite for intelligence. Due to the limited lifetime of a freight object the opportunities for application of learning processes for transport supervision are restricted. The individual quality dynamics are very variable and may not be translated to other freight classes. Already at the beginning of a transport this knowledge has to be completely present. A learning process is only viable on a meta-layer by incrementally building a knowledge base concerning specifics of certain freight classes. Furthermore it is possible that the internal wireless communication network incrementally adapts to communication disturbances caused by the spatial distribution of the freight items within the transport medium.

Examples of intelligent sensor systems

The application of agent-oriented architectures for the control of embedded systems and sensor networks has already been researched in the past. Two examples from literature are summarized in the following section.

Target tracking by radar sensors

One approach to control and coordinate a distributed sensor network was presented by the group of Lesser (Lesser et al. 2003; Mailler 2005). They describe a sensor network for discovering and tracking moving targets. Each sensor node has three independent radar sensors that cover an angle of 120°. In their research they selected a setting that forces local processing and cooperation:

- At least 3 sensor nodes have to cooperate for triangulation.
- The node has to decide in which direction it looks for new targets, because only one of the three radar sensors can be activated at the same time.
- A very limited communication bandwidth prevents central data interpretation and sensor control.
- All sensor data have to be processed in real-time before distance measurements become obsolete.

To reduce communication the sensors are organized in clusters. Best results were achieved with fixed clusters of 5 to 10 sensors (Mailler 2005, p. 11). Architectures for dynamic coalitions forming were also considered (Lesser et al. 2003, pp. 110f). A dynamic coalition is formed in response to an event like the detection of a new target and dissolved when the event no longer exists.

Each sensor is represented by an agent that can take on different roles. One agent in each sector acts as the sector manager that disseminates a schedule to each sensor with frequencies to scan for new targets. When a new target is detected, the sector manager selects a track manager that is responsible for tracking the target as it moves through the environment. The track manger requests and coordinates other sensors and fuses the data they produce (Mailler 2005, p. 6).

The target tracking utilizes the same JADE framework to execute the agents as our transport planning system. But it moves the agent platform to an external computer. The radar nodes are only equipped with a simple processor to control the communication unit and the hardware of the sensor elements.

Fuzzy agent architecture

Human operators are often better at control of complex and non-linear technical processes. Zadeh (1965) who introduced the theory of "fuzzy sets" proposed that the reason for the human superiority is that they are able to make effective decisions on the basis of imprecise linguistic information. Fuzzy logic has become an increasingly popular approach to convert qualitative linguistic descriptions into non-linear mathematical functions. *Fuzzy rules provide an attractive means for mapping sensor data to appropriate control actions* (Hagras et al. 1999 p. 324). Hagras et al. combine Brooks subsumption architecture with fuzzy logic controllers (FLC). Each layer or behaviour of the subsumption architecture is represented by one FLC. An additional FLC is used to combine the output of the different layers. The parameterization of the FLCs is performed by a patented genetic learning mechanism. This hierarchical fuzzy agent was tested for the control of robots. In another implementation the fuzzy learning technique is used to adapt an "intelligent dormitory" that is located at the University of Essex to the personal preferences of a guest (Hagras et al. 2002). The fuzzy agents run on a Motorola 68030 processor with a computation power that is about ten times lower than those of the embedded platform for the intelligent container.

Self-configuration as autonomous cooperative process

Wireless sensor networks are a perfect indicative of autonomous cooperation. As an example the network formation process is discussed with a focus on autonomous cooperation.

The most intuitive way to network WSN devices is to use a fully-meshed network topology. This means that all the nodes in the network are interconnected. This network topology is the optimal representation of heterarchy, as any node in the network may communicate with any other node at once as shown in figure 4.31. However, in wireless sensor networks there are restrictions regarding communication in the network.

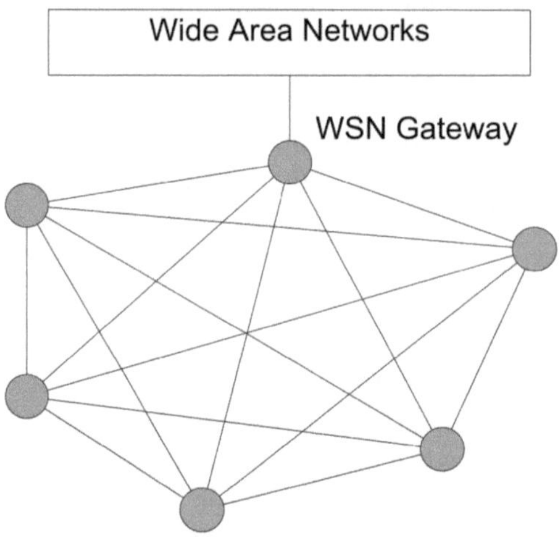

Fig. 4.31 Fully-meshed network topology

A very important aspect that comes into play when designing WSN systems is energy consumption. This is mainly due to dependence on batteries. When taking a closer look at the energy consumption of the three parts (microcontroller, RF-transceiver and sensors) as shown in Table 4.12. it is clear to see that in common WSN systems, the RF-transceiver consumes around 20 to 40 times the energy of the microcontroller. The energy consumption of sensors should not be neglected, but is highly dependable on the application and is not taken into consideration here. Therefore, communication among the nodes in the WSN has to be kept as minimal as possible in order to enable longer system lifetimes. However, this is a contradiction towards the paradigm of autonomous cooperation. Therefore, pure heterarchy has to be traded-of against energy-efficiency.

Table 4.12. Power consumption of selected WSN platforms

	Telos B	Mica2[16]	MicaZ	iMote2[17]
CPU sleep, Radio off[18]	0.0153 mW	0.054 mW	0.054 mW	0.1 mW
CPU on, Radio off	5.4 mW	36 mW	36 mW	>100mW
CPU on, Radio Tx/Rx	58.5 mW	117 mW	75 mW	>150mW
CPU on, Sensors active[19]	7.2 mW	37.8 mW	37.8 mW	>100mW

The power consumption of the WSN platform that was developed within CRC637 almost matches the power consumption of the Telos B platform.

Taking these necessary limitations for WSN systems into consideration, another type of topology has to be found. One possibility is to use hierarchical concepts like clustering. The basic idea of clustering is that a group of network nodes form a cluster (figure 4.32). One device is elected as cluster head of this cluster for a certain period of time. The cluster head manages the communication with any other device that addresses a node inside the cluster, so it acts as a gateway to the cluster. Using this method, the workload is distributed among the network nodes and communication and energy consumption are reduced.

Thus, network lifetime can be prolonged by a factor of up to 4 (Younis and Fahmy 2005), while this cluster-based-topology also allows very simple collection and aggregation of the data.

[16] Both Mica2 and MicaZ were developed by Crossbow Inc. For more information refer to www.xbow.com

[17] The iMote2 platform has been developed by Intel. Compared to the other platforms in the table it features a PDA-class CPU which has more computing power, but also incorporates increased power consumption.

[18] CPU is in sleep mode while the RF unit is switched off. This is usually valid for more than 99% of the operation time of a sensor node.

[19] Here values for a commonly used humidity/temperature sensor (Sensirion SHT15, www.sensirion.com) are shown.

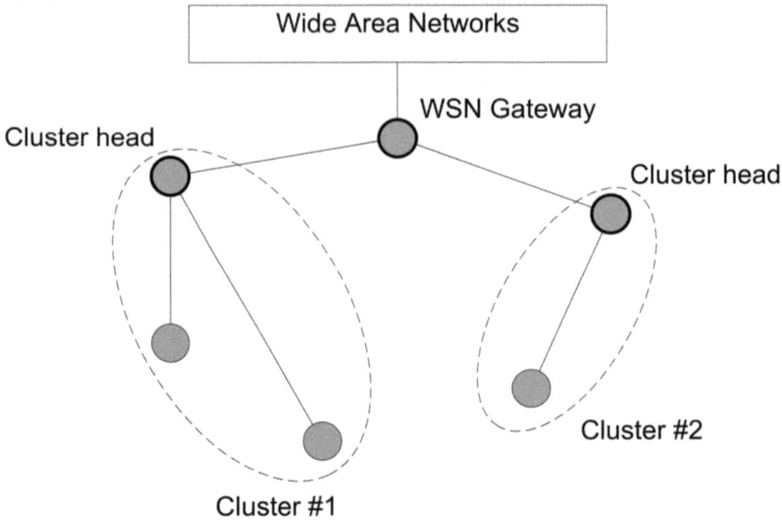

Fig. 4.32 Cluster-based network topology

After expiration of the cluster head period all devices start to compete for the next election. If a node that is currently cluster head fails, the election process will be restarted among the remaining nodes. If new nodes enter the network during a cluster head period they will enter as a cluster member. Different clusters are identified by an address supplement. This ensures the formation of multiple clusters. The size of the clusters varies with the number of nodes in the network and of the corresponding RF power settings of the nodes. For very large numbers of nodes even multi-tier hierarchical clustering is possible in order to ease the network management.

The cluster head election process is mainly probability-based while the random value is influenced by factors like the number of packets sent and received, number of sensor readings, total time running and how often the device had been elected cluster head before. WSNs provide a mapping of heterarchy in order to prolong system lifetime. All nodes in the network participate autonomously in the topology formation process. Therefore, this process is collaborative and decentrally organized. Various aspects of autonomous decision are also included in WSN systems. E.g. the nodes may autonomously decide on routing of messages, if their energy level falls below a certain threshold.

The knowledge of sensor about the "world" is mainly represented by tables containing information about the known neighbouring sensors and values for the quality of corresponding communication links.

All these features show that the topology formation process is one example of autonomous cooperation in wireless sensor networks. Other examples of autonomous cooperation in wireless sensor networks are e.g. routing of messages, data aggregation, and cross-network correlation of data.

After consideration of the above aspects WSN systems imply several aspects that can be seen as autonomous cooperating process.

The scope of an extended decision system

An appropriate distribution of the measurement task could lead to large reductions of the power consumption of the sensor nodes. But available solutions fall behind the crafted configuration of the sensor network. Standard software packages merely check thresholds or calculate mean values. The energy consumption mainly depends on the number of communication and measurement cycles. An intelligent distribution and control could prolong battery lifetime by omitting redundant cycles.

Decision alternatives

To minimize the effort for communication and measurements the system has to choose between different possible alternative actions. The scope of the decision system covers the following fields:

- **Distribution of the measurement task:** The supervision task has to be distributed among the sensor nodes with regard to their tolerances, resolution and individual power consumption per measurement. The battery reserves should also be taken into consideration.
- **Forwarding of measurements:** If a sensor node observes a deviation of a value the node has to weigh up whether it should spend energy to transfer the value to the transport planning. Before an expensive multi hop communication to a central system is initiated the node contacts his immediate neighbours. Cooperative decisions include measurements of several nodes to avoid duplicate or unnecessary notifications. Crucial deviations have to be distinguished from measurement noise.
- **Requests for additional measurements:** For conspicuous sensor values the node could request additional measurements from its neighbours for confirmation. The costs per measurement vary by a factor of more than thousand for different environmental parameters. Semiconductor temperature sensors spend 0.1 mJ (0.0001 Joule) per cycle. While gas sensors require heating to 200°C or 300°C of the

device with an energy consumption of more than 100 mJ. The frequency of such measurements should be reduced as much as possible. Measurements of other 'cheaper' sensors could be an indication for the decision whether the activation of a high-energy sensor is necessary.

- **Plausibility checking:** Plausibility checking is not directly linked with energy. But the distinction between measurement errors and spatial confined or time limited deviation of the surrounding conditions is very important for the robustness of the system. The decision could be based on comparisons with other nodes or on a record of former local measurements.

Boundary conditions and demands on the decision system

Decisions have to be made from a local point of view because the communication costs increase with the distance or the number of network hops. In a cluster topology decisions are made in cooperation through agreement or local voting. The choice between different possible reactions requires complex decision processes. The results of the possible alternatives are often uncertain. Some influence factors are unknown or cannot be measured with sufficient exactness. Because of the non-deterministic system behaviour the alternatives cannot be reduced to a wrong / right decision. Advantages and risks of the alternatives have to be compared instead.

The aim of autonomous cooperation is to quickly find a stable solution where a concise calculation of the optimal solution is not adequate. The decision processes of the sensor system have to be designed for robustness against failure of other systems or breakdown of communication links. Sudden changes in the environment or faulty information should not lead to instability of the overall system.

The power consumption of the processor that controls the sensor nodes is almost proportional to the computational load. Merely averaging of measurement values requires much less energy than communication. Different decision algorithms have to be selected by their power consumption. New simplified solutions for miniaturized microcontrollers have to be developed. The power consumption of the decision system has to be lower than the amount of energy that is saved by selection of a better reaction alternative. The required computation power per decision (Joule per decision) has to be compared against the consumption of measurement and communication cycles (Joule per cycle).

Intelligent freight objects

The developments towards autonomous cooperation, which were presented at the example of sensor networks, could be considered as signpost for further research. In future implementations the freight items could be equipped with miniaturized microcontrollers and ultra-low-power sensors. They connect themselves to neighbour items to extend their communication and measurement facilities. The "intelligent package" makes decisions about its transport route, negotiates with different vehicles for transport and detects quality risks on its own.

Another aim is to support these items over an electro-magnetic field comparable to passive RFID tags. But unlike pure identification tags, batteries are still required to make the system capable of measurement and planning even in the absence of a field. The "VarioSens" data loggers from KSW-Microtec[20] are an example of such a semi-passive system. The temperature logging is powered by a paper-thin battery, but the energy that is required for communication is provided by the electro-magnetic field of the reader.

4.6.7 Conclusions and outlook

To design a concise supervision for transports of perishable goods several requirements have to be taken into account like efficiency of mobile communication, measurement and assessment of spatial distributed sensor data, stability of the solution in case sub-system or communication failure, just in time capabilities as well as the implementation on embedded systems and their networking. In this article we presented autonomous cooperation as a robust solution to handle the vast amount of spatial scattered sensor data.

Wireless sensor networks and RFIDs are supporting technologies to supply the necessary information to autonomous processes. Furthermore, the wireless sensor networks feature ad hoc networking capabilities while providing sufficient means of communication security and robustness.

The agent framework JADE was implemented on a high-performance embedded processor to provide a platform for local pre-processing that reduces the communication volume and avoids overheads caused by central planning.

---

[20] KSW-Microtec AG, Dresden, Germany http://www.ksw-microtec.de

The system features like permanent access to the freight state, instant notifications on quality problems and the option for automated route planning provide important advantages for the huge application field in food logistics. The keeping quality model was introduced as method to evaluate sensor data.

The application of autonomous cooperation at the level of the sensor nodes may also be increased beyond the implementation of self-configuration mechanisms. To extend the sensor node's battery lifetime by reduction of communication it is necessary to move the decision-making ability into the sensor network. Decisions have to be made from a local instead of a bird's eye view. The translation of this approach into application requires further research. Expectable increases in the performance of microcontrollers promote these developments. But the anticipated growing of calculation power at same or lower level of energy consumption reaches not far enough to replace the development of algorithms that are specially adapted to low-power microcontrollers. An agent framework like JADE is probably not feasible on low-cost sensor nodes. The solution would rather be a combination of reactive behaviours and logical reasoning. The organisation of the nodes in clusters is expected to deliver the best results.

The fitness of a solution is finally judged by its energy balance. The energy that is necessary to calculate a decision has to be less than the saved costs for measurement and communication.

Information technologies are of growing importance for transport planning and supervision. Autonomous cooperation will be one of the key concepts in systems that go beyond nowadays remote temperature monitoring.

References

Bellifemine F, Caire G, Poggi A, Rimassa G (2003) Jade – a white paper. In: TILAB "EXP in search of innovation", Vol 3, Italy

Benenson Z, Freiling FC (2005) On the Feasibility and Meaning of Security in Sensor Networks. 4th GI/ITG KuVS Fachgespräch "Drahtlose Sensornetze", Zurich, Switzerland

Bergmann K (1998) Seminar „Lernalgorithmen in der Robotik“, Technische Universität Graz, Austria

Bigus J (2001) Intelligente Agenten mit Java programmieren. Addison-Wesley, München

Bonabeau E, Dorigo M, Theraulaz G (1999) Swarm Intelligence - From Natural to Artificial Systems. Oxford University Press, New York

Brooks RA (1986) A robust layered control system for a mobile robot. IEEE Journal of Robotics and Automation, 2(1): 14-23

Dannenberg A (2006) Fruchtlogisik: Gobale Beschaffung – Lokale Verteilung, Vortrag am 21.03.06, Bremen, Atlanta AG

Gorecki CA (2006) Beiträge zur sicheren Kommunikation mobiler Systeme. Ph.D. thesis University Bremen, ISBN 3-8325-1160-1, Logos Verlag, Berlin, Germany

Hagras H, Callaghan V, Colley M (1999) An embedded-agent technique for industrial control environments. Assembly Automation, Vol 19, Number 4, 1999, pp 323-331

Hagras H, Colley M, Callaghan V, Clarke GS, Duman H, Holmes A (2002) A Fuzzy Incremental Synchronous Learning Technique for Embedded-Agents Learning and Control in Intelligent Inhabited Environments , Proceedings of the 2002 IEEE International Conference on Fuzzy systems, Hawaii, USA

Handziski V, Polastre J, Hauer JH, Sharp C, Wolisz A, Culler D (2005) Flexible Hardware Abstraction for Wireless Sensor Networks. In: Proceedings of the 2nd European Workshop on Wireless Sensor Networks (EWSN 2005), Istanbul, Turkey

Institute of Electrical and Electronics Engineers Ed (2003) IEEE Standard for Information Technology – Telecommunication and information exchange between Systems – Local and metropolitan area networks – Specific requirements. Part 15.4: Wireless Medium Access Control (MAC) and Physical Layer (PHY) Specification for Low-Rate Wireless Personal Area Networks (LR-WPANs). IEEE Computer Society, New York, USA

Jedermann R, Lang W (2006a) Mobile Java Code for Embedded Transport Monitoring Systems. In: Grote, C. and Ester, R. (eds.): Proceedings of the Embedded World Conference 2006, February 14-16, Nuremberg, Germany. Vol 2., pp. 771-777. Franzis Verlag, Poing

Jedermann R, Behrens C, Westphal D, Lang W (2006b) Applying autonomous sensor systems in logistics; Combining Sensor Networks, RFIDs and Software Agents. In: Sensors and Actuators A (Physical) Vol .132, Issue 1, 8 November 2006, pp. 370-375 (http://dx.doi.org/10.1016/j.sna.2006.02.008)

Jedermann R, Schouten R, Sklorz A, Lang W, van Kooten O. (2006c) Linking keeping quality models and sensor systems to an autonomous transport supervision system. In: 2nd intern. Workshop on „Cold-Chain-Management", 8th and 9th May 2005, Bonn, Germany

Jedermann R, Gehrke JD, Lorenz M, Herzog O, Lang W (2006d) Realisierung lokaler Selbststeuerung in Echtzeit: Der Übergang zum intelligenten Container. In: 3. Wissenschaftssymposium Logistik, Bundesvereinigung Logistik, Dortmund, 30.-31 May 2006. Pfohl HC and Thomas Wimmer T (eds) Wissenschaft und Praxis im Dialog. Steuerung von Logistiksystemen - auf dem Weg zur Selbststeuerung, Wirtschaft and Logistik, pp. 145-166, Hamburg, BVL, Deutscher Verkehrs-Verlag

Kesselmann C, Foster I (2003) The Grid. Morgan Kaufmann Publishers, USA

Mattern F (2005) Die technische Basis für das Internet der Dinge. In: Fleisch E und Mattern F (eds) Das Internet der Dinge. Springer, Berlin, pp 39-66

Lesser V, Ortiz CL Jr, Tambe M. eds (2003) Distributed sensor networks - a multiagent perspective. Boston, Mass., Kluwer

Madden SR, Franklin MJ, Hellerstein JM, Hong W (2003) The Design of an Acqusitional Query Processor for Sensor Networks. In: Proceedings of SIGMOD, San Diego, USA

Mailler R, Horling B, Lesser V, Vincent R (2005) The Control, Coordination, and Organizational Design of a Distributed Sensor Network. Under Review

Pöppe C (2005) Führerpersönlichkeit und Herdentrieb. Spektrum der Wissenschaft, September 2005, pp 22-23,

Punt H, HuysamerM (2005) Supply Chain Technology and Assessment - Temperature Variances in a 12 m Integral Reefer Container Carrying Plums under a Dual Temperature Shipping Regime. In: Acta horticulturae, Vol 687, pp 289-296

Rogers A, David E, Jennings NR (2005) Self-Organized Routing for Wireless Microsensor Networks. In: IEEE Transactions on Systems, Man, and Cybernetics – Part A: Systems and Humans, Vol. 35, No. 3

Tanner DJ, Amos ND (2003) Heat and Mass Transfer - Temperature Variability during Shipment of Fresh Produce. In: Acta horticulturae, Vol 599, pp 193-204

Tijskens LMM, Polderdijk J J (1996) A generic model for keeping quality of vegetable produce during storage and distribution. In: Tijskens 2004, pp 171-185

Tijskens LMM (2004) Discovering the Future. Modelling Quality Matters. PhD thesis, Wageningen Universiteit, (Promotor Olaf van Kooten)

Walter KD (2005) Embedded Grid Computing. In: Elektronik, Fachzeitschrift für industrielle Anwender und Entwickler, 9/2005, pp 76-79, Poing, Germany

Wooldridge M, Jennings NR (1995) Intelligent Agents: Theory and Practice. In: The Knowledge Engineering Review 10(2), pp 115-152

Woo A, Culler D (2001) A Transmission Control Scheme for Media Access in Sensor Networks. In: Proceedings of Mobicom 2001, Rome, Italy

Younis O, Fahmy S (2005) An Experimental Study of Routing and Data Aggregation in Sensor Networks. In: Proceedings of the International Workshop on Localized Communication and Topology Protocols for Ad hoc Networks (LOCAN), held in conjunction with The 2nd IEEE International Conference on Mobile Ad Hoc and Sensor Systems (MASS-2005)

Zadeh L (1965) Fuzzy sets. In: Informat. Conf., Vol. 8, pp 338-53

4.7 Transport Scenario for the Intelligent Container

Reiner Jedermann[1], Jan D. Gehrke[2], Markus Becker[3], Christian Behrens[4], Ernesto Morales-Kluge[5], Otthein Herzog[2], Walter Lang[1]

[1] Institute for Microsensors, -Actuators and -Systems (IMSAS), University of Bremen

[2] Center for Computing Technologies (TZI), University of Bremen

[3] Working Group Communication Networks (ComNets), University of Bremen

[4] Institute for Electromagnetic Theory and Microelectronics (ITEM), University of Bremen

[5] Bremen Institute of Industrial Technology and Applied Work Science (BIBA), University of Bremen

4.7.1 Scenario setting

Previous chapters described among others the application of autonomous cooperation on embedded systems, in sensor networks, transport planning and communication systems. For practical demonstration of the implications of the described studies the prototype of the intelligent container was linked to an agent system for transport coordination including communication gateway and vehicle location. We arranged a demonstration scenario that illustrates the cooperation of these system components by displaying the processes that are related to one selected freight item and one transport vehicle.

The scenario describes the automated supervision and management of a transport of perishable goods like foodstuffs. The location of the freight and the vehicles are traced by two RFID systems. The transport coordination automatically initiates reactions to transport disturbances.

The logistic entities and services that are involved in the transport scenario are represented by software agents. The agent system comprises means of transport, freight items and storage facilities as well as secondary services like traffic information, route planning and service brokerage. For our demonstration setup the agents are pooled in two notebooks. The first one depicts the site of the manufacturer or sender of goods who defines the transport order and receives notifications about the current state of his consignment. The second laptop provides a platform for the transport coordination. The sensory supervision related to specific freight items is executed on the embedded platform inside the means of transport. This quality supervision process is realised as a mobile agent that accompanies the fright item along its way through the supply chain.

The required hardware for the demonstration setting is summarised in figure 4.33. A map with the related cities, motorways and the current location of our selected transport vehicle together with status information is displayed on a screen. This electronic roadmap is connected with a model truck that drives around in our shop floor. A network of RFID readers that are mounted under the floor represent the cities and motorway crossings. If they detect a passive RFID tag, which is mounted at the model truck, they send the updated position information to the transport coordination. For easier demonstration we use a second immovable model truck or container to explain the supervision of loading processes and the internal sensor network.

Beside the Ultra High Frequency RFID system at 866 MHz for the location of the truck a second RFID system is used for tracking of the freight items and the loading process. The second system operates in the High Frequency range at 13.56 MHz. A passive RFID tag is attached to the freight that is used to control the transfer of the quality assessing agent. Because of the low data transfer rate of HF-tags only few information are stored on the tag like the unique identification number and the network address of the last platform or server that hosted the freight agent. When the freight is loaded to a new platform like a storage room or a transport container, the local processor reads the information from the RFID tag and requests the corresponding supervision agent under the given address as shown in figure 4.34. After being transferred to the new platform the agent connects itself to the local sensor facilities and continues its supervision task in the same cargo hold as the physical object was loaded to.

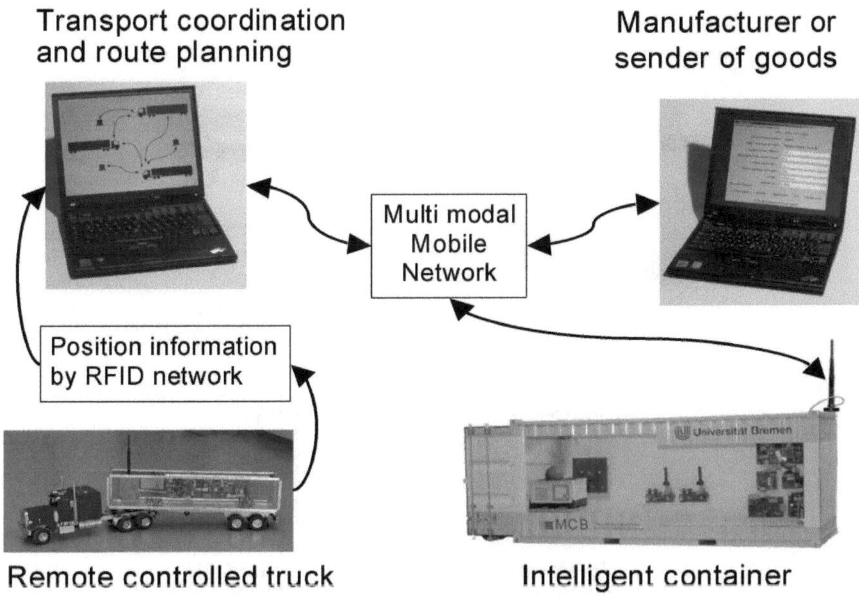

Fig. 4.33 Hardware setting of the demonstrator. Two computers are used to display the processes of the transport coordination and the monitoring by the freight owner. The model container is linked over multiple mobile networks with the information infrastructure. Temperature and other environmental parameters are provided by a sensor network inside the model container. A second remote controlled model truck drives to points in the demonstration hall that represent cities of Germany. Position data are delivered by a network of RFID readers mounted in the floor.

The effects of deviations from the optimal transport conditions are calculated by the concept of shelf life that predicts the remaining time-span until the quality falls below an acceptance limit. The freight specific models take different dynamic effects of temperature into account like 'chilling injuries' by storage below the recommended temperature and accelerated decay processes by temperature rises. The nonlinear dependency of shelf life from these processes is calculated according to the laws of reaction kinetics.

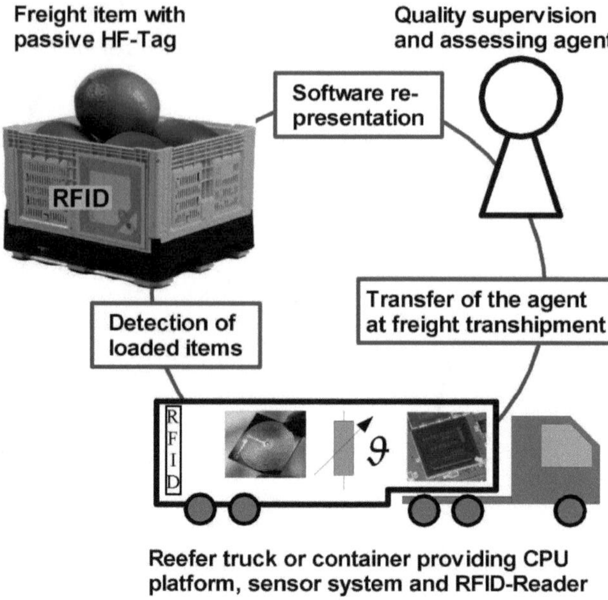

Fig. 4.34 Control of the transfer of the mobile quality supervision agent by an RFID tag attached to the transport packing. After reading the address information from the RFID tag the new means of transport sends a request for the agent that represents the freight item. On the new platform the agent links itself to the sensor facilities provided by the means of transport.

4.7.2 Steps of the transport demonstration

During the course of our example transport scenario the system has to cope with various disturbances like failure of one mobile communication network, traffic congestions and quality losses due to deviations from the recommended transport temperature. The system recognises these problems and handles them automatically by selection of alternative communication paths and means of transport, dynamic route re-planning, and by putting the freight temporarily into stationary cold storage for transhipment.

The following section describes the steps of the transport demonstration scenario, which show how autonomous cooperating processes could improve the supply chain management. The demonstration starts with the definition and placement of the transport order and ends with arrival at the final destination. As example transport order we selected the shipment of a consignment of fish from Bremerhaven to Frankfurt. This freight requires cold storage at a recommended temperature of 1 °C.

Creation of the transport order

The transport packing or freight object is marked with a standard passive RFID tag. In the first step this 'empty' tag has to be linked to the transport order and to a supervision agent. The RFID tag is scanned in the warehouse of the manufacturer or sender of goods. A dialog box (figure 4.35) supports the sender for definition of the transport order. First he selects the kind of good. A knowledge base proposes supervision parameters for the specific product. The user defines which types of sensors are required for supervison. He can choose between different mathematical or bio-chemical models for prediction of quality changes. Finally he enters the destination and other transport parameters. After closing the dialog box two freight specific agents are created. The first agent is responsible for the local sensor supervision of the environmental parameters and assessing their effects on the freight quality.

This supervision or assessing agent accompanies the freight along its way through the supply chain. The agent is started first at the warehouse of the sender. His computer or IP address is stored on the RFID tag. Different agents that are running on the same platform are discerned by a read-only unique identification number given by the RFID tag.

As the second agent that is related to the freight item, the load attendant is started on the transport coordination platform. He is responsible for transport planning and coordination with different means of transport on behalf of the freight or his sender, respectively. The load attendant could run on different platforms. In our test scenario it stays together with the route planning and information systems on the second laptop.

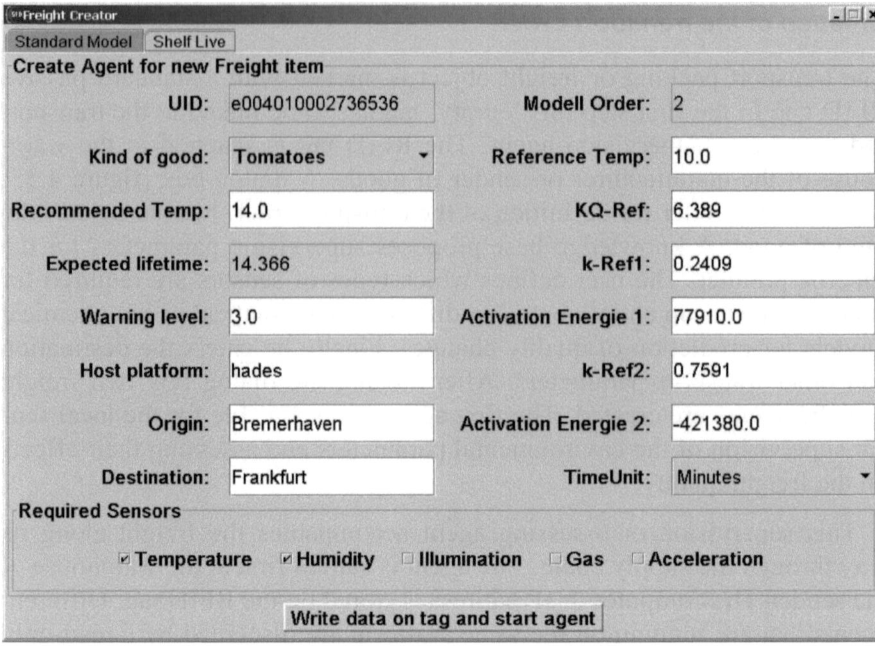

Fig. 4.35 Dialog box for definition of the transport order. The sender selects the kind of good, type of quality model and transport destination. The product specific model parameters in the right column are retrieved from a knowledge base.

Transport Coordination and Route Planning

As soon as the load attendant agent is started, the agent looks for appropriate means of transport (referred to as *trucks* in the following) to get to his intended destination. The needed transport information is given by an electronic consignment note that is passed to the attendant agent on start-up. Transport information includes, e.g., current location and destination as well as technical properties and requirements. In order to get an appropriate truck the agent consults a logistics services broker agent. This service broker provides a managed virtual marketplace for transport capacities and freight. The load attendant registers at the broker with his location, destination, and technical requirements (figure 4.36).

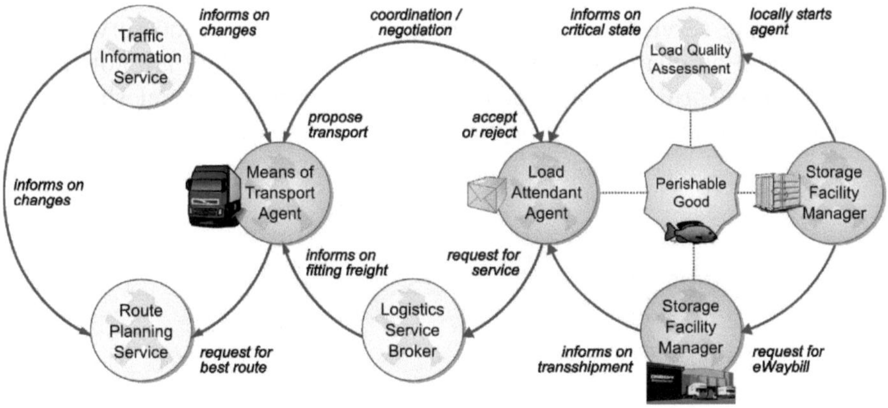

Fig. 4.36 Software agents (as circles) participating in a autonomous transportation process and selected interactions between them.

Trucks that again are represented by software agents also register at the service broker if they want to be informed on new fitting freight. In order to determine which freight meets a truck's requirements he may inform the broker on his technical features he is willing to disclose. In our scenario, these features are, e.g., the ability of cooling and online sensor monitoring. The scenario includes multiple trucks with only two of them providing the features needed to transport fish. One is located in Osnabrück; the other is located in Frankfurt. The truck in Osnabrück virtually hosts the intelligent container.

Both trucks only transport perishable commodities which require temperature-controlled transportation. When the load attendant agent of the consignment of fish in Bremerhaven registers at the broker both trucks are informed on the new matching load. In the following, both agents deliberate about whether to propose to this load to take over its transport to Frankfurt. The truck in Frankfurt does not make a proposal because he would have to accept too much empty miles to get to Bremerhaven. Though, the truck in Osnabrück with our intelligent container proposes to perform the transport.

Thus, the load agent and the truck agent negotiate the conditions of cooperation including, e.g., shipping costs. This interaction follows standardised protocols defined by the Foundation for Intelligent Physical Agent (FIPA). The load attendant agent finally accepts the truck's proposal and the truck plans its way to Bremerhaven. Route planning is done by consulting a route planning service agent that determines the fastest available route given the current traffic situation. Both, truck agent and the route planning service, are registered at a traffic information service that informs them on changes in flow of traffic as soon as they are noted. For tracking of the truck agent's behaviour the monitoring panel shows his current route on the road map (figure 4.37).

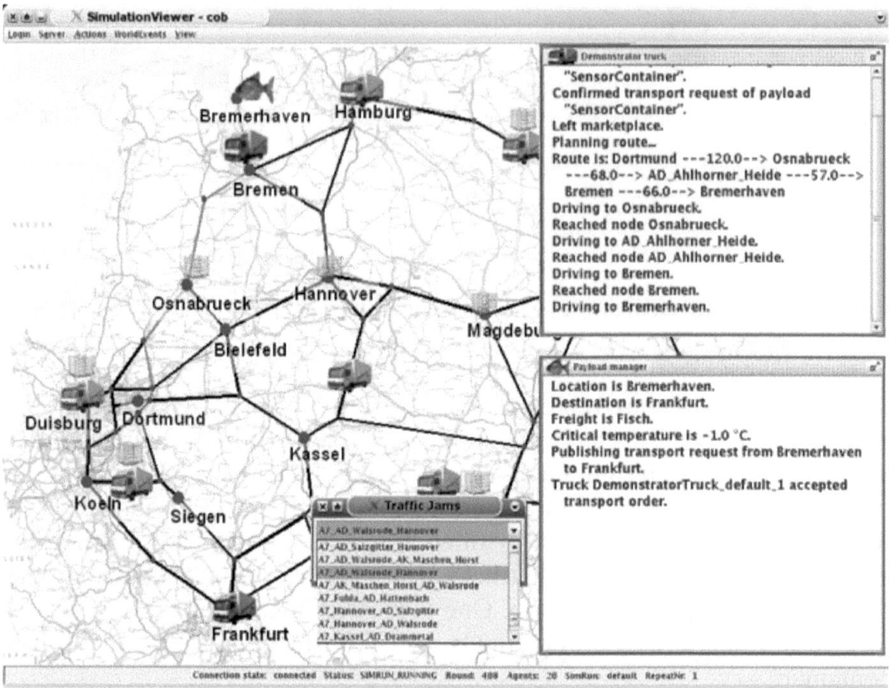

Fig. 4.37 Scenario monitoring GUI showing the traffic network, trucks and loads with their respective positions as well as status information of selected agents.

Loading and transfer of the supervision agent

After arriving of the truck in Bremerhaven the freight can be loaded. Our palette of fish is detected by an RFID reader mounted at the door of the means of transport. In order to transfer the mobile supervision or assessing agent the manager of the means of transport sends a request to the computer platform that was responsible for the package before the transfer. The necessary address information is retrieved form the RFID tag. After being transferred the supervision agent links itself to the sensor facilities that are provided by the means of transport and continues the quality assessment.

Sensor configuration

The first task of the supervision agent is to test whether all required sensors are available inside the means of transport. In our setting the missing of a requested humidity sensor is detected. The agent sends a warning message to the transport operator who adds an additional sensor node to the container. The self-configuration of the sensor network automatically integrates the new node. After completion of the sensor network a message is send to the transport coordination that the vehicle is ready for departure.

Route Re-planning Based on Flow of Traffic

The truck agent again plans his route now directed to freight destination Frankfurt. The currently fastest available route leads via Hanover following German motorway A7. Shortly after departure in Bremerhaven a traffic congestion from Bremerhaven to Bremen is triggered. The traffic information service is informed on the new congestion and immediately reports it to all registered agents including the truck with the intelligent container and the route planning service. Then, the truck agent detects that this congestion affects his current route and requests for an updated route at the routing agent. Given the new traffic information the routing agent now proposes a route via Dortmund following motorway A1. The truck agent changes his plan correspondingly and informs the truck driver on its PDA.

Communication failure

For external communication the embedded platform of the container is connected with an automated gateway that selects between different mobile networks like UMTS, GPRS and WLAN according to their availability, bandwidth and costs. In practical application WLAN would be selected as long as the truck is inside the range of a hotspot. The loss of the WLAN connection can be simulated during the demonstration by removing the WLAN device form the PCMIA slot of the gateway. The gateway immediately switches to UMTS communication. The embedded processor and the notebooks are connected via VPN tunnels that provide a permanent IP-connection with fixed addresses independent of the currently used mobile network.

Quality supervision

During transport the freight owner only receives notifications about important changes of the freight state like transhipment and crucial quality losses (figure 4.38). To retrieve the history of an alarm he can request a full sensor protocol that is displayed in an oscilloscope view. During normal transport operation the sensor data are pre-processed by dynamic quality models. These models represent the knowledge that is necessary to assess the effects of sub optimal transport conditions. Instead of sensor raw data only information about predicted quality changes have to be transferred over external mobile networks.

To force a quality loss we press a test button at one of the sensor nodes to simulate a temperature rise. This deviation from the recommended transport parameters leads to an accelerated decay of the loaded good. Only if this disruption is continued for a certain period a crucial quality loss is predicted by the dynamic model. This process can be observed in the oscilloscope view. If the quality drops below a defined warning threshold or if it becomes foreseeable that the quality will fall below an acceptance limit before the freight arrives its destination a warning is send to the transport coordination agent.

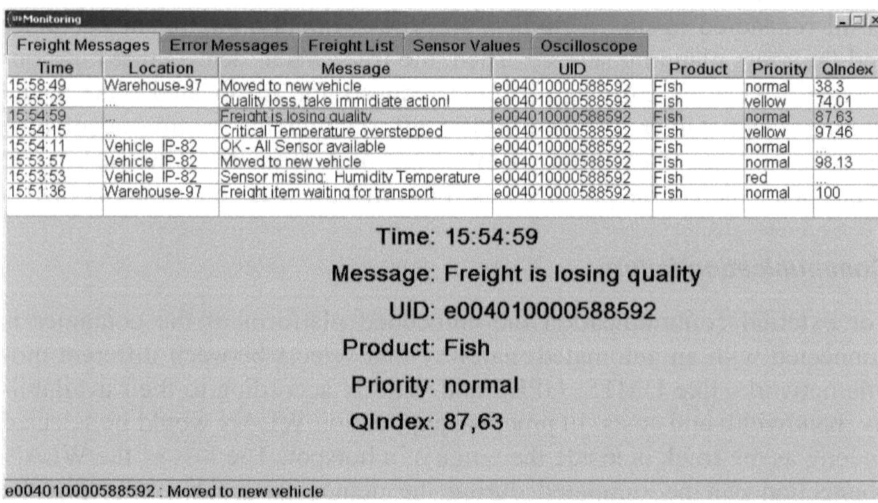

Fig. 4.38 Log of the messages that are received by the owner of the freight. He is informed on transhipments, state of the sensor system, overstepping of the recommended temperature and changes in quality. The messages contain time stamp (time-lapse mode), current location of the package, an unique identification number of the freight object, the kind of good and an index value for the current quality.

Reactions to quality loss and re-planning

The freight's transport planning and coordination agent assesses the quality warning message by estimating the prospective freight quality when reaching the freight destination, i.e., Frankfurt. In the given scenario the agent receives the warning message with the truck located on German motorway A1 between the cities Osnabrück and Dortmund. The prospected quality in Frankfurt is assessed as too low to continue the transport. Thus, the agent searches for alternatives to reach his goal. As one possibility, the freight could be transhipped to another truck that has a properly working cooling system. Although there is such a truck located in Frankfurt the transhipment must not interrupt the cold chain. Thus the agent searches for refrigerated warehouses. This information is provided by the agent's local knowledge base and the logistics service broker agent. In this case, the agent finds an appropriate refrigerated warehouse near Kassel as the shortest possible detour. Subsequently, the agent coordinates with the truck agent and the agent representing the warehouse to organise transhipment with intermediate storage in Kassel.

The truck again re-plans its route to the Kassel warehouse. The truck driver is notified by his PDA display that destination has changed to Kassel. Furthermore, the freight agent informs the truck agent that rising temperatures were responsible for change of destination. This causes the truck to schedule his maintenance because the cooling system may have been damaged.

When reaching Kassel the freight agent requests for a compensatory truck by consulting the service broker. This time the simulated truck in Frankfurt with cooling abilities proposes to take over the transport from Kassel to Frankfurt. The freight agent accepts this proposal and finally the transport successfully reaches Frankfurt within time and acceptable quality. Despite all disturbances, local control strategies implemented by software agents and supported by communication and sensor technologies managed to perform the process successfully. Therewith, this rather simple transportation scenario demonstrated some of the current possibilities of autonomous logistic processes.

4.7.3 Institutional cooperation

The scenario was developed inside the CRC637 by cooperation of different institutes form the fields of electrical engineering, computer science and applied work science.

- The Institute for Microsensors-, actuators and systems (IMSAS) developed the embedded agent platform and the prototype for the 'intelligent container' including software agents for dynamic quality assessment and RFID controlled agent transfer.
- The TZI Intelligent Systems division developed the distributed PC-based agent software platform including simulation and visualisation components. Additionally, they contributed agents running on this platform and managing the transport process by autonomous coordination.
- The Institute for Electromagnetic Theory and Microelectronics (ITEM) developed the wireless sensor nodes and the software for the self-configuration of the sensor network.
- ComNets implemented a self-organized selection of communication networks in a device called Autonomous Communication Gateway. The device has the ability to communicate using three different communication networks, namely *Wireless Local Area Network* (WLAN), *Universal Mobile Telecommunication System* (UMTS) and *General Packet Radio Service* (GPRS).
- The Bremen Institute of Industrial Technology and Applied Work Science at the University of Bremen (BIBA) provided the indoor location system and the facilities to 'run' the remote controlled model trucks.

The theoretical background of the demonstrator was discussed in the previous articles of this volume:

- Chapter 4.6 "Intelligent containers and sensor networks" describes the background of the embedded platform, the layout of the sensor network and future perspectives to incorporate autonomous processes into miniaturized devices with limited resources.
- Chapter 3.5 "Distributed Knowledge Management in Dynamic Environments" discusses the theoretical backgrounds of an agent-based distributed service framework for knowledge exchange in autonomous logistic processes.

We additionally thank Farideh Ganji who worked on the RFID based location system for the trucks, Martin Lorenz who supported agent development, and our students that were committed to the project, namely Javier Antunez-Congil and Christian Ober-Blöbaum.

Index

D

E

I

J

K

L

M

MIX
Papier aus verantwortungsvollen Quellen
Paper from responsible sources
FSC® C105338

If you have any concerns about our products,
you can contact us on
ProductSafety@springernature.com

In case Publisher is established outside the EU,
the EU authorized representative is:
Springer Nature Customer Service Center GmbH
Europaplatz 3, 69115 Heidelberg, Germany

Printed by Libri Plureos GmbH
in Hamburg, Germany